Muin S.A. Tuffaha

Phenotypic and Genotypic Diagnosis of Malignancies

Related Titles

Meyers, R. A. (ed.)

Cancer

From Mechanisms to Therapeutic Approaches

2007

ISBN: 978-3-527-31768-4

Kumar, C. S. S. R. (ed.)

Nanomaterials for Cancer Diagnosis

2007

ISBN: 978-3-527-31387-7

Knudsen, S.

Cancer Diagnostics with DNA Microarrays

2006

ISBN: 978-0-471-78407-4

Kiaris, H.

Understanding Carcinogenesis

An Introduction to the Molecular Basis of Cancer

2006

ISBN: 978-3-527-31486-7

zur Hausen, H.

Infections Causing Human Cancer

2006

ISBN: 978-3-527-31056-2

Kumar, C. S. S. R. (ed.)

Nanomaterials for Cancer Therapy

2006

ISBN: 978-3-527-31386-0

Debatin, K.-M., Fulda, S. (eds.)

Apoptosis and Cancer Therapy

From Cutting-edge Science to Novel Therapeutic Concepts

2006

ISBN: 978-3-527-31237-5

Muin S.A. Tuffaha

Phenotypic and Genotypic Diagnosis of Malignancies

An Immunohistochemical and Molecular Approach

WILEY-VCH Verlag GmbH & Co. KGaA

The Author

Dr. Muin S. A. Tuffaha
German Jordan Center for Laboratory Medicine
P.O. Box 925705
Amman, 11190
Jordan

All books published by Wiley-VCH are carefully produced. Nevertheless, authors, editors, and publisher do not warrant the information contained in these books, including this book, to be free of errors. Readers are advised to keep in mind that statements, data, illustrations, procedural details or other items may inadvertently be inaccurate.

Library of Congress Card No.: applied for

British Library Cataloguing-in-Publication Data
A catalogue record for this book is available from the British Library.

Bibliographic information published by the Deutsche Nationalbibliothek
Die Deutsche Nationalbibliothek lists this publication in the Deutsche Nationalbibliografie; detailed bibliographic data are available on the Internet at <http://dnb.d-nb.de>.

© 2008 WILEY-VCH Verlag GmbH & Co. KGaA, Weinheim

All rights reserved (including those of translation into other languages). No part of this book may be reproduced in any form – by photoprinting, microfilm, or any other means – nor transmitted or translated into a machine language without written permission from the publishers. Registered names, trademarks, etc. used in this book, even when not specifically marked as such, are not to be considered unprotected by law.

Composition Thomson Digital, Noida, India
Printing betz-druck GmbH, Darmstadt
Bookbinding Litges & Dopf GmbH, Heppenheim
Cover Design Adam-Design, Weinheim

Printed in the Federal Republic of Germany
Printed on acid-free paper

ISBN: 978-3-527-31881-0

Dedicated to the light of my life
My father and mother
Sami and Haya
For their love and support

Contents

Phenotypic and Genotypic Diagnosis of Malignancies. Muin S.A. Tuffaha
Copyright © 2008 WILEY-VCH Verlag GmbH & Co. KGaA, Weinheim
ISBN: 978-3-527-31881-0

Preface

Recent years have seen an explosive increase in new methods in diagnostic tumor histopathology, especially in the immunohistochemical, genetic and molecular profiling of tumors. The idea for this book began with the first edition in 2002, when I published my personal notes and experience in immunohistochemistry and molecular pathology, which I frequently use in my daily work as a histopathologist. The continuous development of new antibodies and molecular diagnostic methods make it necessary to constantly update the information in this book; and the second updated edition was published in 2005.

In this edition the majority of the chapters have been reviewed, the immunohistochemical profile of the different tumor groups is updated and the most recent WHO classification of tumors is considered. Also, new chapters in immunohistochemistry, molecular protocols and molecular diagnosis have been added.

The book is designed to be a practical, easy to use bench reference in immunohistochemistry and molecular pathology, useful for the diagnosis and differential diagnosis of tumors. This book should be of value to histopathologists, oncologists, hematologists and laboratory technicians.

Amman, December 2007 *Muin S.A. Tuffaha*

Phenotypic and Genotypic Diagnosis of Malignancies. Muin S.A. Tuffaha
Copyright © 2008 WILEY-VCH Verlag GmbH & Co. KGaA, Weinheim
ISBN: 978-3-527-31881-0

1
Introduction to Immunohistochemical and Molecular Methods in Tumor Diagnosis, and the Detection of Micrometastases and Circulating Tumor Cells

Solid and hematological tumors make a large heterogeneous group with different histogenesis and pathogenesis, different biological and clinical behaviors and consequently different responses to various therapeutic methods, which make accurate diagnosis and precise tumor classification essential for tumor management.

During recent years, classic surgical pathology has rapidly developed and now, beside conventional microscopy and electron microscopy, there are a number of additional highly sensitive diagnostic tools, including immunohistochemistry, cytogenetics and molecular pathology. These methods provide further objectives and reproducible criteria for diagnosis, classification and follow-up of tumors.

Since the late 1980s, immunohistochemistry has presented a sensitive and uncomplicated method to screen the immunophenotypic profailes of different tumors to determine their histogenesis by immunological detection of a specific cellular antigens on tissue sections prepared from frozen tissue or formalin-fixed paraffin-embedded tissue blocks. Nowadays we have several hundreds of monoclonal and polyclonal antibodies specific of cellular and extracellular structures. The use of these antibodies gives us the possibility to classify many undifferentiated tumors including solid tumors and hematological neoplasia, in addition to the possibility to determine the sensitivity of some tumors to specific therapeutic agents.

In the past decade, the rapid development of molecular biology provided other sensitive and powerful molecular tools for the recognition and classification of tumors. Examples are tumors derived from tissue types characterized by unique tissue-specific antigens. These antigens or tissue-specific markers are encoded by the cellular genome and translated into complementary mRNA. The presence of these tissue-specific mRNAs can be detected by RT-PCR-based methods, representative examples are tumors derived from thyroid, prostate and neural tissue. Other tumors are associated with recurrent specific chromosomal and genetic aberrations or rearrangements, which can be used as specific genotypic markers. The detection of tissue-specific transcripts or tumor-specific genetic aberrations by molecular methods offers the basis for precise tumor diagnosis. This method is now used for the diagnosis of tumors exhibiting similar morphology and non-specific immunohistochemical phenotype such as synovial sarcoma, PNET and melanoma.

Phenotypic and Genotypic Diagnosis of Malignancies. Muin S.A. Tuffaha
Copyright © 2008 WILEY-VCH Verlag GmbH & Co. KGaA, Weinheim
ISBN: 978-3-527-31881-0

Table 1.1 Stages of metastatic process.

Tumor development and evasion of immune response
Separation of the tumor cells from the original tumor
Erosion of basement membrane overcoming of the physiological barriers
Invasion of blood and lymphatic vessels
Adhesion and invasion at different distant sites
Organization of metastatic tissue with the induction of tumor stroma

It is also used for the diagnosis of many hematological malignancies such as chronic myeloid leukemia, follicular lymphoma and myelodysplastic syndrome.

Another major problem facing the modern oncology beside the diagnosis and classification of tumors is the metastatic spread of malignant tumors and tumor relapse after therapy. Metastases can be the first manifestation of some tumors and represent an important factor determining patient's final outcome as they cause about 90% of deaths from solid tumors.

Additionally, the presence of micrometastases or individual tumor cells in surgical margins is the major cause of tumor relapse after surgical excision of tumors. The detection of such submicroscopic micrometastases or disseminated tumor cells in surgical margins, lymph nodes, bone marrow or peripheral blood is difficult, impractical or impossible by conventional microscopic examination and needs more sensitive and specific methods like immunohistochemistry and molecular methods. Furthermore, a large number of patients with distant metastases are clinically asymptomatic, which means that more sensitive methods are needed to monitor tumor progression.

Metastatic spread is an important aspect in tumor biology and usually determines the clinical behavior. The metastatic process is a long complicated multistep process which begins with tumor development and evasion of immune response and ends with the organization of metastatic tissue and the induction of tumor stroma and neovascularization (Table 1.1).

An additional cause of metastatic spread is assumed to be the dissemination of tumor cells due to tumor manipulation during surgical intervention or diagnostic assays. Many reports suggest that the manipulation of tumors could provoke shedding of tumor cells into blood or lymphatic vessels causing metastatic spread. To prevent this risk factor the "no-touch isolation technique" with primary lymphovascular ligation was suggested many years ago.

One of the first steps to go behind the metastatic cascade is the examination of peripheral blood to detect potential circulating malignant cells. This method is useful to follow-up malignancies spreading via the blood stream such as sarcomas, melanoma, neuroblastoma, prostatic, thyroid and hepatocellular carcinomas. PCR- and RT-PCR-based molecular techniques are found to be the most efficient assay to detect circulating tumor cells in peripheral blood, and allow the detection of one tumor cell in up to10^5 background nucleated blood cells. However, it is important to consider that the presence of tumor cells in peripheral blood is only one step in a multistep process and not all tumor cells circulating in peripheral blood are able to attach and proliferate to develop metastases. Many studies show that the majority of circulating tumor cells are killed by different mechanisms including immunological and mechanical ways. Experimental studies demonstrate that approximately 0.01% of circulating tumor cells arising from solid tumors are able to establish metastatic colonies after extravasation. Animal experiments show that only about 0.01% of cancer cells injected into the circulation are able to form metastatic colonies with tumor stroma. In contrast, the presence of tumor cells in peripheral blood is considered to be a negative prognostic factor correlating with tumor progression and an essential step to the development of organ metastases. Many clinical observations suggest that tumor cells detected by sensitive methods during remission may reflect a latent tumor stadium under the control of the immune system and tumor cells may be reactivated after suppression of the immune system.

The next station in the metastatic spread after invasion of the lymphatic vessels is the locoregional lymph nodes, which act as clearing sites for foreign antigens and tumor cells, mainly for tumors originated from epithelial tissue. Among many prognostic factors, the status of the locoregional lymph nodes is one of the most important and useful indicators for most types of carcinomas, such as breast, lung, esophageal and gastrointestinal carcinomas, and generally the presence of lymph node metastases is a significant predictor of poor prognosis in all cancers. Many studies showed that 30–60% of lung, breast, and gastrointestinal cancers already have lymph node or distant metastases at the time of diagnosis. Recent studies showed that breast cancer with minimal involvement of a single axillary node, even in form of disseminated tumor cells, notably correlates with worse prognosis compared with cases with no axillary node involvement. Regional lymph nodes are subjected to routine study by conventional microscopy by examination of H&E stained sections from each removed lymph node, but many studies show that about 20% of the cases of gastrointestinal carcinoma with histologically negative lymph nodes (pN0) suffer tumor relapse in less than five years. Immunohistochemical staining of lymph nodes performed on frozen or formalin-fixed paraffin-embedded tissue are found to be more sensitive than the conventional microscopy alone in detection of lymph node metastases. Recently, molecular methods based on PCR and RT-PCR proved to be very sensitive tools for the detection of submicroscopic metastases and disseminated tumor cells with a sensitivity ratio up to one target cell in 10^6 background cells. Because molecular methods are costly and time-consuming to be done on all removed lymph nodes, the alternative examination of the sentinel lymph node has been suggested. The sentinel lymph node is defined as the first locoregional lymph node receiving lymph from the tumor mass and can be detected by dye or radioactive scanning. This method is effective in the management of malignant melanoma and breast carcinoma.

Another important station for tumor cells is bone marrow, which also acts as a clearing and filtration site for antigens and malignant cells. Many studies found a marked correlation between the presence of micrometastases or disseminated tumor cells in bone marrow and the progress of malignant tumors. Several studies showed that even the presence of a small number of tumor cells in bone marrow is associated with poor prognosis in several tumor types such as carcinomas of GIT, breast, prostate and neuroblastoma and considered as an independent indicator of tumor relapse.

The detection of malignant cells in different body fluids is also important for tumor diagnosis and tumor staging. Examination of liquor can be useful in the therapy of various malignancies such as lymphoma/leukemia, melanoma, breast carcinoma or primary brain tumors. Examination of pleural effusion and sputum is also essential in cases of lung and breast carcinoma or primary mesothelioma. The study of ascitic fluid also gives valuable information in pancreatic and hepatocellular carcinomas.

In conclusion, the modern approach in the diagnosis of tumors is a multistep process, which includes conventional histology, immunohistochemistry, cytogenetics and molecular methods. The early detection of micrometastases and circulating tumor cells before metastatic disease becomes clinically evident gives a better chance for the eradication of residual tumor cells to prevent tumor relapse in addition to its prognostic importance.

2
Diagnostic Immunohistochemistry

2.1
Introduction

Immunohistochemistry is a sensitive method to detect tissue- and tumor-specific antigens using highly specific monoclonal or polyclonal antibodies against different cellular and extracellular antigens. The antigen–antibody complex is detected by various sensitive, direct and indirect detection systems such as the peroxidase-anti-peroxidase (PAP) system, the alkaline phosphatase–anti-alkaline phosphatase (APAAP) system and the avidin (streptavidin)–biotin horseradish peroxidase (ABC) system. The immunohistochemical reaction is performed on tissue sections prepared from formalin-fixed paraffin-embedded tissue blocks or frozen tissue, allowing the study of tumor morphology combined with the immunohistochemical phenotype and the pattern of antigen expression. At the present time a large number of monoclonal and polyclonal antibodies are used in immunohistochemistry, covering all types of histogenetic differentiation essential for the diagnosis of tumors and also helpful for determining the strategies of tumor therapy.

An additional important benefit from immunohistochemical methods is the detection of minimal residual tumor cells, since the immunohistochemical stain of tumor cells is more sensitive than conventional histopathology, allowing the possibility to detect small groups of tumor cells or even a single tumor cell in tissue sections or smears. For example, cytokeratin-positive tumor cells can be visualized as isolated tumor cells or micrometastases in lymph nodes or as disseminated tumor cells in bone marrow in up to 50% of patients with stage I and stage II breast cancer. In addition to the diagnostic value and the prognostic information obtained by immunophenotyping of tumors, immunohistochemical methods can give valuable information to evaluate the sensitivity of different tumors to various therapeutic agents used for tumor therapy (so-called pharmacopathology), discussed in the following section.

Phenotypic and Genotypic Diagnosis of Malignancies. Muin S.A. Tuffaha
Copyright © 2008 WILEY-VCH Verlag GmbH & Co. KGaA, Weinheim
ISBN: 978-3-527-31881-0

2.2 Pharmacopathology

Modern immunohistochemical methods play an important role in the therapy management of many types of malignancies. Using highly specific antibodies it is possible to detect the expression of specific epitopes which are specific targets for immunotherapy, such as HER-2, EGRF-1 and CD20. Specific antibodies are also used to detect the expression of hormone receptors and enzymes which are targets for receptor antagonists and enzyme inhibitors. Nowadays the semi-quantitative estimation of steroid receptors, HER-2, EGFR-1, CD20 and CD117 are widely used for cancer therapy.

2.2.1 Evaluation of Steroid Hormone Receptor Status by Semi-Quantitative Estimation of Receptor Expression (Estrogen, Progesterone and Androgen Receptors)

The aim of this assay is to evaluate the expression of steroid receptors (receptor status) in hormone-dependent tumors, since these receptors (estrogen, progesterone and androgen receptors) are essential for the proliferation and differentiation of tumors derived from hormone-dependent tissues such as breast, ovarian and uterine tissue.

2.2.1.1 Estrogen Receptor

Estrogen receptor (ER) is a member of the steroid family of ligand-dependent transcription factors. There are two isoforms of the estrogen receptor (ER), the alpha form (ER α) and beta form (ER β). ER α and ER β isoforms have a ~95% homology in the DNA-binding domain but only ~60% homology in the ligand-binding domain. These isoforms show different expression in different tissue types: the ER α isoform is mainly expressed in both epithelial and stromal cells of the breast, uterus, placenta, liver, CNS, and bone, whereas the ER β isoform is mainly expressed in the prostate, testes, ovary, spleen, thymus, skin and endocrine glands, including the thyroid and parathyroid glands, adrenal glands and pancreas. Both isoforms are expressed on the nuclear membrane of normal and tumor cells and can be detected by immunohistochemical staining using specific monoclonal antibodies (such as the clones 1D5, 6F11) on tissue sections prepared from frozen or paraffin-embedded tumor tissue. Most antibodies used in immunohistochemistry are directed to the alpha isoform of the estrogen receptor. Antibodies directed to the beta form of the receptor molecules are rarely used but overexpression of ER β is noted during the progression of breast tumors and is thought to be a sign of tamoxifen resistance. This assay is essential for the therapy management of breast and endometrial cancers to determine the response of these tumors to the anti-estrogen therapy (i.e. tamoxifen) or oophorectomy. Statistical observations reveal that about 70% of the patients with estrogen receptor-positive tumors showed a significant clinical response to anti-estrogen therapy, whereas about 85% of the patients with estrogen receptor-negative tumors showed no or little response to this therapy.

Table 2.1 Scoring system according to Remmele.

	Percentage of positive cells		Intensity of stain
0	No positive cells	0	No detectable stain
1	<10%	1	Weak nuclear stain
2	10–50%	2	Moderate nuclear stain
3	51–80%	3	Strong nuclear stain
4	>80%		

Since the estrogen receptor is an unstable molecule with a short half-life, biopsies must be handled to preserve the receptor structures. Rapid fixation of tumor tissue after surgical excision using buffered neutral formalin for 16–24 h is essential for optimal staining.

Few semi-quantitative scoring systems were suggested, whereas the modified scoring system suggested in 1987 by Remmele, the modified scoring system suggested in 1985 by McCarty and the Allred scoring system proved to be the most practical and simple systems. The three systems depend on the evaluation of the stain intensity and the percentage of positive cells.

Remmele Scoring System This simple scoring system has a 12-point scale (0–12). To calculate the score one of the numbers 0, 1, 2 or 3 is given according to the intensity of the nuclear stain and one of the numbers 0, 1, 2, 3 or 4 is given according to the percentage of positive tumor cells (Table 2.1). The score is calculated by multiplying the number reflecting the dominant stain intensity by the number reflecting the percentage of these positive tumor cells with a maximum score of 12 (3×4). Tumors with a score of less than 3 show usually a poor response to the anti-estrogen therapy corresponding to 10–20 fmol/mg of cytosol tumor protein using the biochemical assay. Clinical observations demonstrate a good correlation between the results of this scoring system and the tumor behavior, especially the response to anti-estrogen therapy.

McCarty Scoring System This scoring system has a 300-point scale (0–300), calculated by multiplying the percentage of positive cells (0–100) by the number reflecting the intensity of the immunohistochemical stain (0: no detectable staining; 1: weak nuclear staining; 2: moderate nuclear staining; 3: strong nuclear staining). The total score is explained as follows:

- Negative (−): 50 or less
- Weakly positive (+): 51–100
- Moderately positive (++): 101–200
- Strongly positive (+++): 201–300

Allred Scoring System The Allred scoring system has an eight-point scale (0–8). This scoring system is calculated by adding the number representing the proportion of positive cells 0, 1, 2, 3, 4 or 5 to number reflecting the intensity of the nuclear stain

Table 2.2 Allred scoring system.

	Percentage of positive cells		Intensity of stain
0	No positive cells	0	No detectable stain
1	<1%	1	Weak nuclear stain
2	1–10%	2	Moderate nuclear stain
3	10–33%	3	Strong nuclear stain
4	33–66%		
5	>66%		

0, 1, 2 or 3 (Table 2.2). Tumors with a score of less than 3 show usually a poor response to the anti-estrogen therapy.

Different breast cancer types exhibit different expression levels of steroid receptors depending on the histological subtype of the tumor, differentiation and proliferation grade in addition to other cytogenetic features. Papillary, tubular, invasive lobular and mucinous breast carcinomas are usually associated with a high expression level of estrogen receptors, whereas medullary, apocrine and metaplastic carcinomas lack estrogen receptors or show very low expression levels. Note that breast cancers associated with BRCA-1 and BRCA-2 mutations usually lack the expression of estrogen receptors. Breast cancer in men is more frequently positive for estrogen and progesterone receptors and is supposed to be more sensitive to anti-hormonal therapy. Furthermore the expression level of bcl-2 oncoprotein is markedly associated with the expression level of steroid receptors, whereas there is an inverse relationship between the expression of HER-2, EGFR, mammaglobin and Ki-67 (MIB-1) proliferation index and the expression of steroid receptors.

2.2.1.2 **Progesterone Receptors**

Progesterone is a steroid hormone involved in the differentiation of breast parenchyma and endometrium in addition to milk protein synthesis. Progesterone receptor (PR) is another member of the steroid family expressed on the nuclear membrane. The progesterone receptor molecule is a 946-amino-acid protein, which has two homologous isoforms PR-A and PR-B. Both isoforms are expressed in breast parenchyma, but PR-A is the dominant isoform in endometrial stroma and PR-B is the dominant isoform in endometrial glands. Progesterone receptors can be detected by immunohistochemistry using specific polyclonal and monoclonal antibodies (such as the 1A6 clone). Progesterone receptor status is one of the important prognostic factors in breast, endometrial and ovarian cancers. A high expression level of both estrogen and progesterone hormone receptors is a positive prognostic factor for breast and endometrial cancers and predicts a good response to anti-estrogen therapy. Several clinical observations show that estrogen and progesterone receptor-positive breast cancers show a good response to tamoxifen therapy in about 77% of cases, whereas estrogen-positive but progesterone-negative breast cancers are more likely to be poorly differentiated, with more aggressive behavior. The estrogen-positive, progesterone-negative breast cancers show a better response to aromatase inhibitors but are more resistant to tamoxifen therapy, with an optimal response to the tamoxifen therapy in only 27% of cases; and the response to this therapy is usually shorter than estrogen-positive, progesterone-positive tumors. It is interesting that estrogen-negative but progesterone-positive breast cancers show a good response to tamoxifen therapy in about 45% of cases, while breast cancers negative for both estrogen and progesterone receptors show an optimal response to the tamoxifen therapy only in about 10% of cases.

2.2.1.3 Androgen Receptors

Androgen receptor (AR) is also a member of the steroid family of ligand-dependent transcription factors, which can be detected by immunohistochemistry using specific antibodies. Androgen receptor is expressed in the majority of estrogen-positive and in about 50% of estrogen-negative breast carcinomas in addition to DCIS, furthermore the majority of metastatic breast tumors are immunoreactive for androgen receptor. Clinical studies show also that ER-negative/AR-positive tumors have a better prognosis and longer disease-free survival than ER- and AR-negative tumors, which suggests a possible role of the androgen receptors in the breast cancer growth and differentiation. Androgen receptor is also highly expressed in the prostatic tissue and in the majority of prostatic tumors. The expression of the androgen receptor is an important prognostic factor for the cancer of the prostate and correlates with the grade of prostatic tumors and can give additional information useful for the treatment of prostatic carcinoma, namely the response to the anti-androgens or LHRH-analogs.

2.2.2 Estimation of Human Epidermal Growth Factor Receptor-2 Expression

*H*uman *e*pidermal growth factor *r*eceptor-*2* (HER-2 neu) is a member of the type 1 receptor tyrosine kinase family with domains on the cell surface, functioning as growth factor receptors. The type 1 receptor tyrosine kinase family includes the receptors HER-1 (EGFR or c-erbB1), HER-2 (c-erbB-2), HER-3 (c-erbB-3) and HER-4 (c-erbB-4).

The HER-2 neu oncoprotein, also known as p185 or c-erbB-2 (*c*hicken *er*thro*b*lastic virus) is a 185-kDa transmembrane glycoprotein encoded by the HER-2 proto-oncogene located on chromosome 17 q12–21 consists of an extracellular domain (P105), a single transmembrane segment, and a cytoplasmic tyrosine kinase domain. In mammals, the HER-2 receptor plays an important role in the development of neural tissue and cardiac muscle. HER-2 is also normally expressed on the membrane of normal epithelial cells and 20×10^3 to 50×10^3 receptors are generally found on the surface of normal breast epithelial cells, nevertheless some types of epithelial tumors are associated with HER-2 overexpression and 2×10^5 to 3×10^6 receptors may be expressed on the cell membrane of these tumor cells. HER-2 overexpression is found in various human carcinomas, mainly breast carcinoma in addition to some other cancer types such as non-small cell carcinoma of the lung, salivary gland tumors and gastrointestinal tumors. Remarkable is the HER-2 overexpression in the epithelioid components of synovial sarcoma but this phenomenon is not associated with gene amplification.

About 30% of unselected breast carcinomas are associated with HER-2 overexpression; most of them are invasive ductal carcinoma, inflammatory breast cancer, Paget's disease as well as breast cancer arising in pregnant women, whereas invasive lobular carcinoma, mucinous carcinoma and medullary carcinoma of the breast are rarely accompanied by HER-2 overexpression. HER-2 overexpression is also found in a significant percentage of male breast carcinomas (about 30%).

Another considerable observation is the relationship between the expression of HER-2 oncoprotein and the expression of steroid hormone receptors. Tumors with HER-2 overexpression show usually a low expression level of estrogen and progesterone receptors, while tumors with a high expression level of estrogen and progesterone receptors often lack overexpression of the HER-2 oncoprotein. Moreover, there is an inverse relationship between the expression of the bcl-2 oncoprotein and the expression of both HER-2 and EGFR. It is also interesting to mention that BRCA-1- and BRCA-2-related breast carcinomas are rarely associated with HER-2 overexpression.

The overexpression of the HER-2 oncoprotein is an important prognostic factor for breast cancer and the following aspects must be considered:

- Breast cancers with HER-2 overexpression have a poor prognosis with a median survival rate of 3 years compared with tumors lacking the HER-2 overexpression having a median survival rate of 6–7 years.
- Tumors with HER-2 overexpression are frequently poorly differentiated and a high percentage of the tumor cells is in the S phase.
- A high percentage of tumors with HER-2 overexpression is associated with P53 mutations and the tumor cells are usually bcl-2-negative.
- Tumors with HER-2 overexpression are usually resistant to anti-estrogen therapy and CMF (cyclophosphamide/methotrexate/5-fluorouracil) chemotherapy.
- Tumors with HER-2 overexpression are usually highly sensitive to the anthracycline chemotherapy (Doxorubicin, Epirubicin) and have an enhanced response to the taxane chemotherapy (paclitaxel).
- Tumors with HER-2 overexpression are sensitive to the specific immunotherapy and specific tyrosine kinase inhibitors.

The extracellular domain of the HER-2 neu molecule is the therapeutic target for the specific IgG1κ humanized chimeric monoclonal antibody Trastuzumab (Herceptin®), approved 1998 and found to be effective in malignancies with HER-2 overexpression. An important aspect of this specific immunotherapy includes the efficiency against non-proliferating disseminated tumor cells (dormant cells), which are usually in the G0 phase of the cell cycle, where conventional chemotherapy is less effective. The HER-2 neu molecule is also the target for specific tyrosine kinase inhibitors such as Tykerb (lapatinib ditosylate), which block the kinase-substrate interaction and the extracellular tyrosine kinase receptors on tumor cells.

For optimal therapy management and to verify the sensitivity of tumors to the specific antibody a three-point semi-quantitative scoring system to determine the HER-2 overexpression is used. The scoring is estimated after immunohistochemical staining of tumor sections. This immunohistochemical reaction is applicable on routinely processed sections from formalin-fixed paraffin-embedded tissue using specific polyclonal (HercepTest™, Dako) or monoclonal (such as the clones PN2A, CB11, SP3, TAB 250 and B10) antibodies. The antibody included in the HercepTest™ and other clones such as SP3 and CB11 are directed to the extracellular domain of the HER-2 molecule, while some other clones are directed to the intracellular domain. Because of the quick damage of the HER-2 molecules due to oxidation or other factors (slide aging), fresh prepared tissue slides must be used. Tissue after frozen section is not suitable for this reaction. Critical for accurate evaluation is the standardization of the immunohistochemical reaction, which includes standard tissue pretreatment, adequate antigen unmasking (40 min in a water bath at 95–100 °C) and the use of certified antibodies. The hallmark of the three-point scoring system is the interpretation of the intensity and continuity of the membrane staining and the percentage of positive cells (Table 2.3).

For precise scoring the following factors must be considered:

- The interpretation of the immunohistochemical stain must begin with the evaluation of standardized control slides with the scores 0, 1+ and 3+.
- Only membrane staining should be evaluated. Cytoplasmic or nuclear stain must be neglected. Staining caused by edge artifacts must also be ignored.
- Only invasive tumor components must be considered.
- Primary tumors with heterogeneous cell populations exhibiting different expression intensity of the HER-2 oncoprotein and consequently different scores, must be particularly documented. In such cases, the assessment of metastatic tissue to determine the metastasizing cell population is recommended. This phenomenon is described in about 5% of breast tumors.

Tumors with scores of 0 or 1+ have no gene amplification and no HER-2 overexpression and consequently show a poor sensitivity to the specific immunotherapy. Tumors with a score of 3+ are associated with HER-2 overexpression and show a good

Table 2.3 The three-point scoring system.

Score	HER-2 overexpression	Staining result
0	Negative No gene amplification	No detectable staining or membrane staining in less than 10% of tumor cells
1+	Negative No gene amplification	A faint partial membrane staining in more than 10% of tumor cells
2+	Positive	A weak to moderate staining of the entire membrane in more than 10% of tumor cells
3+	Positive High gene amplification	A strong staining of the entire membrane in more than 10% of tumor cells

response to the specific antibody therapy, whereas tumors with a score of 2+ needs a further confirmation to estimate the number of gene copies in the tumor cells to justify this type of therapy because of possible side-effects and high cost. The confirmation of the HER-2 overexpression in tumors with a score of 2+ can be achieved by genetic or chromosomal studies such as fluorescent *in situ* hybridization (FISH), chromogenic *in situ* hybridization (CISH) and real-time PCR assays. The presence of <2 gene copies in the examined cells indicates no gene amplification, whereas 2–6 copies signifies low-level amplification and >6 gene copies signifies strong gene amplification. Many studies show that the concordance between immunohistochemical score 0, 1+ and 3+ and the FISH method is more than 96%, whereas only 10–33% of tumors with a score of 2+ show HER-2 gene amplification in the FISH assay, which emphasize the necessity of further confirmation of HER-2 overexpression in tumors with a score of 2+.

2.2.3
Estimation of Epidermal Growth Factor Receptor Expression

*E*pidermal growth *f*actor *r*eceptor-1 (EGFR, Erb1) is another member of type 1 receptor tyrosine kinase family mentioned in the previous section functioning as growth factor receptors. EGFR is a transmembrane 170-kDa glycoprotein encoded by the c-erbB1 proto-oncogene and composed of the three following major domains:

- An extracellular ligand-binding domain, which binds to different growth factors (see below);
- A transmembrane lipophilic region;
- An intracellular domain characterized by tyrosine kinase activity initiated in response to ligand binding.

Epidermal growth factor receptor-1 (EGFR) binds to epidermal growth factor (EGF), transforming growth factor-α (TGF-α), heparin-binding epidermal growth factor (HB-EGF), amphiregulin, betacellulin and epiregulin. It plays an essential role in the development of many normal tissue types and is normally expressed on the membrane of normal epithelial cells and up to 25×10^4 receptors can be found on the surface of non-neoplastic epithelial cells such as endometrial cells. The activation of EGFR pathways is important for cell cycle progression and cell proliferation in addition to cell differentiation, but the expression or overexpression of EGFR in tumors is usually a sign of aggressive behavior and poor prognosis. The deregulation of cell cycle coupled with the overexpression of EGFR has been observed though the growth and the progression of different tumors of different histogenesis, mostly epithelial malignancies including carcinoma of the breast, non-small cell carcinoma of the lung, squamous cell carcinoma of head and neck, renal cell carcinoma, colon and pancreatic adenocarcinoma, ovarian and bladder carcinoma, however some other non-epithelial tumors such as glioblastoma multiforme are also associated with EGFR overexpression. Note that the EGFR overexpression in glioblastoma multiforme is a favorable prognostic factor.

The inhibition of EGFR activity causes the disruption of EGFR pathways and the blockage of the cell cycle progression leading to cell apoptosis, inhibition of tumor angiogenesis and suppression of the metastatic activity. EGFR is now the therapeutic target of different EGFR tyrosine kinase inhibitors such as lapatinib ditosylate and specific monoclonal antibodies such as the chimeric IgG1κ antibody Cetuximab (Erbitux®), the human IgG4κ antibody Panitumumab (Victibix®) and the humanized IgG1 antibody Nimotuzumab (TheraCIM®). The specific antibodies are now approved and used for the therapy of EGFR positive tumors such as colorectal and head and neck carcinomas.

The overexpression of EGFR can be estimated by the immunohistochemical staining using different specific polyclonal or monoclonal primary antibodies such as the clones E30, EGFR.25 and 31G7. For optimal immunohistochemical staining, unmasking of EGFR antigen by enzymatic digestion is needed (optimally 5 min digestion with 0.1% proteinase K). For the immunohistochemical staining and the evaluation of EGFR overexpression, we have to take the same cautions listed in the previous HER-2 section.

Two scoring systems were proposed for the semi-quantitative evaluation of EGFR overexpression in order to estimate the response of related tumors to the specific immunotherapy:

- The first system is identical to the three-point scoring system used for the estimation of HER-2 overexpression.
- The second system is a modification of the scoring system suggested by Remmele used for the evaluation of the estrogen receptor status. This scoring system has a nine-point scale (0–9) and calculated by multiplying the number reflecting the dominant intensity of the membrane stain 0, 1, 2 and 3 by the number reflecting the percentage of these positive tumor cells 0, 1, 2 and 3, as in Table 2.4.

According to this scoring system tumors having a score of less than 4 considered to have low EGFR expression and show a poor response for the specific therapy. The presence of cells showing cytoplasmic stain must also be mentioned.

Other similar pharmacopathologic diagnostic systems are now developed for other growth factor receptors, which can be the target for immunotherapy agents or specific

Table 2.4 The nine-point scale (0–9) is calculated by multiplying the number reflecting the dominant intensity of the membrane stain (0, 1, 2, 3) by the number reflecting the percentage of these positive tumor cells (0, 1, 2, 3).

	Percentage of positive cells		**Intensity of stain**
0	No positive cells	0	No detectable stain
1	<10%	1	Weak nuclear stain
2	<50%	2	Moderate nuclear stain
3	>50%	3	Strong nuclear stain

inhibitors such as the vascular endothelial growth factor receptor (VEGFR) and the platelet-derived growth factor receptor (PDGFR).

The vascular endothelial growth factor receptor binds to the vascular endothelial growth factor stimulating the proliferation of the vascular endothelium and increasing the vascular permeability, which may cause tumor progression. VEGFR is the therapeutic target for the humanized IgG1-specific antibody Bevacizumab (Avastin®) found to be effective in metastatic colorectal carcinomas.

The platelet-derived growth factor receptor is a further growth factor receptor, which binds to the platelet-derived growth factor causing the proliferation of mesenchymal cells. PDGFR is also the target of specific therapeutic antibodies.

2.2.4
Detection of CD 20

CD20 is a non-glycosylated 33-kDa phosphoprotein encoded by a gene located on chromosome 11q12–13, next to the site of the t(11; 14)(q13; q32) translocation associated with mantle cell lymphoma. This transmembrane antigen is expressed in the late pre-B cell stage after CD19/CD10 expression but before CD21/CD22 and surface immunoglobulin expression. CD20 remains as a constant antigen throughout the course of B-lymphocyte maturation but is absent in plasma cells.

CD20 (also known as L26) acts as a receptor playing role in the differentiation, activation and proliferation of B-lymphocytes. CD20 is also expressed on neoplastic B-lymphocytes and found almost in all lymphomas/leukemias of B-lymphocyte derivation. The extracellular domain of this antigen can be a suitable target for specific antibodies for the treatment of malignancies carrying this antigen. Rituximab (Rituxan®) a chimeric monoclonal IgG1κ antibody, Y-90 radiolabeled Ibritumomab (Zevalin®) murine monoclonal IgG1κ antibody and the I-131 radiolabeled Tositumomab (Bexxar®) murine monoclonal IgG2aλ antibody are examples for CD20-specific antibodies. These specific antibodies are used for the treatment of various types of CD20-positive B-cell lymphoma especially low-grade lymphomas mostly follicular lymphoma. This antigen can be detected on tissue sections using specific monoclonal antibodies (such as clone L26) to determine the sensitivity of these malignancies to the specific immunotherapy.

Many other lymphocyte-, leukocyte- and plasma cell-specific antigens are also now the target for specific therapeutic antibodies and can be detected by immunohistochemistry or flow cytometry such as CD22, CD 33, CD38 and CD52.

CD22 CD22 (also called as B-lymphocyte cell adhesion molecule or BL-CAM) is another B-cell-associated antigen, a type 1 integral membrane glycoprotein expressed in early B-lymphocytes differentiation, almost in the same stage as CD19 expressed. CD22 is also highly expressed in many types of B-cell lymphomas such as hairy cell lymphoma and expressed in about 90% of high-grade B-cell lymphoma. CD22 can be detected by immunohistochemistry using specific antibodies. CD22 is the therapeutic target of specific antibodies such as the humanized IgG1 antibody HLL2

(Epratuzumab). The treatment with the anti-CD22 antibodies is found to produced a high response rate in refractory hairy cell lymphoma.

CD33 CD33 is a 67-kDa transmembrane glycoprotein, a classic myeloid antigen and a member of the sialic acid-binding immunoglobulin-like lactic family. CD33 appears very early on myelomonocytic precursors but not on hematopoietic pluripotent stem cells. It is expressed in both myeloid and monocytic cell lineages and it is expressed in about 90% of blasts in acute myeloid leukaemia. Anti-CD33 antibodies like the humanized IgG4κ antibody Gemtuzumab ozogamicin (Mylotrag®) are developed for the therapy of acute myeloid leukaemia.

CD52 CAMPATH-1 antigen, clustered as CD52 (also called as epididymal secretory protein E5) is a further target for immunotherapy. This antigen is glycoprotein expressed by B- and T-lymphocytes and monocytes in addition to germ cells. The humanized IgG4κ CD52-specific antibody Alemtuzumab (CAMPATH®) is 2001 approved and used for the treatment of chronic lymphocytic leukemia (B-CLL) and for long-term depletion of T-lymphocytes from donor marrow to prevent graft versus host disease. CD52 can be detected by immunohistochemistry or flow cytometry using specific antibodies such as the monoclonal antibody clone YTH34.5 and HI86.

CD38 CD38 is a transmembrane molecule acting as a receptor in adhesion and signaling pathways. It belongs to a family of proteins involved in the production of calcium mobilizing compounds. CD38 is highly expressed on the surface of plasma cells; it is also expressed on early and activated B- and T-lymphocytes and on the majority of CD34+ hematopoietic cells. A specific antibody (HuMax-CD38) shows to be effective against plasma cell neoplasia including multiple myeloma.

Antibodies to CD5, CD19, CD22, CD30, CD33, CD56 CEA (CD66e) and other targets are now under development for the therapy of related malignancies.

2.2.5
Detection of C-Kit Oncoprotein (CD117)

CD117 or c-kit (also known as stem cell factor receptor, mast cell growth factor or steel factor) is a transmembrane tyrosine kinase receptor type III for stem cell factor (SCF), homologue to the PDGF, VEGFR and CSF-R1 receptors. CD117 is composed of an extracellular ligand-binding domain, a transmembrane segment and an intracellular tyrosine kinase domain. CD117 is encoded by the c-kit proto-oncogene, located on chromosome 4q11–21.

CD117 is involved in the development, differentiation and proliferation of hematopoietic progenitors, intraepithelial lymphocytes, mast cells, germ cells, Cajal cells of intestinal nerve plexus and melanocytes. It is normally expressed in various non-neoplastic cells such as Cajal cells of intestinal nerve plexus, mast cells, melanocytes, germ cells, breast epithelium, salivary glands, thyroid follicular cells, renal tubular epithelium and sweet glands. Significant overexpression of CD117 is found in different tumors, mainly tumors derived from the above-mentioned cells such as gastrointestinal stromal tumor (GIST) derived from Cajal cells, acute and chronic myeloid leukemia, seminoma (dysgerminoma) and malignant melanoma. A Variable degree of CD117 overexpression is also reported in many other tumors including small cell carcinoma of the lung, mastocytosis, angiosarcoma, adenoid cystic carcinoma, thyroid carcinoma, chromophobe renal cell carcinoma and Wilms tumor, Merkel cell carcinoma, Ewing's sarcoma, synovial sarcoma, myelofibrosis, gliomas and osteosarcoma.

The overexpression of CD117 in related tumors seems to be a result of "activating gain of function", mutations taking place in different sites of the c-kit gene. Mutations in the region encoding the extracellular domain (exon 2 and 8) are mainly observed in association with CML and myelofibrosis, mutations in the region encoding the juxtamembrane domain (exon 11) are frequently found in association with GIST, whereas mutations in the region encoding the intracellular domain (exon 17) are mostly observed in association with germ cell and mast cell tumors. Interesting observations found that GISTs associated with mutations within exon 11 of the c-kit gene are frequently malignant with increased mitotic activity. The non-random association of different c-kit gene mutations with various tumors indicates that the activation of tyrosine kinase receptor pathway seems to play an active role in the pathogenesis of these tumors.

Different specific monoclonal and polyclonal antibodies to CD117 are used to detect the expression of this antigen in order to predict the sensitivity of related tumors to tyrosine kinase inhibitors. The staining pattern of CD117 is usually cytoplasmic (diffuse or granular) with membrane accentuation. The specific tyrosine kinase inhibitors are able to block the oncogenic pathway activated by tyrosine kinase (signal transduction inhibitors or STI). These therapeutic inhibitors are found to be effective therapeutic agents for BCR-ABL-positive CML and ALL, gastrointestinal stromal tumor and occasionally in lung cancer and adenoid cystic carcinoma. Imatinib mesylate (2-phenylaminopyrimidine, also known as STI 571 or Glivac®) is an effective and selective BCR-ABL-tyrosine kinase signal transduction inhibitor, which is also able to inhibit other homologous tyrosine kinase receptors such as c-kit and PDGFR is now used now for the treatment of related tumors.

2.3 Immunophenotypic Profiles of Tumors

This section provides an overview of the immunophenotypic profiles of different tumor groups, which can be used as a general guide for tumor diagnosis in surgical pathology. It is essential to mention that antibodies should be used as a panel of antibodies taking into consideration the morphologic appearance of examined tumor, the differential diagnoses and the specificity of the used antibodies. It is necessary for the pathologist to have bench-level experience in addition to theoretical and technical knowledge in immunohistochemistry, since this is essential in troubleshooting problems as they occur. It is important to remember that the interpretation of immunohistochemical results is not the description of positive or negative stains but the characteristics of each antibody and the expression pattern of targeted antigens must be considered besides the results of internal positive and negative controls, which may be found in the sections of examined tissue.

Key to symbols used throughout Section 2.3:

+	Usually positive, or positive in more than 90% of cases
+/−	Positive in more than 50% of cases
−/+	Positive in less than 50% of cases
−	Usually negative, or positive in less than 10% of cases

2.3.1 Soft Tissue Tumors

2.3.1.1 Fibrous and Myofibroblastic Tumors

–Desmoid fibromatosis (abdominal and extra-abdominal fibromatosis):	Vimentin +, Actin +/−, Desmin −/+, S100 −/+, CD34, CD117 −, Pan-Cytokeratin −, EMA −
–Nodular fasciitis:	Vimentin +, Actin +/−, CD68 +/−, Desmin −, S100 −, Pan-Cytokeratin −
–Proliferative fasciitis:	Vimentin +, Myoglobulin +, Actin +/−
–Angiomyxoid fibroma:	Vimentin +, sm Actin +
–Giant cell angiofibroma:	Vimentin +, CD34 +, CD31 −, S100 −
–Cellular angiofibroma:	Vimentin +, CD34 −/+, Actin −/+, Desmin −/+
–Dermatomyofibroma:	Vimentin +, Actin +, Calponin −/+, h-Caldesmon −, Desmin −, CD34 −, S100 −
–Myofibroblastoma:	Vimentin +, Actin +, Desmin +
–Angiomyofibroblastoma:	Vimentin +, Desmin +, Actin −/+, CD34 +/−, ER +/−, Pan-Cytokeratin −, S100 −
–Solitary myofibroma (myofibromatosis):	Vimentin +, Actin +, Desmin +, Pan-Cytokeratin −, S100 −
–Intranodal myofibroblastoma:	Vimentin +, Actin +, S100 −
–Infantile myofibromatosis:	Vimentin +, Actin +, Desmin −/+

–Inflammatory myofibroblastic tumor (inflammatory pseudotumor):	Vimentin +, ALK (p80) +/–, Cyclin D1 +/–, Actin +/–, Desmin –/+, CD68 –/+, bcl-2 –/+, Pan-Cytokeratin –/+, EMA –, CD34 –, CD117 –
–Congenital and infantile fibrosarcoma:	Vimentin +, Actin +/–, Desmin –/+, S100 –/+, CD34 –/+, Myoglobin –
–Solitary (localized) fibrous tumor (primarily of the pleura and peritoneum):	Vimentin +, CD34 +, F XIIIa +, bcl-2 +/–, CD99 +/–, CD10 +/–, Actin –/+, TLE1 –/+, β-Catenin (nuclear stain) –/+, Desmin –, S100 –, Pan-Cytokeratin –, EMA–, CD56 –, CD68 –
–Low-grade fibromyxoid sarcoma:	Vimentin +, Actin –/+, Desmin –/+, bcl-2 +/–, CD34 –/+, Pan-Cytokeratin –/+, S100 –, EMA –
–Sclerosing epithelioid fibrosarcoma:	Vimentin +, EMA +/–, Pan-Cytokeratin –/+, S100 –/+
–Acral myxoinflammatory fibroblastic sarcoma (inflammatory myxohyaline tumor):	Vimentin +, CD34 –/+, CD68 –/+
–Superficial acral fibromyxoma:	Vimentin +, CD34 +, CD117 +/–, EMA +/–
–Infantile and congenital fibrosarcoma:	Vimentin +, CD34 +/–
–Fibrosarcoma (adult):	Vimentin +, No other positive markers

2.3.1.2 Fibrohistiocytic Tumors

–Fibrous histiocytoma (dermatofibroma):	Vimentin +, FXIIIa +, α-1 antitrypsin +, Actin +/–, Desmin –/+, CD34 –/+, S100 –
–Localized giant cell tumor of tendon sheath:	Vimentin +, CD68 + (in multinucleated cells)
–Dermatofibrosarcoma protuberans:	Vimentin +, CD34 +, PDGF +, p53 +, bcl-2 +/–, CD63 +/–, Calponin –/+, Actin –, h– Caldesmon –, Desmin –, CD31 –, CD56 –, F VIII –, Pan-Cytokeratin –, EMA –
–Giant cell fibroblastoma:	Vimentin +, CD34 +, PDGF +, Actin –/+, Desmin –, FVIII –, CD31 –, S100 –, Pan-Cytokeratin –
–Atypical fibroxanthoma:	Vimentin +, CD68 +/–, Actin +/–, TLE1 –/+, Pan-Cytokeratin –, Desmin –

2.3.1.3 Lipomatous Tumors

–Hibernoma:	Vimentin +, Estrogen receptors +, aP2 (P422) +
–Lipoma:	Vimentin +, S100 +, aP2 –

–Spindle cell lipoma: Vimentin +, CD34 +, bcl-2 +, S100 +, aP2 –, MDM2 –

–Chondroid lipoma: Vimentin +, S100 +, CD68 –/+, aP2 –, MDM2 –

–Spindle cell liposarcoma: Vimentin +, CD34 +, aP2 +, CDK4 +, MDM2 +, Ki-67 (clone K-2) +, S100 –

–Myxoid liposarcoma: Vimentin +, aP2 +, CDK4 +, MDM2 +, Ki-67 (clone K-2) +, S100 –/+ (~40%)

–Pleomorphic liposarcoma: Vimentin +, S100 +, aP2 +, Ki-67 (clone K-2) +, MDM2 +/–

–Dedifferentiated liposarcoma: Vimentin +, aP2 +, CDK4 +, MDM2 +, Ki-67 (clone K-2) +, S100 +/–

2.3.1.4 Smooth Muscle Tumors

–Leiomyoma: Vimentin +, sm Actin +, Desmin +/–, Calponin +/–, h – Caldesmon –/+, bcl-2 –/+

–Leiomyosarcoma: Vimentin +, sm Actin +, Desmin +/–, CD146 +/–, Calponin +/–, D2-40 +/–, CK8 –/+, CK18 –/+, h– Caldesmon –/+, bcl-2 –/+

2.3.1.5 Skeletal Muscle Tumors

–Fetal rhabdomyoma: Vimentin +, sr Actin +, Desmin +, Myosin +, Myoglobin +/–, MyoD1 +/–, GFAP –/+

–Adult rhabdomyoma: Vimentin +, Desmin +, Actin +, Myoglobin +, Myosin +/–, Myotilin +/–, S100 –

–Rhabdomyosarcoma (embryonal, alveolar and pleomorphic types): Vimentin +, Desmin +, sr-Actin +, Myoglobin +, Myogenin +, Myosin +/–, Myotilin +/–, MyoD1 +, Myf-3 +, Myf-4 +, Myf-5 +, Tropomyosin +, bcl-2 +/–, PLAP –/+

2.3.1.6 Endothelial Tumors

–Epithelioid hemangioendothelioma: Vimentin +, CD31 +, CD34 +/–, F VIII +, Fli-1 +/–, Actin –/+, Pan-Cytokeratin –/+, CK8 –/+, ER –/+, Melan A –/+, HMB-45 –/+, EMA –, S100 –

–Angiosarcoma: Vimentin +, CD31 +, CD34 +, CD105 (endoglin) +, Factor VIII +, CD141 (thrombomodulin) +, Fli-1 +, Laminin +/–, CK1 +/–, Pan-Cytokeratin –/+, MDM2 –/+, CD117 –/+, D2-40 –, CD56 –, CDK4 –

–Epithelioid angiosarcoma: Vimentin +, CD31 +, CD34 +, F VIII +, Fli-1 +, Pan-Cytokeratin +/–, EMA –, S100 –

–Kaposiform hemangiosarcoma:	Vimentin +, CD31 +, CD34 +, CD105 +, F VIII +, HHV-8 +, D2-40 +, Fli-1 +, bcl-2 +/–, MDM2 –/+, CDK4 –
–Hemangioblastoma:	CD31 +, CD34 +, CD10 +, CD 133 +, Inhibin A +/–, CD117 –/+
–Malignant endovascular papillary angioendothelioma (Dabska tumor):	Vimentin +, CD31 +, D2-40 +, F VIII +, CD34 +/–, Pan-Cytokeratin –, EMA –, S100 –

2.3.1.7 **Perivascular Tumors**

–Solid glomus tumor/ glomangiosarcoma:	Vimentin +, sm-Actin +, Myosin +, Calponin +, Laminin +, Collagen IV +, h– Caldesmon +/–, CD34 +/–, CD56 –, Desmin –, F VIII –, S100 –, EMA –, Pan-Cytokeratin –
–Myopericytoma:	Vimentin +, Actin +, h– Caldesmon +/–, Desmin –/+, S100 –, Pan-Cytokeratin –, EMA –
–Sinonasal hemangiopericytoma:	Vimentin +, Actin +/–, F VIII –/+, Desmin –, CD31 –, CD34 –, F VIII –, EMA –, Pan-Cytokeratin –
–Hemangiopericytoma:	Vimentin +, CD34 +/–, CD57 –/+, bcl-2 +/–, TLE1 –/+, CD31 –, CD99 –, F VIII –, S100 –, sm-Actin –, Desmin –, Pan-Cytokeratin –

2.3.1.8 **Peripheral Nerve Tumors**

–Neurofibroma:	Vimentin +, CD34 +, S100 +/–, bcl-2 +/–, NF +/–, CD56 –, Calretinin – Proliferation index [Ki-67 (MIB-1) index]: 1–13% (main ~5%; for atypical neurofibroma ~8%)
–Neurolemoma (Schwanoma):	Vimentin +, S100 +, CD56 +/–, leu7 (CD57) +/–, Calretinin +, D2-40 +, GFAP +/–, NGFR (gp75) +/–, bcl-2 +/–, CK1 +/–, CD34 –/+ (in Antoni B areas), TLE1 –/+, CK5/6 –, CK7 –, CK 8 –, CK18 –, CK20 –
–Perineurioma:	Vimentin +, Claudin-1 +, EMA +/–, CD34 –/+, S100 –/+
–Paraganglioma:	NSE +, Synaptophysin +, S100 + in benign and –/+ in malignant paraganglioma, CD56 +, PGP9.5 +/–, Chromogranin +/–, VIP +/–, Serotonin +/–, Somatostatin +/–, Bombesin +/–, Vimentin +/–, GFAP –/+, Pan-Cytokeratin –
–Neurothekoma (dermal nerve sheet myxoma):	S100 +, Vimentin +, NGRF +, GFAP +, Col IV +, CD34 +/–, EMA –/+, CD57 (leu7) –/+, Actin –, NSE –, Calponin –/+, Neurofilaments –, Pan-Cytokeratin –

–Cellular neurothekoma:	Vimentin +, CD63 (NK1-C3) +, Actin +/–, NSE +/–, Desmin +/–, S100 –
–Granular cell tumor:	Vimentin +, S100 +, NSE +, Laminin +, Nestin +, Calretinin +/–, CD56 –/+, CD68 +/–, GFAP –, Neurofilaments –, EMA –, Pan-Cytokeratin –
–Malignant nerve sheet tumor:	Vimentin +, Myelin basic protein +, PGP9.5 +, CD57 (leu7) +/–, NGFR (gp75) +/–, CD99 +/–, S100 +/– (~50%), bcl-2 +/–, c-MET +/–, CD34 –/+, GFAP –/+, EMA –/+, bcl-2 –/+, CD56 –/+, Pan-Cytokeratin – Proliferation index [Ki-67 (MIB-1) index]: 5–38% (main ~18%)

2.3.1.9 **Primitive Neuroectodermal Tumors and Related Lesions**

–Neuroblastoma/olfactory neuroblastoma (esthesioneuroblastoma):	CD56 +, CD57 +, NSE +, Neurofilaments +, PGP9.5 +, NB84 +, Bombesin +/–, S100 +, Synaptophysin +/–, Chromogranin +/–, Fli-1 +/–, Pan-Cytokeratin –/+, EMA –, WT-1 –, CD99 –
–PNET:	CD99 +, Vimentin +, Fli-1+/–, Synaptophysin +, NSE +/–, Chromogranin +/–, S100 +/–, Bombesin +/–Pan-Cytokeratin –/+, CD56 –, WT-1 –
–Ewing's sarcoma:	CD99 +, Fli-1+/–, Vimentin +/–, NSE –/+, Pan-Cytokeratin –/+, CD117 –/+, Synaptophysin –, CD56 –, WT-1 –
–Merkel cell carcinoma:	CK20 (perinuclear) +, EMA +, NSE +, CD56 +/–, Fli-1 +/–, Chromogranin +/–, Neurofilaments –/+, CK8 +/–, CK18 +/–, Pax-5 +/–, CK7 –/+, S100 –, HMB45 –, CEA –

2.3.1.10 **Mesothelial Tumors**

–Mesothelioma and adenomatoid tumor:	CK5 +, CK7 +, CK19 +, CK18 +, Mesothelin +, Calretinin +, Thrombomodulin (CD141) +/–, Podoplanin +/–, D2-40 (M2A) +/–, EMA +/–, Vimentin +/–, WT-1 +/–, h-Caldesmon +/– (in epithelioid mesothelioma), ESA –/+, N-Cadherin –/+, E-Cadherin –/+, Actin –/+, p63 -, Myoglobin –, Myosin –, Myogenin –, CEA –, CK20 –, TTF-1 –

2.3.1.11 Synovial Tumors

–Tenosynovial giant cell tumor:	Vimentin +, CD45 +, CD68 +, CD31 +/–, CD34 +/–, Desmin –/+, Actin –, h-Caldesmon –, F VIII –, S100 –

2.3.1.12 Extraskeletal Osseous and Cartilaginous Tumors

–Chondroblastoma:	Vimentin +, S100 +, NSE +, EMA +, Pan-Cytokeratin +/–
–Mesenchymal chondrosarcoma:	Vimentin +, S100 +, CD99 +/–, CD57 +/–, Desmin –, Actin –, Chromogranin –, Pan-Cytokeratin –, EMA –, Osteonectin –
–Extraskeletal osteosarcoma:	Vimentin +, Osteonectin +, Osteocalcin +/–, Androgen receptors +/–, CD117 –/+, S100 –

2.3.1.13 Melanocytic Tumors

–Melanoma:	Vimentin +, HMB45 +, S100 +, MAGE1 +, MITF (microphthalmia transcription factor) +, Melan-A +, Tyrosinase +, CD63 (NK1-C3) +, PNL2 +, Nestin +/–, CD117 –/+, Pan-Cytokeratin – Proliferation index [Ki-67 (MIB-1) index]: high in melanoma but very low in nevus cells

2.3.1.14 Miscellaneous Tumors and Tumors of Uncertain Differentiation

–Synovial sarcoma:	In epithelioid components (monophasic epithelioid): TLE1 +, bcl-2 +, SYT +/–, CK7 +/–, CK14 +/–, CK19 +/–, EMA +/–, HER-2 +/–, KL1 –/+, CEA –/+, Vimentin –/+, Calponin –/+, CD34 – In spindle cell components (monophasic fibrous): TLE1 +, Vimentin +, bcl-2 +, SYT +/–, Calponin +, Calretinin +/–, CD99 +/–, CD56+/–, CD57+/–, E-Cadherin –/+, CD34 –/+, S100 –/+, CD117 –/+, Pan-Cytokeratin –/+, EMA –/+, Actin –/+, h-Ch-Caldesmon –, Desmin – Demonstration of specific t(X; 18) translocation is recommended to confirm the diagnosis.
–Clear cell sarcoma:	Vimentin +, HMB45 +, S100 +, MITF +, NSE +/–, Melan-A +/–, Leu-7 +/–, MDM2 –/+, Desmin –, Actin –, Pan-Cytokeratin –, EMA –, CDK4 –
–Epithelioid sarcoma:	Vimentin +, Pan-Cytokeratin +, CK1 +/–, CK8 +/–, CK18 +/–, CK19 +/–, EMA +/–, CD34 +/–, CK7 –/+, S100 –/+, CK5/6 –, CD31 –, F VIII –, CK20 –, Fli-1 –, CEA –

–Desmoplastic small round cell tumor:	Vimentin +, Pan-Cytokeratin +, CK 8/18 +, EMA +, NSE +/−, Desmin (perinuclear stain) +/−, CK19 +/−, WT1 +/−, CD57 +/−, CD15 +/−, CD99 −/+, CD34 −, CD117 −, Actin −, MyoD1 −, Myoglobin −, S100 −, CK5/6 −, CK20 −
–Extraskeletal myxoid chondrosarcoma:	Vimentin +, S100 +/−, Leu 7 +/−, NSE +/−, Synaptophysin −/+, EMA −/+, Chromogranin −, CD68 −, Pan-Cytokeratin −, CEA −, Actin −, Desmin −
–Rhabdoid tumor:	Vimentin +, Pan-Cytokeratin +, CK8 +/−, EMA +/−, CD99 +/−, NSE +/−, Synaptophysin −/+, Actin −/+, Desmin −/+, Myoglobin −, CD34 −, S100 −
–Alveolar soft part sarcoma:	Vimentin +/−, TEF3 +/−, Desmin +/−, sm Actin −/+, S100 −/+, NSE −/+, Pan-Cytokeratin −/+, CD34 −/+, Synaptophysin −, Chromogranin −, Myoglobin −, Myogenin −, MyoD1 −, HMB45 −, EMA −, CD 31 −, CD117 −
–Pleomorphic hyalinizing angiectatic tumor:	Vimentin +, CD34 +/−, CD31 −, Pan-Cytokeratin −, S100 −
–Myxoma (cutaneous and intramuscular):	Vimentin +, Calretinin +/−, CD34 +/−, Actin −/+, Desmin −/+, Pan-Cytokeratin + (only in epithelial components if present), CD68 −/+, S100 −
–Chordoma:	Vimentin +, NSE +, CK8 +, CK19 +, EMA +, S100 +, CK5 +/−, CK8 +/−, CK10 +/−, CK14 +/−, CK18 +/−, E-Cadherin +/−, β-Catenin +/−, Desmin −, CK7 −/+, CK20 −
–Myxoma of the jaw:	Vimentin +, S100 +, Pan-Cytokeratin −, Desmin −
–Aggressive angiomyxoma:	Vimentin +, Actin +/−, PgR −/+, CD34 −/+, ER +/−, PgR +/−, Desmin −/+, S100 −, Pan-Cytokeratin −
–Myoepithelioma (mixed tumor of soft tissue):	Pan-Cytokeratin +/−, CK8/18 +/−, EMA +/−, S100 +/−, Calponin +/−, CK14 +/−, Desmin −/+, Actin −/+, GFAP −/+
–Ossifying fibromyxoid tumor:	Vimentin +, S100 +/−, Desmin +/−, Actin −/+, GFAP −/+, Pan-Cytokeratin −, EMA −

2.3.2
Central Nervous System Tumors

–Astrocytoma and glioblastoma:	Vimentin +, GFAP +, S100 +, NSE +, CD99 +/–, CD56 +/–, HER-2 +/–, Synaptophysin –/+, Pan-Cytokeratin –/+, NF –, Chromogranin – Proliferation index [Ki-67 (MIB-1) index]: Low-grade astrocytoma: <5% Anaplastic astrocytoma: 5–10% Glioblastoma: >10%
–Oligodendroglioma:	GFAP +, S100 +, NSE +, CD56 +/–, Synaptophysin +, Chromogranin –/+, Pan-Cytokeratin –/+, Vimentin –/+, NF –
–Neurocytoma:	NF +, Synaptophysin +, S100 –/+, GFAP –, Pan-Cytokeratin –, Chromogranin –
–Ependymoma:	Podoplanin +, GFAP +, S100 +, Vimentin –/+, Synaptophysin –/+, CD99 –/+, Pan-Cytokeratin –/+, Chromogranin –
–Pineocytoma and Pineoblastoma:	Synaptophysin +, NSE +, NF +, S100 +/–, GFAP +/–
–Choroid plexus papilloma/ carcinoma:	Pan-Cytokeratin +, CK8 +, Transthyretin +, S100 +, Vimentin +/–, CD44 +/–, GFAP –/+, EMA –/+, Synaptophysin +, Chromogranin –
–Medulloblasoma:	Vimentin +, S100 +, CD56 +, Synaptophysin +/–, PGP9.5 +/–, GFAP –/+, NF –/+, bcl-2 –/+
–Retinoblastoma:	CD56 +, Synaptophysin +, NSE +, S100 +, GFAP +, Pan-Cytokeratin –
–Meningioma (intra and extracranial):	Vimentin +, S100 +/–, D2-40 +/–, CD141 +/–, NSE +/–, Nestin +/–, Pan-Cytokeratin +, CK8/18 +/–, EMA +/–, CD99 +/–, Osteonectin –/+, bcl-2 –/+, CD34 –/+, CK7 –/+, GFAP –, Synaptophysin –, Chromogranin –, NF –, CD56 –, CK5/6 –, CK20 –

2.3.3
Respiratory Tract and Lung Tumors

2.3.3.1 Upper Respiratory Tract Tumors

–Sinonasal (undifferentiated) carcinoma:	CK8 +, CK7 –/+, CK19 –/+, EMA –/+, CK4 –, CK5 –, CK6 –, CK10 –, CK13 –, CK14 –, CK20 –
–Nasopharyngeal (undifferentiated) carcinoma:	CK5/6 +, CK8 +, CK13 +, CK19 +, EMA +, bcl-2 +/–, EBV +/–, HLA-DR +/–, CK4 –, CK7 –, CK10 –, CK14 –
–Squamous cell carcinoma:	CK5/6 +, CK8 +, CK13 +, CK19 +, EMA +, bcl-2 –, EBV –, HLA-DR –, CK4 –, CK7 –, CK10 –, CK14 –

2.3.3.2 **Lung Tumors**

–Squamous cell carcinoma:	CK 34E12 +, CK1 +, CK5/6 +, CK14 +, CK10/13 +, CK 17 +, CK18 +, CK19 +, CK8 +, p63 +, Calretinin –/+, TTF-1 –, CK4 –, CK7 –, CK20 –
–Adenocarcinoma:	CK7 +, CK8 +, CK18 +, CK19 +, CEA +, Villin +/–, TTF-1 +/–, Surfactant +/–, CK12+/–, CK14 –/+, Mesothelin –/+, Thrombomodulin –, Calretinin –, CK5/6 –, CK20 –, CK 34E12 –, CDX2 –, p63 –
–Bronchioloalveolar carcinoma:	CK7 +, CK8 +, CK18 +, CK19 +, CEA +, TTF-1 + (negative in mucinous type), Surfactant + (negative in mucinous type), CK12 +/–, CDX-2 +/–, CK14 –/+, Thrombomodulin –, Mesothelin –/+, Calretinin –, CK5/6 –, CK20 –, CK 34E12 –, P63 –
–Large cell carcinoma:	CK7 +, CK8 +, CK14 +, CK18 +, CK19+, EMA +, TTF-1 –/+, Surfactant –/+, CK5/6 –, CK20 –
–Large cell neuroendocrine carcinoma:	CK7 +, CK8 +, CK18 +, CK19 +, CK20 -, CD56 +, TTF-1 –/+, Chromogranin +, Synaptophysin +
–Small cell carcinoma:	KL1 +, CK-MNF +, CK8 +, CK18 +, CK19 +/–, CD56 +, NSE +, Synaptophysin +, S100 +/–, Leu7 +, Chromogranin +, Neurofilaments +/–, Vimentin +/–, TTF-1 +/–, CD117 –/+, CK7 –, CK14 –, CK20 – Proliferation index [Ki-67 (MIB-1) index]: >90%
–Atypical carcinoid tumor:	KL1 +, CK-MNF +, CK8 +, CK18 +, CD56 +, Chromogranin +, NSE +, Synaptophysin +, S100 +/–, Leu7 +, PGP 9.5 +, TTF-1 +/–
–Typical carcinoid tumor:	KL1 +, CK-MNF +, CK8, CK18+, CD56 +, Chromogranin +, NSE +, Synaptophysin +, Leu7 +, PGP 9.5 +, S100 +/–, TTF-1 +/–, E-Cadherin +/– Proliferation index [Ki-67 (MIB-1) index]: >1%
–Clear cell tumor (sugar tumor):	HMB45 +, Cathepsin B +, CD63 +, Vimentin +/–, S100 –/+, CD57 (leu7) –/+, Synaptophysin –/+, NSE –/+, CD34 –/+, EMA –, CD56 –
–Pulmonary sclerosing hemangioma (inverting alveolar pneumocytoma):	Polygonal clear cells in solid portions (stromal cells): EMA +, TTF-1 +, Vimentin +/–, Estrogen and progesterone receptors +/–, CK7 –/+, CK5/6 –, CK20 –, CD31 –, CD34 –, Surfactant –, Calretinin –, MIB-1 + (membrane and cytoplasmic staining pattern) Surface lining cells: CK 7 +, EMA +, TTF-1 +, Surfactant +, CD15 +/–, CK5/6 –, CK20 –, Vimentin –, Calretinin –, Estrogen and progesterone receptors –

–Epithelioid hemangioendothelioma:	CD31 +, CD34 +, Vimentin +, Pan-Cytokeratin –, Calretinin –
–Inflammatory pseudotumor (pulmonary inflammatory myofibroblastic tumor):	Vimentin +, Cyclin D1 +/–, ALK (p80) +/–, Actin (in spindle cells) +, Desmin –/+, bcl-2 –/+, Pan-Cytokeratin –, EMA –, CD56 –
–Pulmonary histiocytosis X:	S100 +, CD1a +, CD11c +/–, CD68 +/–, CD31 +/–, HLA-DR +

2.3.4 Pituitary Gland Tumors

·Pituitary adenoma:

–General markers:	Pan-Cytokeratin +/–, EMA +/–, NSE +, Synaptophysin +, Chromogranin +/–, Vimentin –, CEA –
–Somatotrope adenoma:	STH +, Prolactin +/–, TSH +/–, FSH +/–, LH +/–, α-subunit +/–
–Lactotrope adenoma:	Prolactin +, α-subunit +/–, Galectin-3 +/–
–Corticotrope adenoma:	ASTH +, α-subunit +/–
–Gonadotrope adenoma:	LH +, FSH +, α-subunit +/–
–Thyrotrope adenoma:	TSH +, STH +/–, Prolactin +/–, α-subunit +/–
–Oncocytoma:	Nonfunctional (No hormone secretion)
–Null cell adenoma:	Nonfunctional (No hormone secretion)
–Plurihormonal adenoma:	STH +/–, TSH +/–, LH +/–, FSH +/–, Prolactin +/–
–Craniopharyngioma:	CK5/6 +, CK7 +, CK17 +, CK19 +, β-Catenin +, CK18 –/+, CK10 –, CK20 –, EMA –, GFAP –, Vimentin –

2.3.5 Thyroid Tumors

–Follicular thyroid adenoma:	Thyroglobulin +, Thyroid peroxidase +, TTF-1 +, CK18 +, CK7 +/–, CK19 –/+, CK5/6 –, CK20 –, Calcitonin –, CD44V6 –, Galectin-3 –
–Follicular thyroid carcinoma:	Thyroglobulin +, Thyroid peroxidase +, TTF-1 +, CK7 +, CK8 +, CK18 +, CD44V6 +, S100 +, Galectin-3 +/–, Vimentin +/–, CK19 +/–, HBME1 +/–, Calcitonin –, CK5/6 –, CK20 –, HER-2 –
–Papillary thyroid carcinoma:	Thyroglobulin +, Thyroid peroxidase +, CK1 +, CK7 +, CK8 +, CK18 +, CK19 + (negative in benign thyroid tissue), CK34BE12 +, TTF-1 +, Galectin-3 +, CD44V6 +, HBME-1 +, CK5/6 +/–, CK14 +/–, EMA +/–, Vimentin +/–, CD15 –/+, CD34 –/+, CK20 –, Calcitonin –

–Poorly differentiated thyroid carcinoma: Thyroglobulin +, Thyroid peroxidase +, CK18 +, TTF-1 +, CK18 +, Galectin-3 +, CD44V6 +, Vimentin +/–, bcl-2 +/–, CK5/6 –, CK19 –, CK20 –, Calcitonin –

–Medullary thyroid carcinoma: Calcitonin +, Chromogranin +, Synaptophysin +, TTF-1 +, CD56 +, Leu7 +, S100 +, NSE +, CEA +, Vimentin (in spindle cell components) +, CK7 +, CK8 +, CK18 +, CK19 +, Synapsin I +, HER-2 +, bcl-2 +/–, Galectin-3 –/+, CK5/6 –, Thyroglobulin –, CK20 –

–Anaplastic thyroid carcinoma: Pan-Cytokeratin +, CK8 +, CK18 +, CK19 +/–, EMA –/+, CEA +/–, TTF-1 –/+, Galectin-3 –/+, Vimentin +/–, bcl-2 –/+, CK20 +, Thyroglobulin –, Calcitonin –

2.3.6 Parathyroid Tumors

–Parathyroid adenoma: Parathyroid hormone (PTH) +, Chromogranin +, Synaptophysin +, Neurofilaments +, CK8 +, CK18 +, CK19 +/–, Vimentin +/–, Cyclin D1 –/+, RCC (gp200) +/–, Calcitonin –/+, CK7 –/+, TTF-1 –, Thyroglobulin –, CD56 –, CK5/6 –, CK20 –/+
Proliferation index [Ki-67 (MIB-1) index]: <5%

–Parathyroid carcinoma: Parathyroid hormone (PTH) +, Chromogranin +, Synaptophysin +, Neurofilaments +, Vimentin +/–, Pan-Cytokeratin +, CK8 +, CK7 –/+, CK19 +/–, Cyclin D1 +/–, Calcitonin –/+, Galectin-3 –/+, CK5/6 –, CK20 –, TTF-1 –, Thyroglobulin –, CD56 –
Proliferation index [Ki-67 (MIB-1) index]: >6%

2.3.7 Thymic Epithelial Tumors

–Thymoma (type A and B): Neoplastic epithelial component: CK5/6 +, CK7 +, CK8 +, CK18 +, CK19 +, CD57 (leu7) –/+, EMA +/–, CD20 +/– (in type A only), CD5 –, bcl-2 –, CD117 –, HER-2 –
Tumor-associated lymphocytes: predominantly immature T-lymphocytes: TdT +, CD1a +, CD3 +, CD99 +

–Thymic carcinoma: Neoplastic epithelial component: CD5 +, bcl-2 +, CD117 +, CK6 +/–, CK8 +/–, CK18 +/–, CK19 +/–, CK7 –/+, EMA+, HER-2 –/+
Tumor-associated lymphocytes: predominantly mature T- and B-lymphocytes (TdT) –

2.3.8 Salivary Gland Tumors and Odontogenic Tumors

2.3.8.1 Salivary Gland Tumors

–Pleomorphic adenoma: In epithelial components: CK8 +, CK14 +, CK18 +, CK19 +, CK7+/−, EMA +, CEA +/−, CK14 −, CK20 −, Vimentin −, GFAP −
In myoepithelial components: Vimentin +, S100 +, Calponin +, Actin +, GFAP +, CK5/6 +, CK14 +, EMA −, CEA −
Proliferation index [Ki-67 (MIB-1) index]: >2%

–Oncocytoma: CK7+, CK8 +, CK18 +, EMA +, CEA +, Actin −

–Myoepithelial adenoma/carcinoma: Vimentin +, GFAP +/−, S100 +, Pan-Cytokeratin +, CK5/6 +, CK14 +, Calponin +, Actin +/−, CK19 +/−, EMA +/−, CEA −

–Basal cell adenoma/carcinoma: In epithelial luminal components: Pan-Cytokeratin +, CK8 +, CK18 +, EMA +, CEA +/−, CD43 −, Vimentin −
In myoepithelial components: S100 +, Actin +, Calponin +, Vimentin +, GFAP +, CK5/6 +, CK14 +

–Mucoepidermoid carcinoma: In mucus-secreting cells: CK8 +, CK17 +, CK18 +, CK19 +, EMA+, CK7 +/−
In epidermoid cells: CK5/6 +, CK8 +, CK10/13 +, CK14 +, CK7 −

–Acinic cell carcinoma: CK7 +, CK8 +, CK18 +, EMA +, CK19 +/−, CEA +, NSE +/−, bcl-2 −/+, Amylase −/+, CK14 −

–Adenoid cystic carcinoma: CK5/6 +, CK7 +, CK8 +, CK14 +, CK18 +, CK19 +, EMA +, bcl-2 +, CD43 +, CEA +/−, S100 +/−, CD117 (c-kit) +/−, p63 +/−, CK20 −, Vimentin −
Proliferation index [Ki-67 (MIB-1) index]: >20%

–Polymorphous low-grade adenocarcinoma (terminal duct carcinoma): CK7 +, CK8 +, CK18 +, CK19 +, EMA +, S100 +, Vimentin +, bcl-2 +, CK20 −, CD43 −, GFAP −, Actin −
Proliferation index [Ki-67 (MIB-1) index]: 1.5–7.0%

–Salivary duct carcinoma: CK7 +, CK8 +, CK14 +, CK18 +, CK19 +, EMA +, Androgen receptors +/−, CEA +/−, GCDFP15 +/−, HER-2 +/−, S100 −/+, PSA −/+, CK20 −
Proliferation index [Ki-67 (MIB-1) index]: mean 21%

–Epithelial–myoepithelial carcinoma: In epithelial components: CK8 +, CK18 +, EMA +, CEA −/+
In myoepithelial components: S100 +, Vimentin +, Actin +, CK5/6 +, CK14 +

–Hyalinizing clear cell carcinoma: CK7 +, CK8 +, CK18 +, EMA +, CEA +/−, S100 −, Actin −, Calponin −

2.3.8.2 Odontogenic Tumors

–Ameloblastoma:	Pan-Cytokeratin +, CK 5 +, CK 14 +, Vimentin +, Calretinin +/–
–Clear cell odontogenic carcinoma:	Pan-Cytokeratin +, CK19 +, EMA +, Vimentin –, Desmin –, Actin –, S100 –

2.3.9 Gastrointestinal Tract Tumors

2.3.9.1 Esophageal and Gastric Tumors

–Squamous cell carcinoma of esophagus:	CK5/6 +, CK8 +, CK14 +, CK19 +, β-Catenin +/–, p63 +/–, Cyclin D1 +/–, CK7 –, CK20 –
–Adenocarcinoma of esophagus:	CK7 +, CK8 +, CK18 +, CK19 +, E-Cadherin +/–, CDX-2 +/–, Cyclin D1 +/–, Villin +/–, CK20 –/+, CK5/6 –
–Adenocarcinoma of stomach:	CK8 +, CK18 +, CK19 +, Villin +, EMA +, CK20 +/–, CEA +/–, Glicentin +/–, CDX-2 +/–, CK7 –/+, CK5/6 –, CK14 –, CK17 –, CA125 –

2.3.9.2 Intestinal Tumors

–Small bowel adenocarcinoma:	CK8 +, CK18 +, CK19 +, CDX-2 +, Villin +, CK20 +/–, CK7 –/+
–Ampulla of Vater adenocarcinoma:	CK8 +, CK18 +, CK19 +, CK7 +, CK20 +/–, CDX-2 –/+
–Colorectal adenocarcinoma:	CK8 +, CK18 +, CK19 +, CK20 +, CEA +, Villin +, MUC2 +, CDX-2 +, CK7 +/–, β-Catenin (nuclear stain) +/–, CK5/6 –, CK14 –, Thrombomodulin (CD141) –
–Colorectal mucinous adenocarcinoma:	CK20 +, CDX-2 +, Villin +, β-Catenin + (nuclear stain), CK7 –/+
–Cloacogenic carcinoma:	CK1 +, CK5 +, CK8 +, CK15 +, CK17 +, CK18 +, CK19 +, CK6 +/–, CK10 +/–, CK7 –/+, CK20 –
–Recto-anal squamous cell carcinoma:	CK5 +, CK6 +, CK10 +, CK17 +, CK18 +, CK19 +, CK7 –, CK20–
–Anal Paget's disease:	CK7 +, CK8 +, CK18 +, EMA +, MUC-2 +, CEA +/–, CK20 –/+, GCDFP-15 –/+, MUC-1 –

2.3.9.3 Gastrointestinal Neuroendocrine Tumors

–Carcinoid of small bowel and colon:	KL1 +, CK-MNF +, CK19 +, CD57 +, Chromogranin +, NSE +, Synaptophysin +, leu7 +, PGP 9.5 +, Serotonin +, Villin +, CD56 +/–, Somatostatin +/–, CDX-2 +/–, CEA +, PP (pancreatic polypeptide) –/+, CK7 –/+ (in colonic carcinoids), CK20 –/+, CK7 – (in small bowel carcinoids), E-Cadherin –, β-Catenin –

–Appendiceal carcinoid (nongoblet):	KL1 +, CK-MNF +, CK19 +, Chromogranin +, NSE +, Synaptophysin +, leu7 +, PGP 9.5 +, Serotonin +, E-Cadherin +, β-Catenin +, CD56 +/–, Somatostatin +/–, PP –/+, CEA –, CK20 –
–Goblet cell carcinoid of the appendix (adenocarcinoid):	KL1 +, CK-MNF +, CK20 +, Chromogranin +, NSE +, Synaptophysin +, Leu7 +, PGP 9.5 +, CEA +, E-Cadherin +, β-Catenin +, CD56 +/–, Somatostatin –/+, Serotonin –, CK7 –
–Small-cell and high-grade neuroendocrine carcinoma of GIT:	KL1 +, CK-MNF +, CK8 +, CK18 +, CK19 +/–, NSE +, Synaptophysin +, S100 +/–, leu7 +, CD56 +/–, Vimentin +/–, CDX-2 +/–, Chromogranin –/+, TTF-1 –/+, CK7 –/+, CK20 –

2.3.9.4 **Gastrointestinal Stromal Tumors**

–Gastrointestinal stromal tumor (GIST):	Vimentin +, CD117 (c-Kit) +, CD34 +/–, CD99 +/–, Nestin +/–, Tau +/–, bcl-2 +/–, sm-Actin –/+, D2-40 +/–, h-Caldesmon –/+, S100 –/+, CK8 –/+, CK18 –/+, PDGFR-α –/+ (positive in CD117 negative GISTs), Synaptophysin –, Chromogranin –, Desmin –, PGP9.5 –, Calponin –, β-Catenin – GISTs with epithelioid morphology are frequently CD117 negative
–Gastrointestinal autonomic nerve tumor (plexosarcoma; GANT as subtype of GIST):	Vimentin +, CD117 +, CD34 +/–, NSE +/–, Synaptophysin +/–, β-Catenin +/–, PGP9.5 +/–, Chromogranin –/+, S100 –/+, Neurofilaments –/+, h-Caldesmon –/+, Desmin –, Actin –, Calponin –
–Inflammatory fibroid polyp of the gastrointestinal tract:	Stromal cells: CD34 +, Fascin +, Cyclin D1 +, Calponin +/–, CD35 +/–, sm Actin –/+, CD117 –, S100 –, Desmin –, h-Caldesmon –, bcl-2 –
–Mesenteric fibromatosis:	Vimentin +, β-Catenin + (nuclear and cytoplasmic stain), sm Actin +/–, Desmin –/+, CD117 (c-Kit) –/+, Calponin –, Pan-Cytokeratin –, S100 –

2.3.10
Hepatobiliary Tumors

–Hepatocellular carcinoma:	CK8 +, CK18 +, AFP +/– (–/+ in fibrolamellar hepatocellular carcinoma), CD10 +/–, EMA +/–, Hepatocyte-specific antigen (Hep Par1) +/–, Glypican-3 + (negative in all non-malignant liver tissue), MAGE-1 +/–, Osteonectin +/–, CD66a +/–, CD56 +/–, CK7 –/+ (+ in fibrolamellar hepatocellular carcinoma), CD34 + (in sinusoidal lining cells), CK19 –/+, Vimentin –/+, ER –/+, PR –/+, CK5/6 –, CK14 –, CK20 –, EMA –, CEA –, Inhibin –, Melan A –

–Hepatoblastoma:	Hep Par1+, CK8 +, Glypican-3 +, CK18 +/–, AFP +/–, CEA +/–, CD34 +/–, EMA +/–, Chromogranin +/–,Vimentin +/–, S100 –/+
–Cholangiocellular carcinoma:	CK7 +, CK8 +, CK18 +, CK19 +, CEA +, EMA +, CK17 +/–, CDX-2 +/–, CD5 +/–, CK20 + (in tumors from extrahepatic bile ducts), CK20 – (in tumors from inrahepatic bile ducts), Vimentin –/+, AFP –, CK5/6 –
–Adenocarcinoma of gallbladder:	CK7 +, CK18 +, CK19 +, EMA +, CEA +, CK5/6 –, CK20 –/+

2.3.11 Pancreas Tumors

2.3.11.1 Tumors of Exocrine Pancreas

–Serous cystadenoma:	CK7 +, CK8 +, CK18 +, CK19 +, EMA +, CK20+/–, CEA –
–Intraductal papillary mucinous carcinoma:	CK7 +, CK8 +, CK19 +, CK20 +, CEA +
–Ductal adenocarcinoma:	CK7 +, CK8 +, CK18 +, CK 19 +, MUC1 +, CEA +, EMA +, CK17 +/–, CK20 +/–, Mesothelin +/–, CDX-2 +/–, MUC2 –, Lipase –, Calretinin –, Thrombomodulin (CD141) –
–Acinic cell carcinoma:	CK8 +, CK18 +, Lipase +, Trypsin +, CK7 –/+, CK19 –/+, CEA –, MUC1 –, MUC2 –

2.3.11.2 Tumors of Endocrine Pancreas

–General markers for endocrine/neuroendocrine pancreas tumors:	CK8 +, CK18 +, CK19 +, Chromogranin +, NSE +, PGP9.5 +, Synaptophysin +, Leu7 +, S100 +/–, CK5/6 –, CK7 –, CK20 – Proliferation index [Ki-67 (MIB-1) index] in well differentiated neuroendocrine tumor: ≤2% Proliferation index [Ki-67 (MIB-1) index] in well differentiated neuroendocrine carcinoma: >2% Proliferation index [Ki-67 (MIB-1) index] in poorly differentiated neuroendocrine carcinoma: >30%
–Beta cell tumors (insulinoma):	Insulin +, Proinsulin +
–G cell tumors (gastrinoma):	Gastrin +, hPP +/–
–Alpha cell tumors (glucagonoma):	Glucagon +, Glicentin +/–
–Delta cell tumors (somatostatinoma):	Somatostatin +, Calcitonin –/+, ACTH –/+
–D1 cell tumors (VIPoma):	VIP (vasoactive intestinal polypeptide) +
–PP cell tumor:	hPP (human pancreatic polypeptide) +

2.3.12 Adrenal Gland Tumors

–Adrenocortical adenoma/carcinoma: Melan-A +, CD56 +, Inhibin +/–, Vimentin +/–, D11 +/–, Synaptophysin +/–, NSE +/–, bcl-2 +/–, Calretinin +/–, CD56 +/–, CEA –/+, bcl-2 –/+, Pan-Cytokeratin –/+, CK7 –, CK19 –, CK20 –, EMA –, Chromogranin –, RCC (gp200) –, CD10 –
Proliferation index [Ki-67 (MIB-1) index] for adrenocortical adenoma: <2.5%
Proliferation index [Ki-67 (MIB-1) index] for adrenocortical carcinoma: >4%

–Pheochromocytoma: Chromogranin +, Synaptophysin +, CD56 +, NSE +, S100 +/– (+ in sustentacular cells with nuclear stain), bcl-2 +/–, GFAP +/–, Vimentin –/+, Pan-cytokeratin –/+, CK5/6 –, CK7 –, CK19 –, CK20 –, EMA –, D11 –, Melan-A –
Proliferation index [Ki-67 (MIB-1) index] in benign pheochromocytoma: <2%
Proliferation index [Ki-67 (MIB-1) index] in malignant pheochromocytoma: >2%

–Neuroblastoma: Vimentin +, CD56 +, NSE +, Neurofilaments +, PGP9.5 +, NB84 +, Synaptophysin +/–, Chromogranin +/–, S100 +/–, CD117 +/–, Pan-Cytokeratin –/+, CK5/6 –, CK7 –, CK20 –

2.3.13 Renal, Urinary Tract and Urinary Bladder Tumors

2.3.13.1 Renal Tumors

–Oncocytoma: CK8 +, CK14 +, CK18 +, EMA +/–, CD117 +/–, CK7 –, Vimentin –, RCC38 –, PAX2 –

–Metanephric adenoma: CK8 +, CK18 +, Vimentin +, CK7 –, CK19 –, EMA –

–Clear cell carcinoma: CK8 +, CK18 +, MOC31 +, α B-Crystallin +, PAX2 +, Vimentin +/–, RCC (gp200) +/–, CD10 +/–, CK19 +/–, EMA +/–, CD9 –/+, P504S (AMACR) –/+, CK7 –, CK13 –, CK20 –, CD117 –, Inhibin –

–Chromophobe carcinoma: CK8 +, CK18 +, EMA +, CK19 +/–, CK7 +/–, CD9 +/–, CD117 +/–, PAX2 –/+, RCC (gp200) –/+, Vimentin –/+, CD10 –, CK13 –, CK20 –

–Papillary (chromophile) carcinoma: CK8 +, CK18 +, EMA +/–, CK19 +/–, PAX2 +/–, RCC (pg200) +/–, CD9 +/–, CD10 +/–, CK7 +/–, CK19 +/–, Vimentin +/–, P504S (AMACR) +/–, CK5/6 –, CK13 –, CK20 –

–Collecting duct carcinoma: CK8+, CK18 +, CK19 +, EMA +/–, Vimentin +/–, UEA-1 lectin +, CD10 –/+, CK7 –/+, CK5/6 –, CK13 –, CK17 –, RCC –

–Spindle cell (sarcomatoid) carcinoma: CK8 +, CK18+, Vimentin +, EMA +/–, RCC –

–Neuroendocrine carcinoma: CK8 +, CK18 +, CD56 +, S100+, Chromogranin +, Synaptophysin +, NSE +, CK19 –, Vimentin –

–Juxtaglomerular cell tumor: Renin +, CD34 +/–, Actin +/–, Calponin +/–

–Nephroblastoma (Wilms tumor): WT1 +, Vimentin +, CD56 +, Myogenin +/–, PAX2 +/–, S100 +/–, Pan-Cytokeratin +/–, NSE +/–, CD57 –, CK19 –

–Clear cell sarcoma of the kidney: Vimentin +, CK7 –, CK8 –, CK18 –, CK19 –, RCC –

–Angiomyolipoma: HMB45 +, Melan-A +, CD117 +/–, Actin +, Calponin +, MIFT –/+, ER –/+, PgR –/+, EMA –, Pan-Cytokeratin –

–Rhabdoid tumor: Vimentin +, Pan-Cytokeratin +, CK8 +/–, EMA +/–, CD99 +/–, NSE +/–, Synaptophysin –/+, Actin –/+, Desmin –/+, Myoglobin –, CD34 –, S100 –

2.3.13.2 Urinary Tract and Urinary Bladder Tumors

–Transitional cell (urothelial) carcinoma: CK5 +, CK7 +, CK8 +, CK13 +, CK17 +, CK18 +, CK19 +, Thrombomodulin (CD141) +, CK20 –/+, CEA +/–, Uroplakin (Ia and II) +/–, Fascin +/– (negative in normal urothelium), Calretinin –/+, PAX2 –, WT1 –, Vimentin –

–Adenocarcinoma of urinary bladder: CK8 +, CK18 +, Thrombomodulin (CD141) +, CK19 +, CK7 +/–, CK20 +/–, CDX-2 –/+, CK5 –, PAP –, PSA –

–Squamous cell carcinoma of urinary bladder: CK5/6 +, CK8 +, CK14 +, CK19 +,CK7 –, CK20 –

2.3.14 Breast Tumors

–Ductal hyperplasia: Lesion is composed of heterogeneous cell populations:

1. Glandular epithelial cells: CK8 +, CK18 +, CK19 +
2. Intermediate glandular cells: CK5/6 +, CK14 +, CK8 +, CK18 +, CK19 +
3. Myoepithelial cells: CK5/6 +, CK14 +, Ma 903 +, p63 +, Actin +

–Atypical ductal hyperplasia: Clonal proliferation of glandular epithelial cells: CK8 +, CK18 +, CK19 +

	No luminal or only residual CK5/6/14 positive intermediate epithelial cells Estrogen receptors (ER) +, Progesterone receptors (PgR) +, Cyclin D1 −/+, bcl-2 +/−, P53 −, HER-2 −
–Ductal carcinoma *in situ* (DCIS):	
–Low-grade:	CK8 +, CK18 +, CK19 +, Cyclin D1 −/+, ER +, PgR +/−, bcl-2 +/−, HER-2 −/+, P53 −/+ No luminal CK5/6/14 and p63 positive intermediate epithelial cells
–High-grade:	CK8 +, CK18 +, CK19 +, Cyclin D1 +/−, ER −/+, PgR −/+, HER-2 +/−, p53 +/−, bcl-2 −/+ Typically no luminal CK5/6/14 positive intermediate epithelial cells
–Invasive ductal carcinoma:	CK7 +, CK8 +, CK18 +, CK19 +, E-Cadherin +, CD44 +, Maspin +/−, Human milk fat globule +/−, HER-2 +/−, EGFR +/−, bcl-2 −/+, CK10/13 −/+, CK1 −, CK14 −, CK20 − ER + in 55–65%, PgR −/+
–Lobular carcinoma *in situ*:	CK7 +, CK8 +, CK18 +, CK19 +, Cyclin D1 −/+, CK20 −, E-Cadherin − No luminal CK5/6/14 positive intermediate epithelial cells ER and PgR + in ~100%
–Invasive lobular carcinoma:	CK7 +, CK8 +, CK18 +, CK19 +, Maspin +/−, GCPF15 +/−, CEA +/−, Cyclin D1 +/−, HER-2 −/+, EGFR −/+, E-Cadherin −, CK1 −, CK5/6 −, CK14 −, CK20 − ER + in 80–90%, PgR +/−
–Tubular carcinoma:	CK7 +, CK18 +, CK19 +, HER-2 −, CK5/6 −, CK20 − No Actin, CK5/6, CK-Ma903, h-Caldesmon or Calponin-positive myoepithelial cells ER + in 85–95%, PgR +
–Cribriform carcinoma:	CK7 +, CK8 +, CK18 +, CK19 +, Human milk fat globule +/−, CK10/13 −/+, CK14 −, CK20 − ER + in 75–85%, PgR +/−
–Mucinous carcinoma:	CK7 +, CK18 +, CK19 +, CEA +, NSE +, EGFR −/+, HER-2 −, CK20 −, CDX-2 − ER + in 60–70%, PgR +
–Papillary carcinoma:	CEA +, CK18 +, CK19 +, CK5/6 −, CK14 − ER + in ~100%, PgR +
–Medullary carcinoma:	CK 8 +, CK 18+, p53 +/−, Vimentin −/+, S100 −/+, CK5/6 −/+, CK7 −, CK14 −, CK19 −, CK20 −, HER-2 −, GCDFP15 − ER + in 0–10% (usually negative in typical medullary carcinoma), PgR −/+

–Apocrine carcinoma: CK8 +, CK18 +, CK 19 +, Androgen receptors +, GCDFP15 +/–, CEA +/–, HER-2 –, S100 –
ER –, PgR –

–Secretory carcinoma: CK8 +, CK18 +, CK 19 +, S100 +, lactalbumin +, HER-2 –/+, GCDF15 –
ER + in <40%, PgR –/+

–Metaplastic carcinoma: Vimentin +, Pan-Cytokeratin +, MNF +, CK7 +/–, EMA –/+, Actin –/+, S100 –/+
ER –, PgR –

–Paget's disease of the nipple: CK7 +, CK8 +, CK18 +, EMA +, MUC-1 +, CD63 (NK1-C3) +, CEA +/–, GCDFP15 +/–, HER-2 +/–, CK5/6 –, CK20 –, MUC-2 –

–Myofibroblastoma of the breast: Vimentin +, Desmin +, CD34 +, Actin –/+, Pan-Cytokeratin –, S100 –, ER –, PgR –/+

–Phyllodes tumor: Stromal cells: Vimentin +, bcl-2 +/–, CD34 –/+, Actin –/+, Desmin –/+, CD117 –/+, S100 –, Pan-Cytokeratin –, EMA –
Proliferation index [Ki-67 (MIB-1) index] in benign type <20%
Proliferation index [Ki-67 (MIB-1) index] in malignant type: >20%
Epithelial cells: CK 5 +, CK6 +, CK14 +, CK8/18 +, CEA +/–

2.3.15 Female Genital Tract Tumors

2.3.15.1 Tumors of Uterine Cervix and Uterine Corpus

A. Tumors of the Uterine Cervix

–Squamous cell carcinoma: CK1 +, CK5 +, CK6 +, CK13 +, CK17 +, CK18 +, CK19 +, P16+, CK14 +/–, CK7 –, CK20 –

–Endocervical adenocarcinoma: CK7 +, CK8 +, CK18 +, CK19 +, CEA +, EMA +, P16 +, CK20 –/+, Vimentin –/+, ER –, PgR –, CK5/6 –, GFAP –

–Mesonephric adenocarcinoma: CK5/6 +, CK7 +, CK8 +, CK18 +, EMA +, CD15 +, CD10 +, bcl-2 +/–, Vimentin –, CK20 –, CEA –

–Endometrioid adenocarcinoma: CK7 +, CK8 +, CK18 +, CK19 +, EMA +, GFAP +/–, ER +/–, PgR +/–, Vimentin +/–, CD56 –/+, CK20 –, CK5/6 –, CEA –, CDX-2 –

B. Tumors of the Uterine Corpus

–Endometrioid adenocarcinoma: CK7 +, CK8 +, CK18 +, CK19 +, EMA +, GFAP +/–, ER +/–, PgR +/–, Vimentin +/–, CD56 –/+, P53 –/+, P16 –, CK20 –, CK5/6 –, CEA –, CDX-2 –

–Uterine serous carcinoma:	CK7 +, CK8 +, CK18 +, CK19 +, EMA +, ER +/–, PgR +/–, P53 +/–, P16 –/+, CK5/6 –, CK20 –
–Endometrial stromal sarcoma:	Vimentin +, CD10 +, PgR +/–, ER +/–, bcl-2 +/–, Actin –/+, Desmin –/+, h– Caldesmon –, Calponin –, CD34 –, Pan-cytokeratin –, EMA–, Inhibin –
–Perivascular epithelioid tumor of the uterus (PEComa):	HMB45 +, Melan A +, Actin +/–, CD10 –, CD34 –, Pan-cytokeratin –, S100 –
–Uterine leiomyoma:	Vimentin +, Desmin +, Actin +, Calponin +, h-Caldesmon +/–, CD10 –, Pan-cytokeratin –
–Placental site trophoblastic tumor:	βhcG +, Human placental lactogen +, CD146 +, Cytokeratin + Proliferation index {Ki-67 (MIB-1) index}: >15%
–Gestational Choriocarcinoma:	See choriocarcinoma of the ovary

C. Tumors of the Fallopian Tube

–Endometrioid adenocarcinoma:	CK7 +, CK8 +, CK18 +, CK19 +, EMA +, GFAP +/–, ER +/–, PgR +/–, Vimentin +/–, CD56 –/+, P53 –/+, P16 –, CK20 –, CK5/6 –, CEA –, CDX-2 –
–Serous carcinoma:	CK7 +, CK8 +, CK18 +, CK19 +, EMA +, ER +/–, PgR +/–, CK5/6 –, CK20 –

2.3.15.2 Ovarian Tumors

A. Surface Epithelial-Stromal Tumors

–Serous ovarian neoplasia:	CK7 +, CK8 +, CK18 +, CK19 +, EMA +, CA125 +, HAM56 +, WT1 +/–, CK5/6 +/–, ER +/–, p16 +/–, Mesothelin +/–, Vimentin –/+, Calretinin –/+, S100 –/+, CEA –, Villin –, CK20 –, MUC-2 –, CDX-2 –, Inhibin –
–Mucinous ovarian neoplasia:	CK7 +, CK8 +, CK18 +, CK19 +, EMA +, HAM56 +, Villin +, CDX-2 + (in mucinous adenocarcinoma and intestinal type adenoma), CK20 +/– (intestinal type CK20 +), MUC-2 +/–, MUC5AC +/–, CEA +/–, ER +/–, Villin –/+, WT1 –, CK17 –, Vimentin –, Inhibin –
–Endometrioid carcinoma:	CK7 +, CK8 +, CK18 +, CK19 +, EMA +, ER +/–, Vimentin +/–, Mesothelin +/–, CK5 –/+, CK20 –, WT1 –, Inhibin –
–Clear cell carcinoma:	CK 7 +, EMA +, CD15 +/–, CK20 –, CD10 –
–Brenner tumor and ovarian transitional cell carcinoma:	Epithelial components: EMA +, CK7 +, CEA +, WT1 –/+, CK5/6 + (in basal epithelial cells), CK19 –, CK20 –, Thrombomodulin (CD141) –, Vimentin – Fibrous stroma: Vimentin +, bcl-2 +/–, Pan-cytokeratin –

B. Sex Cord-Stromal Tumors

–Granulosa cell tumor: Vimentin +, Inhibin +/–, Calretinin +/–, CD99 +/–, Actin +/–, S100 +/–, WT-1 +/–, Pan-cytokeratin –/+, CK8 –/+, CK18 –/+, EMA –, Desmin –

–Thecoma: Vimentin +, Inhibin +, Calretinin +, sm Actin +/–, WT-1 +/–, ER –/+, PgR –/+, Pan-cytokeratin –

–Sclerosing stromal tumor: Vimentin +, sm Actin +, Inhibin +/–, Calretinin +/–, PgR +, ER –/+, Desmin –/+, Pan-cytokeratin –

–Leydig cell tumor: Vimentin +, Inhibin +, Melan –A +, Calretinin +, CD99 +/–, Pan-cytokeratin –/+, S100 –/+, Synaptophysin –/+, Chromogranin –/+, EMA –, PLAP –, AFP –, CEA

–Sertoli cell tumor: Vimentin +, Inhibin +/–, AFP +/–, CD99 +/–, Pan-cytokeratin +/–, Calretinin +/–, NSE +/–, S100 +/–, Synaptophysin –/+, Chromogranin –/+, EMA –, PLAP –, CEA –

–Steroid cell tumor: Vimentin +, Inhibin +, Melan A +, Calretinin +/–, Pan-cytokeratin –/+

C. Germ Cell Tumors

–Dysgerminoma: PLAP +, Oct-4 +, Vimentin +, CD117 +/–, CK8/18 –/+, AFP –, βhcG –, Inhibin –, S100 –, EMA –

–Embryonal carcinoma: PLAP +, AFP +/–, Pan-cytokeratin +, CD30 +, Oct-4 +, CK19 +/–, NSE +/–, βhcG –, Vimentin –, CEA –, EMA –, CD117 –

–Yolk sac tumor: AFP +, Pan-cytokeratin +, PLAP +/–, βhcG –, Oct-4 –, EMA –, Vimentin –, CD30 –, CK7 –

–Choriocarcinoma: Syncytiotrophoblastic cells: βhcG +, Inhibin +, CD10 +, Pan-cytokeratin +, CK8/18 +, CK19 +, EGFR +, PLAP +/–, Human placental lactogen +/–, EMA +/–, CEA +/–, Vimentin –/+, CD30 –, AFP –, Oct-4 –
Cytotrophoblastic cells: CD10 +, Pan-cytokeratin +, CK8/18 +, CK19 +, CEA +, PLAP +/–, βhcG –, Inhibin –, EMA –, CD30 –, AFP –, Oct-4 –

–Polyembryoma: in embryonal bodies: AFP +, Pan-cytokeratin +, PLAP +/–

–Gonadoblastoma: Germ cells: PLAP +, CD117 +, Oct-4 +, Pan-cytokeratin +/–
Sex cord cells: Vimentin +, Inhibin +, Pan-cytokeratin +

2.3.16
Prostate and Seminal Vesicle Tumors

–Adenocarcinoma of prostate gland:	CK8 +, CK18 +, CK19 +, PSA +, PAP +, hK2 +, P504S (AMACR) +/–, Androgen receptors +/–, CK7 –/+, CK5 –, CK10 –, CK20 –, CEA – No myoepithelial basal cell layer positive for high molecular cytokeratins [CK5/6, CK-34E12 (Ma-903)], p63 or LP34
–Adenocarcinoma of seminal vesicle:	CK8 +, CK18 +, CK19 +, CEA +, CK7 +/–, CA 125 (MUC 16) +/–, CK20 –, PAP –, PSA –

2.3.17
Testicular and Paratesticular Tumors

2.3.17.1 Testicular Tumors

A. Germ Cell Tumors

–Intratubular germ cell neoplasia (IGCN):	PLAP +, sCD143 (angiotensin converting enzyme, ACE) +, NSE +, Oct-4 +, CD117 +/–, Ferritin +/–, AFP –, βhcG –, Inhibin –
–Seminoma/dysgerminoma:	PLAP +, tCD143 +, CD117 +/–, D2-40 (M2A) +/–, Oct-3/4 +/–, AP-2γ +/–, Vimentin +/–, CK8 +/–, CK18 –/+, NSE +/–, CK7 –/+, EMA –, CK19 –, CK20 –, CEA –, AFP –, βhcG –, Inhibin –, CD30 –, Glypican-3 –
–Spermatocytic seminoma:	CK8/18 +, Vimentin +/–, NSE +7-, PLAP –/+, CD117 –/+, AFP –, Oct-4 –, βhcG –, CEA –, EMA –, CD30 –, D2-40 –
–Embryonal carcinoma:	PLAP +, AFP +/–, CD30 (Ki-1) +, Oct-4 +, CK8 +, CK18 +, CK7 +/–, CK19 +/–, NSE +/–, βhcG –, Vimentin –/+, EMA –, CK20 –, CEA –, Glypican-3 –, D2-40 –
–Yolk sac tumor:	AFP +, Pan-Cytokeratin +, Glypican-3 +, PLAP +/–, CD34 +/–, NSE –/+, βhcG –, Oct-4 –, EMA –, Vimentin –, CD30 –, CEA –
–Choriocarcinoma:	Syncytiotrophoblastic cells: βhcG +, Inhibin +, CD10 +, Pan-Cytokeratin +, CK8/18 +, CK19 +, EGFR +, Glypican-3 +, PLAP +/–, Human placental lactogen +/–, EMA +/–, CEA +/–, Vimentin –/+, CD30 –, AFP –, Oct-4 – Cytotrophoblastic cells: CD10 +, Pan-Cytokeratin +, CK8/18 +, CK19 +, CEA +, Glypican-3 +/–, PLAP +/–, βhcG –, Inhibin –, EMA –, CD30 –, AFP –, Oct-4 –
–Polyembryoma:	In embryonal bodies: AFP +, Pan-Cytokeratin +, PLAP +/–

B. Sex Cord-Stromal Tumors

–Leydig cell tumor:	Vimentin +, Inhibin +, Melan-A +, Calretinin +, CD99 +/–, Pan-Cytokeratin –/+, S100 –/+, Synaptophysin –/+, Chromogranin –/+, EMA –, PLAP –, AFP –
–Sertoli cell tumor:	Vimentin +, Inhibin +/–, AFP +/–, CD99 +/–, Pan-Cytokeratin +/–, Calretinin +/–, NSE –/+, S100 –/+, Synaptophysin –/+, Chromogranin –, EMA –, PLAP –
–Granulosa cell tumor:	Vimentin +, Inhibin +/–, CD99 +/–, CK8 –/+, CK18 –/+, Actin –/+, S100 –/+, EMA –, Desmin –

2.3.17.2 Paratesticular Tumors

–Adenomatoid tumor:	Vimentin +, Calretinin +, Pan-Cytokeratin +, CK5/6+, EMA +, Thrombomodulin (CD141) +, CD31 –, CD34 –, CEA –
–Adenocarcinoma of rete testis:	Pan-Cytokeratin +, EMA +, CEA +/–, AFP –, PLAP –
–Melanotic neuroectodermal tumor:	Large pigmented cells: Pan-Cytokeratin +, NSE +, HMB45 +, Synaptophysin +, S100 +/– Small cells: NSE +, HMB45 +, Synaptophysin +, Pan-Cytokeratin –/+, Vimentin –

2.3.18 Lymphoid Neoplasia

2.3.18.1 Evolution of Immunophenotypic Profile of Non-Neoplastic Lymphocytes

A. Evolution of Immunophenotypic Profile of Non-Neoplastic B-Cells at Different Stages of Differentiation

–Pluripotent stem cell:	CD117, CDw123, CD243, CDw338, HLA-DR →
–Lymphoid stem cell:	CD10, CD34, CD38, CD117, CD124, CD127, CD228, TdT, HLA-DR →
–Pro-B-cell:	CD19, CD22, CD24, CD34, CD38, CD72, CD79a, CD79b, CD124, CD127, CD179a, CD179b, TdT, HLA-DR →
–Pre-Pre-B cell:	CD10, CD19, CD20, CD22, CD24, CD34, CD72, CD79a, CD79b, CD124, CD179a, CD179b, TdT, HLA-DR →
–Pre-B-cell:	CD9, CD10, CD19, CD20, CD22, CD24, CD38, CD40, CD72, CD74, CD79a, CD79b, CD124, CD179a, CD179b, CD275, CD316, CD317, TdT, HLA-DR →
–Immature B-cell:	CD19, CD20, CD22, CD24, CD37, CD40, CD72, CD74, CD79a, CD79b, CD124, CD268, CD269, CD275, CD316, CD317, HLA-DR, s-IgM →

–Intermediate B-cell: CD19, CD20, CD21, CD22, CD24, CD32, CD35, CD37, CD40, CD48, CD52, CD72, CD74, CD75, CD75s, CD79a, CD79b, CD99, CD124, CDw210, CD268, CD269, CD275, CD316, CD317, s-IgM →

–Mature B-cell: CD19, CD20, CD21, CD22, CD24, CD32, CD35, CD37, CD40, CD48, CD49b, CD49c, CD49d, CD52, CD72, CD74, CD78, CD79a, CD79b, CD84, CD99, CD102, CD119, CDw121b, CD124, CD185, CD192, CD196, CDw210, CD213a1, CD213a2, CD218a, CD218b, CD268, CD277, CD305, CD316, CD317, CD322, CDw327, HLA-DR, s-IgM, s-IgD →

–Follicle center B-cell: CD10, CD19, CD20, CD21, CD22, CD79a, HLA-DR, s-IgM, s-IgG, s-IgA, bcl-6 →

–Immunoblast: CD10, CD19, CD20, CD22, CD24, CD37, CD40, CD72, CD74, CD79a, CD79b, CD139, CD275, CD316, CD317, HLA-DR, c-Ig, s-IgM, s-IgG, s-IgA →

–Lymphoplasmacytoid cell: CD19, CD20, CD38, CD79a, CD79b, CD275, CD316, CD317 →

–Plasma cell: CD38, CD79a, CD126, CD138, CD269, CD275, CD316, CD317, c-Ig, PCA-1

B. Evolution of Immunophenotypic Profile of Non-Neoplastic T-Cells at Different Stages of Differentiation

–Prothymocyte (Pre-T-cell): CD7, TdT →

–Stage I thymocyte (subcortical): CD2, CD4, CD5, CD7, CD8, CD38, CD200, TdT →

–Stage II thymocyte (cortical): CD1a, CD1b, CD1c, CD2, CD3, CD4, CD5, CD7, CD8, CD38, CD52, CD165, CD200, TdT →

–Stage III thymocyte (medullary)/T-helper/inducer cell: CD2, CD3, CD4, CD5, CD7, CD27, CD28, CD48, CD52, CD69, CD99, CD121a, CD127, CDw150, CD155, CD165, CD200

–Stage III thymocyte (medullary)/T-suppressor and cytotoxic cell: CD2, CD3, CD5, CD8, CD7, CD27, CD28, CD48, CD52, CD69, CD99, CD121a, CD127, CDw150, CD155, CD165, CD200

–Natural killer cell (NK-cell/Null cell/LGL): CD11b, CD11c, CD16, CD48, CD56, CD57, CD69, CD94, CD122, CD158a, CD158b, CD159, CD161, CDw210, CD226, CD244, CD247

2.3.18.2 B-Cell Neoplasia

–Precursor B-lymphoblastic leukemia/lymphoma: TdT +, HLA-DR (CD74) +, CD19 +, CD79a +, PAX-5 +, CD10 (CALLA) +/–, CD22 +/–, CD99 +/–, CD20 +/–, CD22 +/–, CD34 +/–, FLI-1 +/–
Proliferation index [Ki-67 (MIB-1) index]: 50–80%

–B-cell chronic lymphocytic lymphoma: CD5 +, CD19 +, CD20 +, CD22 +, CD23 +, CD43 +, CD74 +, CD79a +, CD22 +/–, bcl-2 +, sIgM +, PAX-5 +, sIgD +/–, CD10 –

	Proliferation index [Ki-67 (MIB-1) index]: ~5%
–B-cell prolymphocytic leukemia:	CD19 +, CD20 +, CD22 +, CD74 +, CD79a +, PAX-5 +, bcl-2 +,CD5 −/+, CD10 −, CD23 −, CD43 −, Cyclin D1 −
–Lymphoplasmacytic lymphoma:	CD19 +, CD20 +, CD22 +, CD74 +, CD79a +, MUM1 +, PAX-5 +, IgM +, CD38 +/−, CD43 +/−, bcl-2 +/−, CD138 +/−, CD5 −, CD10 −, CD23 − Proliferation index [Ki-67 (MIB-1) index]: 5–10%
–Mantle cell lymphoma:	CD5 +, CD19 +, CD20 +, CD22 +, CD37 +, CD43 +, CD74 +, CD79a +, sIgM +, sIgD +, Cyclin D1 (bcl-1) +, PAX-5 +, bcl-2 +/−, CD10 −, CD11c −, CD23 −, bcl-6 − Proliferation index [Ki-67 (MIB-1) index]: 5–50%
–Follicular lymphoma:	CD19 +, CD20 +, CD22 +, CD74 +, CD79a +, CD10 +, bcl-2 + (+ in grade 1 and 2, and +/− in grade 3; −/+ in primary cutaneous follicular lymphoma), sIg +, PAX-5 +, bcl-6 +/−, CD23 −/+, CD5 −, CD43 −, κ/λ light chain restriction −/+ Proliferation index [Ki-67 (MIB-1) index] in bcl-2-positive neoplastic follicles: <15% Proliferation index [Ki-67 (MIB-1) index] in bcl-2-negative reactive follicles: >60%
–Nodal marginal zone B-cell lymphoma:	CD19 +, CD20 +, CD21 +, CD22 +, CD35 +, CD74 +, CD79a +, PAX-5 +, sIgM +, sIgA +/−, sIgG +/−, CD11c +/−, bcl-2 +/−, CD43 −/+, sIgD -, CD5 −, CD10 −, CD 23 −, Cyclin D1 −, bcl-6 −
–Marginal zone B-cell lymphoma of MALT type:	CD19 +, CD20 +, CD21 +, CD22 +, CD35 +, CD74 +, CD79a +, PAX-5 +, sIgM +, sIgD +/−, sIgA +/−, sIgG +/−, CD11c +/−, bcl-2 +/−, CD43 −/+, CD5 −, CD10 −, CD23 −, Cyclin D1 −, bcl-6 −
–Splenic marginal zone B-cell lymphoma:	CD19 +, CD20 +, CD21+, CD22 +, CD35 +, CD74 +, CD79a +, PAX-5 +, bcl-2 +, sIgM +, sIgD +, sIgA +/−, sIgG +/−, CD11c +/−, CD43 −, CD5 −, CD10 −, CD23 −, CD25 −, CD43 −, CD103 −, Cyclin D1 − Proliferation index [Ki-67 (MIB-1) index]: <5%
–Hairy cell leukemia:	CD11c +, CD19 +, CD20 +, CD22 +, CD25 +, CD74 +, CD79a +, CD103 +, Annexin A1 +, PAX-5 +, sIgM +, CD23 +/−, CD25 +/−, CD68 +/− (cytoplasmic dots), PCA-1 +/−, HC1 +/−, HC2 +/−, CD5 −/+, CD10 −, CD23 −, CD43 −, bcl-6 −, TRAP + Proliferation index [Ki-67 (MIB-1) index]: <5%

–Plasma cell myeloma/ plasmacytoma:	CD38 +, VS38c +, CD138 (Syndecan-1) +, PCA-1 +, MUM1 +, Vimentin +, CD43 +/–, CD56 +/–, CD79a –/+, CD45 –/+, EMA –/+, CD19 –, CD20 –, CD22 –, PAX-5 – κ or λ Ig light chain restriction Proliferation index [Ki-67 (MIB-1) index] 50–60%
–Diffuse large B-cell lymphoma:	CD19 +, CD20 +, CD22 +, CD74 +, CD79a +, CD45 +, PAX-5 +, bcl-6 +/–, bcl-2 –/+, CD5 –/+, CD30 –/+, Fascin –/+, CD3 –, CD15 –
–T-cell/histiocyte-rich variant of diffuse large B-cell lymphoma: Neoplastic cells:	CD19+,CD20+,CD22+,CD74+,CD79a+,CD45+, PAX-5 +, bcl-6 +, BOB 1 +, OCT-2 +, CD30 –/+, EMA –/+, CD3 –, CD15 –, bcl-2 –, PU 1 – Non-neoplastic cells: CD3 +, CD8 +, Cytotoxic molecules +, CD20 –/+, bcl-2 –, CD68 + in histiocytes
–Mediastinal (thymic) large B-cell lymphoma:	CD19 +, CD20 +, CD45 +, CD74 +, CD79a +, PAX-5 +, CD23 +/–, CD30 –/+, CD5 –, CD10–
–Intravascular large B-cell lymphoma:	CD20 +, CD79a +, PAX-5 +, CD5 –/+
–Primary effusion lymphoma:	CD45 +, CD79a +/–, CD38 +, CD138 +/–, VS38c +, PAX-5 +, HHV-8 +, CD30 +/–, MUM1+/–, CD20 –/+, CD19 –/+, CD43 –
–Burkitt's lymphoma:	CD10 +, CD19 +, CD20 +, CD22 +, CD74 +, CD79a +, PAX-5 +, sIgM +, CD43 +, bcl-6 +/–, c-myc +/–, EBV+/–, CD5 –, CD23 –, TdT –, bcl-2 – Proliferation index [Ki-67 (MIB-1) index]: >90%
–Burkitt-like lymphoma:	CD19 +, CD20 +, CD22 +, CD74 +, CD79a +, PAX-5 +, CD43 +/–, bcl-6 +/–, CD10 –/+, sIgG +/–, IgM +/–, bcl-2 – Proliferation index [Ki-67 (MIB-1) index]: >90%

2.3.18.3 **T-Cell Neoplasia**

–Precursor T-cell lymphoblastic leukemia/ lymphoma:	TdT +, CD10 (CALLA) +, CD7 +, CD2, CD3 +/– (cytoplasmic), CD99 +, CD1a +/–, CD5 +/–, Fli-1 +/–, CD13 –/+, CD15 –/+, CD33 –/+, CD34 –/+ Proliferation index [Ki-67 (MIB-1) index] 50–80%
–T-cell prolymphocytic leukemia:	CD2 +, CD3 +, CD5 +, CD7 +/–, CD4 +/–, CD8 –/+, CD25 -, CD28 –, CD56 –, TdT –
–T-cell large granular lymphocytic leukemia:	CD2 +, CD3 +, CD5 +, CD8 +/–, CD16 +/–, CD57 +/–, CD4 –/+, CD7 –/+, CD56 –/+, CD25 –
–Adult T-cell lymphoma (HTLV1+):	CD2 +, CD3 +, CD4 +, CD5 +, CD25 +, CD7 –, CD8 –
–Aggressive NK-cell leukemia:	CD2 +, CD16 +, CD8 +/–, CD56 +/–, CD57 +/–, CD3 –, CD4 –

–Peripheral T-cell lymphoma: CD2 +/–, CD3 +/–, CD5 +/–, CD7 –/+, CD25 –/+, CD30 –/+, CD134 –/+, CD15 –

–Angioimmunoblastic T-cell lymphoma: CD2 +, CD3 +, CD4 +, CD5 +, CD7 +, CD28 +, CD30 –/+, CD8 –, CD15 –

–Extranodal NK/T cell lymphoma, nasal and nasal type: CD2 +, CD56 +, Cytotoxic molecules (TIA-1, perforin, granzyme B) +, EBV +, CD43 +/–, CD95 (Fas) +/–, CD3 –/+ (cytoplasmic), CD5 –/+, CD7 –/+, CD30 –/+, CD4 –, CD8 –, CD20 –
Proliferation index [Ki-67 (MIB-1) index]: >80%

–Mycosis fungoides/Sezary syndrome: CD2 +, CD3 +, CD5 +, CD45RO +, CD4 +, CD7 –/+, CD8 –, CD25 –
Proliferation index [Ki-67 (MIB-1) index]: <5%

–Enteropathy-type T-cell lymphoma: CD2+, CD3 +, CD7 +, CD103 +, CD8 +/–, Cytotoxic molecules +/–, CD56 –/+, CD4 –, CD5 –

–Hepatosplenic $\gamma\delta$ T-cell lymphoma: CD2 +, CD3 +, CD7 +, CD45RO +, TIA-1 +, CD56 +/–, CD8 –/+, CD4 –, CD5 –, Perforin –

–Subcutaneous (panniculitis-like) T-cell lymphoma: CD2 +, CD3 +, CD43 +, CD45 +, Cytotoxic molecules (TIA-1, perforin, granzyme B) +, CD5 +/–, CD8 +/–, CD25 +/– CD30 –/+ CD4 –

–Anaplastic large cell lymphoma: CD30 +, Clusterin + (Golgi stain pattern), ALK (p80)+/–,CD2+/–,CD3+/–,CD4+/–,CD25+/–, CD43 +/–, CD45 +/–, CD25 +/–, EMA +/–, Cytotoxic molecules +, Galectin-3 +/–, CD15 –/+, Fascin–/+,CD5–,CD7–,CD20–,CD28–,PAX-5–

–Primary cutaneous anaplastic CD30-positive T-cell lymphoma: CD30 +, CD45 +/–, CD2 +/–, CD3 +/–, CD4 +, Cytotoxic molecules +, Clusterin –, CD8 –, CD15 –, EMA –, CD246 (ALK, p80) –, PAX-5 –

–NK-cell neoplasia: CD16 +, CD56 +, CD94 +, CD 123 +, CD43 +/–, cytoplasmic CD3 +/–, surface CD3 -, CD4 +/–, Cytotoxic molecules +/–, CD5 –, PAX-5 (BSAP) –
No TCR gene rearrangement

2.3.18.4 **Hodgkin's Lymphoma**

–Lymphocyte predominance Hodgkin's lymphoma:

Neoplastic L&H cells (lymphocytic/histiocytic cells) or popcorn cells: CD19 +, CD20 +, CD22 +, CD45 +, CD75 +/–, CD79a +/–, CD40 +/–, CD86 +, PU.1 +, J-chain +, Oct-2 +, BOB.1 +, PAX-5 (BSAP) +, bcl-6 +/–, EMA +/–, CD30 –/+ (–), CD15 –, Fascin –, CD138 –, ALK (p80) –, EBV –
Usually negative IgH and TCR gene rearrangements

–**Classic Hodgkin's lymphoma**:	
–Nodular sclerosis	
–Lymphocyte-rich classic	
–Mixed cellularity	
–Lymphocyte-depleted	
–Unclassifiable	
–Hodgkin and Reed–Sternberg (H&RS) cells in classic subtypes of Hodgkin's lymphoma:	CD30 +, Fascin +, CD15 +/−, CD83 +/−, CD138 +/−, HLA-DR +/−, PAX-5 +/−, MUM1 +/−, CD20 −/+, CD79 −/+, CD22 −, CD45 −, Oct-2 −, BOB.1 −, J-chain −, PU.1 −, EMA −, bcl-6 −, ALK −, EBV (LMP1) +/− Usually negative IgH and TCR gene rearrangements

2.3.19 Myeloid Neoplasia

2.3.19.1 Evolution of Immunophenotypic Profile of Non-Neoplastic Myeloid Cells

–Pluripotent stem cell:	CD117, CD123, CD143, CDw338, HLA-DR →
–Myeloid stem cell:	CD33, CD34, CD117, CD123, CDw131, CD176, CD228, CD280, HLA-DR →
–CFU-G:	CD13, CD15, CD33, CD111, CD112, CD116, CDw123, HLA-DR →
–Myelobast:	CD13, CD33, CD114, CD116 →
–Pro-myelocyte:	CD13, CD33, CD89, CD116, CDw123, MPO →
–Myelocyte:	CD13, CD89, CD91, CD114, CD116, CDw123, CDw131, MPO →
–Neutrophil:	CD10, CD11b, CD13, CD15, CD16, CD17, CD24, CD32, CD35, CD43, CD65, CD66, CD89, CDw92, CD93, CD11, CD112, CD114, CD116, CDw123, CDw128, CD156, CD157, CD162, CD170, CD181, CD281, CD282, CD312, MPO (in activated neutrophil also CD64, CD66)
–CFU-Eo:	CD13, CD33, CD34, CD116, CDw123, CDw131 →
–Myelocyte:	CD11b, CD13, CD32, CD33, CD35, CD114, CD116, CDw131 →
–Eosinophphil:	CD9, CD11b, CD15, CD24, CD32, CD35, CD43, CD114, CD116, CDw125, CDw131, CD193, CDw218, CD294 (in activated eosinophil CD23, CD64, CD66)
–CFU-Bas:	CD34, CDw123 →
–Myelocyte:	CD33, CD114, CDw123 →
–Basophil:	CD9, CD17, CD25, CD33, CD38, CD43, CD114, CDw125, CDw131, CD154, CD192, CD193, CD203c, CD294 (in activated basophil also CD64, CD66)
–CFU-M:	CD13, CD15, CD33, CD111, CD112, CD115, CD116, CDw123, CDw131, HLA-DR →

–Monoblast: CD11c, CD13, CD15, CD33, CD115, CD116, CDw123, CDw131, HLA-DR, MPO →

–Pro-monocyte: CD13, CD14, CD33, CD111, CD112, CD115, CD116, CDw123, CDw131, HLA-DR, MPO →

–Monocyte: CD9, CD11b, CD11c, CDw12, CD13, CD14, CD15, CDw17, CD32, CD33, CD35, CD36, CD38, CD43, CD49b,CD49e, CD49f, CD63, CD64, CD65s, CD68, CD84, CD85, CD86, CD87, CD89, CD91, CDw92, CD93, CD98, CD101, CD102, CD111, CD112, CD115, CD116, CD119, CDw121b, CDw123, CD127, CDw128, CDw131, CD147, CD155, CD156a, CD157, CD162, CD163, CD164, CD168, CD171, CD172a, CD172b, CD180, CD184, CD191, CD192, CD195, CDw198, CD206, CDw210, CD213, CD226, CD277, CD281, CD282, CD300a, CD300c, CD300e, CD302, CD305, CD312, CD317, CD322, CDw328, CDw329, HLA-DR, MPO.

–Macrophage and monocyte-derived dendritic cells: CD11c, CD14, CD16, CD26, CD31, CD32, CD36, CD45RO, CD45RB, CD63, CD68, CD71, CD74, CD87, CD88, CD101, CD119, CD121b, CD155, CD156a, CD204, CD206, CDw210, CD312, HLA-DR, (in activated macrophage also CD23, CD25, CD69, CD105).

–CFU-Meg: CD34, CD110, CDw123 →

–Megakaryoblast: CD41, CD42, CD61, HLA-DR →

–Megakaryocyte: CD41, CD42, CD49f, CD61, CD110, CDw123, CDw131, CD151, CD203c →

–Platelet: CD9, CDw17, CD23, CD31, CD36, CD41, CD42, CD49b, CD49f, CD60a, CD61, CD63, CD84, CD92, CD109, CD147, CD151, CD173, CD226

–BFU-E: CD33, CD34, CDw123, CDw131, CD297, CD324 →

–CFU-E: CD36, CDw123, CDw131, CD175s, CD297, CD324 →

–Erythroblast: CD36, CD71, HLA-DR, Glycophorin A, Glycophorin C →

–Normoblast: CD36, CD71, Glycophorin A, Glycophorin C →

–Reticulocyte: CD71, Glycophorin A →

–Erythrocyte: CD35, CD44, CD55, CD59, CD147, CD173, CD233, CD234, CD235a, CD235b, CD236, CD236R, CD238, CD239, CD240CE, CD240D, CD241, CD242, CD297, Glycophorin A, Glycophorin C

Key to abbreviations used in myeloid cell names:
CFU: Colony-forming unit
G: Granulocyte
M: Monocyte
Meg: Megakaryocyte
E: Erythroid
BFU: Burst-forming unit
Eo: Eosinophil
Bas: Basophil

2.3.19.2 Acute Myeloid Leukemia

–Myeloblastic, minimally differentiated (M0):	CD34 +, CD13 +/−, CD33 +/−, CD65 −/+, CD117 +/−, MPO +/−, HLA-DR +/−
–Myeloblastic, without maturation (M1):	CD13 +, CD15 +/−, CD19 +/−, CD33 +/−, CD34 +/−, CD56 −/+, CDw65 +/−, CD117 +/−, HLA-DR +, MPO +
–Myeloblastic, with maturation (M2):	CD13 +, CD15 +, CD33 +, MPO +, HLA-DR +, CD117 −/+, CD56 −/+, CD34 −/+, CAE +
–Myelocytic (M3):	CD13 +, CD33 +, MPO +, CAE + CD64 +/−, CD15 −/+, CD65 −/+, CD117 −/+, CD56 −/+, CD34 −, HLA −DR −
–Myelomonoblastic (M4):	CD11b +, CD13 +, CD33 +, CD64+, CDw65+, CD68 +, MPO +, HLA-DR +, CD14 +/−, CD15 +/−, CD117 +/−, CD34 −/+
–Monoblastic/monocytic (M5a/M5b):	CD11c +, CD15 +, CD33 +, CD64 +, CD68 +, HLA-DR +, CAE +, HLA-DR+, CD13 +/−, CD14 +/−, CDw65+/−, MPO −/+
–Erythroblastic (M6):	MPO +, Glycophorin +, HLA-DR +/−, CD15 +/−, CD33 +/−, CAE −, CD13 −
–Megakaryoblastic (M7):	CD61 +, CD41 +, CD42 +, CD33 +/−, CDw65 −/+, CD13 −/+, HLA-DR −/+, CD15 −, MPO −, CAE −

2.3.19.3 Chronic Myeloid Neoplasia

–Chronic myeloid leukemia:	CD11b +, CD11c +, CD14 +, CD15 +, CD117 −/+
–Granulocytic sarcoma (myelosarcoma):	CD43 +, Vimentin +, Lysozyme +, CD15 +/−, HLA-DR +/−, MPO +/−, CD34 −/+, CD68 −/+, CD3 −, CD20 −
–Mast cell disease:	Tryptase +, Chymase +, CD45 +, CD33 +, CD117 +, Calretinin +/−, CD2 −/+, CD25 −/+, CD 3 −, CD15 −, CD 20 −

2.3.20
Histiocytic and Dendritic Cell Tumors

–Histiocytic sarcoma:	CD45 +, CD68 +, HLA-DR +, CD163 +/–, Lysozyme +/–, CD11c +/–, CD14 +/–, S100 –/+, CD45 –/+, CD4 –/+, CD15 –/+, CD1a –, CD3 –, CD20 –, CD21 –, CD23 –, CD33 –, CD34 –, CD35 –, CD30 –, Fascin –, MPO –
–Tumors of Langerhans cell type/Langerhans cell histiocytosis (histiocytosis X):	S100 +, CD1a +, CD86 +, Langerin +, CD4 +/–, CD11c +/–, CD163 +/–, CD45RB +/–, HLA-DR +/–, PLAP +/–, CD68 +/–, CD45 –/+, CD2 –, CD3 –, CD20 –, CD21 –, CD30 –, CD34 –, CD35 –, MPO –, EMA –
–Follicular dendritic cell tumor/sarcoma:	CD21 +, CD23 +, CD35 +, KiM4p +, Fascin +, Desmoplakin +/–, HLA-DR +/–, S100 +/–, EMA +/–, CD20 –/+, CD45 –/+, CD68 –/+, Actin –/+, CD1a –, CD2 –, CD3 –, D30 –, CD 34 –, CD35 –, CD79a –, CD163 –, MPO –, Pan-Cytokeratin –
–Interdigitating dendritic cell tumor/sarcoma:	S100 +, CD4 +, CD45RB +, Fascin +, CD68 +/–, CD45 –/+, CD1a –, CD2 –, CD3 –, CD20 –, CD21 –, CD23 –, CD30 –, CD34 –, CD35 –, MPO –, EMA –, Pan-Cytokeratin –

2.4
Common Antibodies used for Tumor Diagnosis on Paraffin Sections and Expression Pattern

The following part provides an overview of the most common antibodies used for tumor diagnosis as well as the expression pattern of targeted antigens detected by the immunohistochemical reaction. The expression pattern of the different cellular antigens is an important factor to consider and essential for the precise interpretation of the immunohistochemical stains and includes the following expression (stain) patterns:

1. Nuclear stain pattern: characteristic for antigens expressed within cellular nuclei or on the nuclear membrane. Good examples for this expression pattern are transcription factors and steroid hormone receptors.
2. Cytoplasmic stain pattern: characteristic for antigens located within the cytoplasm. Common examples are cellular skeletal proteins such as vimentin, actin and desmin.
3. Membrane stain pattern: characteristic for antigens expressed on the cell membrane, typical example is the expression of the lymphoid surface antigens.
4. Extracellular stain pattern: characteristic for extracellular and tissue matrix antigens in addition to the cellular secretion products such as CEA.

It is noteworthy to mention that some antigens have a different expression pattern during the cell cycle or during different differentiation stages such as the immunoglobulin expression in lymphoid tissue. Other antigens have a unique expression pattern in some tumors. The following examples demonstrate this phenomenon, which is also of diagnostic value:

- Paranuclear expression of cytokeratins in different tumors. This phenomenon is reported in variable percentage of different tumors, good example are the following tumors:

1. Merkel cell carcinoma with perinuclear cytokeratin deposits (mainly cytokeratin 20).
2. Small cell carcinoma (mainly cytokeratin 19).
3. Carcinoid tumors and pancreatic endocrine tumors.
4. Renal oncocytoma (with low molecular weight cytokeratins).
5. Medullary thyroid carcinoma.
6. Seminoma (with low molecular weight cytokeratins).
7. Granulosa cell tumor.
8. Rhabdoid tumor.
9. Few mesenchymal tumors including desmoplastic small round cell tumor, leiomyosarcoma and monophasic synovial sarcoma.

- Membrane and cytoplasmic expression of MIB-1 (Ki-67 equivalent) in some types of follicular thyroid neoplasia, namely in the hyalinizing trabecular adenoma and carcinoma. The same phenomenon is also reported is salivary gland pleomorphic adenoma and sclerosing hemangioma of the lung, nevertheless such phenomena can be the result of a cross-reaction with the MIB-1 clone of Ki-67.

As formalin – the most widely used tissue fixative – may destroy or change the structure of many antigens by inter- and intramolecular cross-linking of targeted antigen molecules, unmasking of changed antigen molecules is required to enable the antigen–antibody reaction. For this purpose one of the two following unmasking methods is used:

1. High-temperature unmasking: a high temperature (~95 °C) is able to break the chemical bonds masking the antigens. High-temperature unmasking is usually achieved using a water bath or microwave with different buffers (such as citrate or EDTA) with a pH ranging between 6 and 9. Simultaneous exposure to ultrasound can enhance the immunohistochemical stain.
2. Unmasking by enzymatic digestion using proteolytic enzymes like trypsin, pronase and proteinase K.

Table 2.5 includes the most common used antibodies for tumor diagnosis with the characteristic expression pattern in addition to the recommended unmasking method. Table 2.6 lists the main diagnostic use of different antigens, the expression of these antigens in normal tissue and different tumors.

Table 2.5 Commonly used antibodies for tumor diagnosis, with characteristic expression pattern and recommended unmasking method. *Key*: C: cytoplasmic stain pattern; D: antigen unmasking by digestion with proteolytic enzymes (pronase, trypsin, pepsin, proteinase K); E: extracellular stain pattern; H: high-temperature antigen unmasking (microwave, water bath); M: membrane stain pattern; N: nuclear or perinuclear stain pattern.

Antibody	Expression pattern	Unmasking method
Cytokeratins and epithelial markers		
CK 1	C	H
CK 4	C	H
CK 5	C	H
CK 6	C	H
CK 7	C	H/D
CK 8	C	H
CK 10	C	H
CK 13	C	H
CK 14	C	H
CK 15	C	H

(continued)

Table 2.5 *(Continued)*

Antibody	Expression pattern	Unmasking method
CK 17	C	H
CK 18	C	H
CK 19	C	H/D
CK 20	C	H
KL-1	C	D
CK MNF-116 (CK 5/6/8/17/19)	C	H
34E12 (CK 1/5/10/14)	C	H
CAM 5.2 (CK 8/18)	C	H
AE1 (CK type I:CK 10/13/14/15/16/19)	C	H/D
AE3 (CK type II:CK 1/2/3/4/5/6/7/8)	C	H/D
EMA (epithelial membrane antigen)	M/C	–
ESA (HEA, epithelium-specific antigen)	M	H/D
Villin	M/H	H
EGFR	M/C	H/D
CEA (CD66e)	E	–
Myoepithelial markers		
CK 5	C	H
CK 14	C	H
sm-Actin	C	–
sm Myosin	C	H
Calponin	C	D
P63	N	H
Pan-mesenchymal marker		
Vimentin	C	H
General myogenic markers		
Desmin	C	–
Muscle-specific actin (HHF35)	C	–
Smooth muscle markers		
sm-Actin	C	–
Myosin (heavy chain)	C	H
Calponin	C	D
h-Caldesmon	C	H
Striated muscle markers (skeletal and cardiac muscle)		
Myoglobin	C	–
Myosin (Y32)	C	H
Myotilin	C	H
Rhabdomyosarcoma markers		
Myo D1	N	H
Myf-3	N	H
Myogenin (Myf-4)	N	H
Neural and neuroendocrine markers		
S100	C	–

Table 2.5 *(Continued)*

Antibody	Expression pattern	Unmasking method
NSE	C	–
Neurofilaments	C	H
Chromogranin	C	H
Synaptophysin	C	–
Nerve growth factor receptor (gp75)	C	H
GFAP	C	H
PGP 9.5	C	H
Myelin basic protein	C	H
CD56	M	H
CD99 (MIC-2)	M	H
CD 117 (c-kit)	M/C	H
Leu 7	C	–
Melanoma markers		
Melanosome (HMB45)	C	D
Melan A (MART-1)	C	H
Tyrosinase	C	H
S100	C/N	H
MAGE-1	N	H
CD63 (NK1-C3)	C	H
Microphthalmia transcription factor	N	H
Mesothelial markers		
Calretinin	C	H
Mesothelin	M	H
Thrombomodulin (CD141)	M	H
D2-40 (M2A)	M	H
Podoplanin	M	H
Lymphoid, myeloid, histiocytic and plasma cell markers		
Lymphoid markers		
CD45 (LCA)	M	H
CD246 (ALK, p80)	N/C	H
TdT	N	H
B-cell markers		
CD10 (CALLA)	M	H
CD19	M	H
CD20	M	H
CD21	M	D
CD22 (BL-CAM)	M/C	H
CD23	M	H
CD45RB	M	–
CD74	M	H
CD79a	M	H
Oct-2	N	H

(continued)

Table 2.5 *(Continued)*

Antibody	Expression pattern	Unmasking method
MB2	C	–
PAX-5 (BSAP)	N	H
HLA-DR	M	H
Kappa light chain (Ig κ)	C	D
Lambda light chain (Ig λ)	C	D
Ig M/G/A/D	M/C	D
T-cell markers		
CD2	M	H
CD3	M	H
CD4	M	H
CD5	M	H
CD7	M	H
CD8	M	H
CD25 (IL-2 receptor)	M/C	H
CD43	M	–
CD45RO (Uchl-1)	M	H
CD56 (leu 19)	M	H
Natural killer(NK)-cell markers		
CD2	M	H
CD56 (NCAM, leu 19)	M	H
CD57 (leu 7)	M	D
Cytotoxic molecules		
TIA-1	C	H
Granzyme B	C	H
Perforin	C	H
Plasma cell markers		
VS38c	C	H
CD38	M	H
CD138 (syndican-1)	M/C	H
MUM-1	N/C	H
Kappa light chain (Ig κ)	C	D
Lambda light chain (Ig λ)	C	D
Hodgkin's and Reed–Sternberg cell markers		
CD15	M	H
CD30 (Ber H2, Ki-1)	M	H/D
CD83	M	H
Oct-2	N	H
J-chain	C	H
Fascin	M/C	H
PAX-5 (BSAP)	N	H

Table 2.5 *(Continued)*

Antibody	Expression pattern	Unmasking method
Follicular dendritic cells		
CD21	M	D
CD35	M	H
Fascin	M/C	H
Macrophage and histiocytic markers		
CD1a	M	H
CD68 (KP1, KiM6)	C	H
CD86	M	H
Alpha-1 antitrypsin	C	D
Alpha-1 antichemotrypsin	C	–
Langerin	M/C	H
Lysozyme (Muramidase)	C	D
Myeloid markers		
CD15	M	H
Myeloperoxidase (MPO)	C	H
Megakaryocytic and platelet markers		
CD61 (glycoprotein IIIa)	C	D
Erythroid markers		
Glycophorin A (CD235 A)	M	–
Glycophorin C (CD235 R)	M	–
Organ-specific markers and hormones		
Lung-specific markers		
Thyroid transcription factor-1 (TTF-1)	N	H
Surfactant Apoprotein A	C/N	–
Surfactant precursor protein	C/N	H
Thyroid-specific markers		
Thyroid transcription factor-1 (TTF-1)	N	H
Thyroglobulin	C	–
Thyroid peroxidase	C	–
Calcitonin	C	–
Parathyroid-specific markers		
Parathyroid hormone	C	H
Breast-specific markers		
Estrogen receptor	N	H
Progesterone receptor	N	H
Gross cystic disease fluid protein-15 (GCDFP15)	E/C	H

(continued)

Table 2.5 *(Continued)*

Antibody	Expression pattern	Unmasking method
Human milk fat globule	C	H
Mammaglobin	C	H
Gastrointestine-specific markers		
CDX-1	N	H
CDX-2	N	H
Hepatocellular carcinoma		
Alpha fetoprotein (AFP)	C	–
Hepatocyte-specific antigen (OCH1E5)	C	H
Glypican-3	M	H
MAGE-1	N	H
Kidney-specific markers		
Renal cell carcinoma marker (RCC; gp200)	M	D
Wilms tumor protein (WT-1)	N/C	H/D
CD10	M	H
Transitional cell (urothelium)-specific markers		
Uroplakin (subtypes Ia and II)	M	H
Thrombomodulin (CD 141)	M	H
Prostate-specific markers		
PSA	C	H
PAP	C	–
Glandular kallikrein (hK2)	C	H
Androgen receptor	N	H
P504S (AMACR)	C	H
Germ cell markers		
Alpha fetoprotein (AFP)	C	–
Placental alkaline phosphatase (PLAP)	M	–
β-Human chorionic gonadotrophin (βhcG)	C	–
Placental lactogen	M	–
Oct-3	N	H
Oct-4	N	H
Angiotensin converting enzyme (ACE, CD143)	C	H
AP-2γ	N	H
D2-40 (M2A)	M	H
Glypican-3	M	H
Sex cord-stromal tumor markers		
Inhibin	C	H
Anti-Müllerian hormone	C	H
Bone markers		
Osteonectin	C	H
Osteopontin	E	H

Table 2.5 *(Continued)*

Antibody	Expression pattern	Unmasking method
Cell cycle and proliferation markers		
Ki-67 (MIB-1) Ki-67 positive in G_1, S, G_2 and M and negative only in G_0 cell phase	N	H
Cyclin A (positive in S cell phase)	N	H
Cyclin D1 (positive in G_1 cell phase)	N	H
PCNA	N	–
PCNA positive in G_1, S, G_2 and M phases (mainly in the S phase), negative only in G_0 phase		
Bromodeoxyuridine (in the S phase)	N	–
Oncoproteins and tumor suppressor proteins		
bcl-2	M/C	H
bcl-6	N	H
CD117 (c-kit)	M	H
p53	N	H
HER-2 (c-erb-2)	M	H
EGFR	M/C	H/D

Table 2.6 Diagnostic use and expression of antigens.

Antigen	Main diagnostic use	Possible expression in other tumors	Expression in normal cells
Actin (smooth muscle)	Smooth muscle tumors	Myoepithelial and myofibroblastic tumors	Smooth muscle cells, myoepithelial cells, myofibroblasts
Actin (sarcomeric)	Rhabdomyosarcoma, striated muscle tumors		Striated and cardiac muscle
Alpha fetoprotein (AFP)	Hepatocellular carcinoma, yolk sac tumor	Tumors with hepatoid differentiation	Fetal liver
a-Methylacyl-CoA racemase (AMARC, P504s)	Prostatic adenocarcinoma, high-grade PIN	Gastrointestinal adenocarcinoma, hepatocellular carcinoma, carcinoma of breast, ovarian carcinoma, renal cell carcinoma, urothelial carcinoma, lymphoma	Periurethral glands, liver, salivary glands, renal tubular epithelium, pancreas epithelium

(continued)

Table 2.6 *(Continued)*

Antigen	Main diagnostic use	Possible expression in other tumors	Expression in normal cells
Anaplastic lymphoma kinase (ALK)	Anaplastic large cell lymphoma, inflammatory myofibroblastic tumor	Malignant peripheral nerve sheath tumor, rhabdomyosarcoma, neuroblastoma, Ewing's tumor/PNET, leiomyosarcoma	
Androgen receptors (AR)	Prostatic carcinoma	Osteosarcoma, apocrine breast carcinoma	Prostate, apocrine and sebaceous glands, skin, oral mucosa
Angiotensin converting enzyme (AEC) CD143	seminoma		tCD143 (testicular form): cells of spermatogenesis; sCD143 (somatic form): endothelial cells of lung capillaries, epithelial cells of proximal renal tubules, small intestine epithelium, neural cells, T-cells and macrophage
bcl-2	Follicular lymphoma	DLBCL, solitary fibrous tumor, synovial sarcoma, Kaposiform hemangiosarcoma, neurofibroma, schwannoma, nasopharyngeal carcinoma, dermatofibrosarcoma protuberans, spindle cell lipoma, rhabdomyosarcoma	Small B lymphocytes in mantle zone and cells within T-cell areas, medullary cells in thymus
bcl-6	Large cell lymphoma	Follicular lymphoma (interfollicular area), Burkitt's lymphoma	Germinal centers of lymph nodes
CA125	Serous, endometrioid and clear cell ovarian carcinoma	GIT, breast, uterine and lung adenocarcinoma, mesothelioma	Breast ductal epithelium, epithelium of lung, GIT, biliary tract, pancreas, female genital tract and apocrine glands, meso-thelial cells
CA19-9	GIT and pancreas carcinoma	Ovarian and lung carcinoma, renal cell carcinoma	Breast ductal epithelium, epithelium of lung, GIT and hepatobiliary, kidney, salivary and sweet glands

Table 2.6 *(Continued)*

Antigen	Main diagnostic use	Possible expression in other tumors	Expression in normal cells
β-Catenin	Desmoid fibromatosis	Solitary fibrous tumor, synovial sarcoma, hepatic adenoma, neuroendocrine carcinoma, GIT adenocarcinomas, endometrioid carcinoma, cholangiocellular carcinoma	Fibroblasts, endothelial cells, alveolar macrophages, hepatocytes, cytotropho-blasts in placenta, normal prostate
h-Caldesmon	Smooth muscle tumors	Glomus tumors, GIST, epithelioid mesothelioma	Smooth muscle cells, myoepithelial cells
Calcitonin	Medullary thyroid carcinoma	Neuroendocrine carcinoma	Thyroid parafollicular (C) cells
Calponin	Smooth muscle tumors	Myofibroblastic tumors	Smooth muscle, myoepithelial cells
Calretinin	Mesothelioma, mast cell lesions	Squamous cell carcinoma, transitional cell carcinoma, granular cell tumor, adrenocortical tumors, ovarian sex cord stromal tumors	Central and peripheral neural tissue, mesothelial cells, mast cells
CD1a	Langerhans cell histiocytosis	Myeloid leukemias, mycosis fungoides, cutaneous T-cell lymphomas, T-ALL	Cortical thymocytes, Langerhans cells, immature dendritic cells
CD2	T-cell lymphoma	Mastocytosis	Thymocytes, mature peripheral T-cells, NK-cells
CD3	T-cell lymphomas	NK-lymphoma (cytoplasmic stain)	Thymocytes, peripheral T-cells, activated NK cells, Purkinje cells of cerebellum
CD4	T-cell lymphomas	Histiocytic lymphoma acute myeloid leukemia	Thymocytes, T-helper cells, macrophages, Langerhans cells, dendritic cells, granulocytes
CD5	Mantle cell lymphoma	B-CLL, T-ALL, T-cell lymphoma, prolymphocytic leukemia, atypical thymoma and thymic carcinoma	T-cells, B-cells of mantle zone of spleen and lymph nodes

(continued)

Table 2.6 *(Continued)*

Antigen	Main diagnostic use	Possible expression in other tumors	Expression in normal cells
CD7	T-ALL and T-cell lymphomas	CML, ALL, cholangiocellular carcinoma, pancreas carcinoma	Thymocytes, mature T-cells and NK cells, pre-B cells, monocytes, early myeloid cells
CD8	Mycosis fungoides, subcutaneous panniculitis-like T-cell lymphoma	T-cell large granular lymphocytic Leukemia, CLL, mantle cell lymphoma	Suppressor/cytotoxic T-cells and NK-cells
CD9	Papillary (chromophile) renal cell carcinoma, clear cell renal cell carcinoma, Pre B-ALL, AML (M3)	Astrocytoma, ovarian carcinoma, vascular tumors	B-cells, plasma cells, activated T-cells, megakaryocytes/ platelets, macrophages, granulocytes, endothelial cells, brain, cardiac muscle, GIT epithelium, renal tubular epithelium, liver, ovarian surface epithelium, thyroid
CD10	Burkitt's lymphoma, ALL, angioimmunoblastic lymphoma, endometrial stromal tumors, renal cell carcinoma	Follicular lymphoma, B-cell lymphoma, plasma cell neoplasia, hepatocellular carcinoma, placental site trophoblastic tumor, rhabdomyosarcoma, Ewing's sarcoma	Pre-B- and pre-T-cells, germinal center cells, granulocytes, adrenal cortex, endometrial stroma cells, hepatocytes, renal tubular cells, endothelial cells, myoepithelial cells, fibroblasts, brain, choroid plexus
CD 11b	AML (M1-5)	Hairy cell leukemia	Follicular dendritic cells, promyelocytes and granulocytes, NK-cells, macrophages
CD11c	Hairy cell leukaemia	AML (M4-M5), Langerhans cell histiocytosis, B-cell lymphoma	T-, B- and NK-cells; granulocytes, monocytes, macrophages, dendritic cells

Table 2.6 *(Continued)*

Antigen	Main diagnostic use	Possible expression in other tumors	Expression in normal cells
CD13	AML (M1–M6), CML	ALCL, pre-B- and T-ALL	Myeloblasts, promyelocytes, and myelocytes, macrophages/monocytes, granulocytes, interdigitating dendritic cells, immature mast cells, endothelial cells, fibroblasts, osteoclasts, perineurium of peripheral nerves, intestinal and bile duct epithelial cells
CD14	B-CLL, follicular lymphoma	AML (M4-M5), DLBCL	Macrophages, B-cells, granulocytes, Langerhans cells, dendritic cells
CD15	Hodgkin's lymphoma (Reed–Sternberg cells)	Myeloid leukemia, adenocarcinoma, peripheral T-cell lymphoma, ALCL	Neutrophils and eosinophils, activated B- and T-cells, proximal tubules of kidney
CD19	B-cell lymphoma/ leukemia	AML (M0), blast phase of CML	B-cells, follicular dendritic cells
CD20	B-cell lymphoma/ leukemia		B-cells, follicular dendritic cells
CD21	Follicular dendritic cell sarcoma	Hairy cell leukemia, mantle cell and marginal zone lymphoma	Mature B cells, follicular dendritic cells, pharyngeal and cervical epithelial cells
CD22 (BL-CAM)	Hairy cell leukaemia	Pre-B-ALL, B-cell lymphoma	B-cells
CD23	B-CLL, follicular dendritic cell tumors	Mediastinal large B-cell lymphoma, lymphoplasmacytic lymphoma, hairy cell leukemia, DLBCL, Reed–Sternberg cells	Activated B-cells, subpopulations of T-cells, monocytes, Langerhans cells, follicular dendritic cells, intestinal epithelium
CD25 (interleukin-2 receptor)	Hairy cell leukemia, ALCL	T-cell lymphoma, mast cell tumors	Activated B- and T-cells, macrophages, myeloid precursors, oligodendrocytes

(continued)

Table 2.6 *(Continued)*

Antigen	Main diagnostic use	Possible expression in other tumors	Expression in normal cells
CD30	Anaplastic large cell lymphoma, Reed–Sternberg cells in classic Hodgkin's lymphoma, mediastinal large B-cell lymphoma	Embryonal carcinoma, mesothelioma	Granulocytes, monocytes, activated B-, T-cells, NK-cells, monocytes, decidua
CD31	Vascular tumors	Plasmacytoma, Langerhans cell histiocytosis, granulocytic sarcoma, Ewing's sarcoma	Endothelial cells, megakaryocytes/platelets, macrophages/monocytes, Kupffer cells, osteoclasts, granulocytes, T-, NK- and plasma cells
CD33	AML (M0, M1-M5), CML	AML (M6, M7), B- and T-ALL	Myeloid cells, monocytes
CD34	Vascular tumors, dermatofibrosarcoma protuberans, GIST, solitary fibrous tumor, epithelioid sarcoma, Kaposi's sarcoma, AML (M0), granulocytic sarcoma, neurofibroma, liposarcoma	Pre-B-ALL, AML (M7), alveolar soft part sarcoma, congenital and infantile fibrosarcoma, inflammatory fibrous polyp of GIT, breast fibroadenoma, giant cell fibroblastoma, juxtaglomerular cell tumor, superficial acral fibromyxoma	Hematopoietic progenitor cells, B- and T-lymphocyte precursors, endothelial cells, hepatic sinusoids, interstitial cells of Cajal, endometrial stroma, fibroblasts
CD35	Follicular dendritic cell sarcoma	Mantle cell lymphoma, marginal zone lymphoma	Follicular dendritic cells, red blood cells, populations of B-, T- and NK-cells
CD38	Plasma cell neoplasia, plasmablastic lymphoma	Pre-T-ALL, primary effusion lymphoma, types of B-cell lymphoma	Plasma cells, erythroid and myeloid precursors, early B- and T-cells, pancreatic islets, neurons
CD41	AML (M7)		Megakaryocytes/platelets
CD43	T-/NK-cell lymphomas	B-ALL, Burkitt's lymphoma, mantle cell lymphoma, marginal zone lymphoma, adenoid cystic carcinoma	Activated B-cells, T-cells, NK-cells, plasma cells, granulocytes

Table 2.6 *(Continued)*

Antigen	Main diagnostic use	Possible expression in other tumors	Expression in normal cells
CD45	Lymphoma/ leukemia	Giant cell tumor of tendon sheet, interdigitating dendritic cell and histiocytic tumors	Lymphocytes, monocytes, macrophages, histiocytes, mast cells
CD45RA	B-cell lymphoma/ leukemia	Reed–Sternberg cells in classic Hodgkin's lymphoma	B-cells, small population of T-cells and monocytes
CD45RO, (UCHUL)	T-cell lymphoma/ leukemia		Thymocytes, subpopulation of resting CD4 and CD8 T-cells, activated T-cells
CD52	Lymphoma	Hodgkin's lymphoma (Reed–Sternberg cells)	B- and T-lymphocytes, monocytes, germ cells
CD56	NK-lymphomas, neuroendocrine tumors (small cell carcinoma and carcinoid), pheochromocytoma, neuroblastoma, paraganglioma, Merkel cell tumor	Schwannoma, synovial sarcoma, embryonal and alveolar rhabdomyosarcoma, meningioma, melanoma, chordoma, epithelioid sarcoma	NK-cells, T-cells, cerebellum and brain cortex, neuromuscular junctions, neuroendocrine tissue, neurons, osteoblasts
CD57 (leu 7)	Neuroendocrine tumors	T-cell lymphoma and pre T-ALL, chordoma, MPNST, germ cell tumors, papillary thyroid carcinoma, thymic carcinoma, germ cell tumors, synovial sarcoma, carcinoid tumors, schwannoma, mesothelioma, astrocytoma and glioblastoma	NK-cells, Schwann cells, gastric chief cells, adrenal medulla, pancreatic islets, renal tubular epithelium, epithelial cells of thymic cortex, retina, oligodendrocytes
CD61	AML (M7)		Megakaryocytes/platelets
CD63	Melanoma and melanocytic tumors	Angiomyolipoma, cellular neurothekoma, dermatofibrosarcoma protuberans, renal oncocytoma and chromophobe carcinoma	Endothelial cells, fibroblasts, macrophages, mast cells, osteoclasts

(continued)

Table 2.6 *(Continued)*

Antigen	Main diagnostic use	Possible expression in other tumors	Expression in normal cells
CD66e carcinoembryonic antigen (CEA)	Gastrointestinal and pancreatic adenocarcinoma, cholangiocellular and hepatocellular carcinoma	Breast cancer, cervical adenocarcinoma, lung cancer, ovarian mucinous tumors, medullary thyroid carcinoma	GIT mucosa, hepatocytes, thyroid C cells, granulocytes
CD68	Histiocytic tumors, dendritic cell tumors, AML (M4/M5), giant cell tumors	Fibrous histiocytoma, nodular fasciitis, villonodular synovitis, granular cell tumor, inflammatory myofibroblastic tumor, mast cell disease, hairy cell leukemia	Macrophage, monocytes, osteoclasts, Kupffer cells, mast cells, synovial cells, microglia, dendritic cells, fibroblasts, Langerhans cells, myeloid cells, CD34+ progenitor cells, neutrophils, B- and T-cells
CD74	B-cell lymphoma/ leukemia, Hodgkin's and Reed–Sternberg cells	Renal clear cell and chromophobe carcinoma, histiocytic tumors	B-cells, activated T-cells, null cells, interdigitating dendritic cells, Langerhans cells, macrophages, endothelial cells, thymic epithelium
CD79a	B-cell leukemia/ lymphomas		B-cells, small population of CD3+ T-cells
CD99 (MIC2)	Ewing's sarcoma/ PNET, T- and B-ALL, solitary fibrous tumor	T-cell lymphoma, ALCL, AML, GIST, breast metaplastic carcinoma, ovarian sex cord stromal tumors, synovial sarcoma, Merkel cell carcinoma, endocrine and neuroendocrine tumors, desmoplastic small round cell tumor, melanoma, nephroblastoma, ependymoma, mesenchymal chondrosarcoma, extrarenal malignant rhabdoid tumor, rhabdomyosarcoma, meningeal hemangiopericytoma	Cortical thymic lymphocytes, ovarian granulosa cells, pancreatic islets, Sertoli cells, endothelial cells, T-cells and activated B-cells, ependymal cells

Table 2.6 *(Continued)*

Antigen	Main diagnostic use	Possible expression in other tumors	Expression in normal cells
CD103	Hairy cell leukaemia	Enteropathy-type T-cell lymphoma, AML	Intraepithelial CD8 + lymphocytes, intestinal T-lymphocytes
CD117 (c-kit)	GIST	Clear cell sarcoma, adenoid cystic carcinoma, small cell lung carcinoma, Ewing sarcoma, synovial sarcoma, osteosarcoma, angiosarcoma, AML, CML, mast cell disease, renal oncocytoma, renal chromophobe carcinoma, thymic carcinoma, seminoma, gliomas	Interstitial cells of Cajal, hematopoietic progenitor cells, melanocytes, germ cells, glial and Purkinje cells, endothelial cells, mast cells, renal tubular cells, ovarian stroma and corpus lutum
CD138 (syndecan-1)	Plasma cell tumors (myeloma, plasmacytoma)	Primary effusion lymphoma, thyroid carcinoma, breast carcinoma, adrenal cortical tumors, keratoacanthoma0	B-cell precursors, plasma cells, stratified squamous epithelium
CD141 (thrombo-modulin)	Mesothelioma, transitional cell carcinoma	Squamous cell carcinoma, trophoblastic tumors, vascular tumors, synovial sarcoma	Endothelial cells, urothelium, mesothelial cells, keratinizing epithelial cells, meningeal lining cells, platelets/ megakaryocytes, monocytes, neutrophils, smooth muscle cells, syncytiotrophoblasts, synovial lining cells
CD235a (glycophorin A)	AML (M6)		Erythroid cells
CD246 (anaplastic lymphoma kinase; ALK1; p80)	Anaplastic large cell lymphoma	Inflammatory myofibroblastic tumor	Normal small intestine, T-cells, brain, colon, prostate
CDK4	Liposarcoma	Adrenal oncocytoma, rhabdomyosarcoma,	

(continued)

Table 2.6 *(Continued)*

Antigen	Main diagnostic use	Possible expression in other tumors	Expression in normal cells
CDX-2	Colorectal adenocarcinoma	Gastric adenocarcinoma, GIT carcinoids, islet pancreas tumors, mucinous adenocarcinomas of ovary, bronchoalveolar carcinomas of lung, adenocarcinomas of urinary bladder	Intestinal epithelium, pancreatic epithelial cell
Chromogranin A	Neuroendocrine tumors: pituitary adenomas, medullary thyroid carcinoma, pheochromocytoma, islet cell tumors, Merkel cell tumors, small cell carcinoma, carcinoid and neuroendocrine carcinoma	Oligodendroglioma, neuroblastoma, PNET, paraganglioma	Neuroendocrine cells
Clusterin (apolipoprotein J)	ALCL, follicular dendritic cell tumors	DLBCL, carcinomas of breast, colon, pancreas and prostate	Germ cells
c-myc	Burkitt's lymphoma	breast carcinoma	
Cyclin D1 (bcl-1)	Mantle cell lymphoma	Inflammatory pseudotumor (myofibroblastic tumor), parathyroid adenoma/carcinoma, hairy cell leukemia	Cells in the G1 phase of cell cycle
Cytokeratin 1 (CK1)	Cloacogenic carcinoma		Keratinizing epithelium, suprabasal cells
Cytokeratin 4 (CK4)	Squamous cell carcinoma	Adamantinoma	Non-keratinizing stratified epithelium, suprabasal cells
Cytokeratin 5 (CK5)	Squamous cell carcinoma, mesothelioma, myoepithelial tumors	Detection of myoepithelial cells in prostatic carcinoma	Squamous epithelium, basal epithelial cells, myoepithelial cells, mesothelial cells, cornea
Cytokeratin 6 (CK6)	Squamous cell carcinoma	Poorly differentiated breast carcinoma	Suprabasal cells

Table 2.6 *(Continued)*

Antigen	Main diagnostic use	Possible expression in other tumors	Expression in normal cells
Cytokeratin 7 (CK7)	Adenocarcinoma of lung, GIT, pancreas, biliary tract, breast, endometrium, transitional cell carcinoma, serous ovarian tumors	Merkel cell carcinoma, cloacogenic carcinoma	Epithelium of GIT, salivary glands, biliary tract, pancreas, lung, female genital tract, renal collecting ducts transitional epithelium, mesothelial cells, thyroid follicle cells
Cytokeratin 8 (CK8)	Adenocarcinoma of lung, GIT, pancreas, biliary tract, breast, endometrium, transitional cell carcinoma, hepatocellular carcinoma, renal cell carcinoma, neuro-endocrine carcinoma	Leiomyosarcoma	Epithelium of GIT, salivary glands, biliary tract, pancreas, lung, female genital tract, hepatocytes, proximal renal tubules, transitional epithelium, mesothelial cells, smooth muscle cells, myofibroblasts, arachnoid cells
Cytokeratin 10 (CK10)	Squamous cell carcinoma	Breast ductal carcinoma	Keratinizing epithelium
Cytokeratin 13 (CK13)	Squamous cell carcinoma		Mature non-keratinizing squamous epithelium, basal and intermediate cells of transitional epithelium
Cytokeratin 14 (CK14)	Squamous cell carcinoma, basal cell carcinoma, Hürthel cell tumors	Detection of myoepithelial cells in prostatic carcinoma	Keratinizing and non-keratinizing squamous epithelium, basal and myoepithelial cells in salivary glands, breast prostate and uterus, Hürthel thyroid cells
Cytokeratin 15 (CK15)		Squamous cell carcinoma, basal cell carcinoma	Basal keratinocytes, fetal epidermis
Cytokeratin 17 (CK17)	Squamous cell carcinoma, basal cell carcinoma, transitional cell carcinoma, bile duct and pancreas adenocarcinoma	Carcinoma of breast lung, ovary and thyroid, mesothelioma	Myoepithelial and basal cells, hair shaft epithelium

(continued)

Table 2.6 *(Continued)*

Antigen	Main diagnostic use	Possible expression in other tumors	Expression in normal cells
Cytokeratin 18 (CK18)	Adenocarcinoma of lung, GIT, pancreas, biliary tract, breast, endometrium, transitional cell carcinoma, hepatocellular carcinoma, renal cell carcinoma, neuro-endocrine carcinoma	Leiomyosarcoma	Epithelium of GIT, salivary glands, biliary tract, pancreas, lung, female genital tract, hepatocytes, proximal renal tubules, transitional epithelium, mesothelial cells, smooth muscle cells, myofibroblasts, arachnoid cells
Cytokeratin 19 (CK19)	Adenocarcinoma of lung, GIT, pancreas, biliary tract, breast, endometrium, transitional cell carcinoma	Neuroendocrine tumors and papillary thyroid carcinoma, mesothelioma	Epithelium of GIT, salivary glands, biliary tract, pancreas, lung, female genital tract, transitional epithelium, mesothelial cells, thyroid follicle cells, basal squamous epithelium
Cytokeratin 20 (CK20)	Adenocarcinoma of GIT, transitional cell carcinoma, mucinous ovarian tumors	Merkel cell carcinoma	Gastric and colorectal epithelium, transitional epithelium
D2-40	Lymphangioma and other tumors of lymphatic vessels	Germ cell tumors, Kaposi's sarcomas, schwannoma, meningioma, GIST, synovial sarcoma, leiomyosarcoma, desmoid, malignant peripheral nerve sheath tumor	Lymphatic endothelium
Desmin	Rhabdomyosarcoma and rhabdomyoma, smooth muscle tumors	Desmoplastic small round cell tumor, alveolar soft part sarcoma, tenosynovial giant-cell tumor	Smooth and striated muscle, myoblasts and myofibroblasts
Epidermis growth factor receptor (EGFR)	Endometrial stromal sarcoma, squamous cell carcinoma, detection of EGFR overexpression for immunotherapy in different carcinomas	Colorectal adenocarcinomas, small cell carcinoma of the lung, breast, ovarian and bladder carcinoma, glioblastoma multiforme	Simple and squamous epithelium, endometrial stromal cells

Table 2.6 *(Continued)*

Antigen	Main diagnostic use	Possible expression in other tumors	Expression in normal cells
Epithelial antigen (Ber-EP4)	Epithelial tumors, basal cell carcinomas		Human epithelial cells
Epithelial membrane antigen (EMA; CD227)	Adenocarcinoma of different origin, ALCL	Epithelioid sarcoma, meningioma, chordoma, plasmacytoma	Apical surface of glandular and ductal epithelial cells, activated T-cells, plasma cells, monocytes, follicular dendritic cells
Estrogen receptor (ER)	Breast and endometrium carcinoma	Ovarian serous, mucinous and endometrioid carcinoma, transitional cell carcinoma, hepatocellular carcinoma	Breast and endometrial epithelium, endometrial stromal cells and myometrium
Factor VIII (FVIII)	Vascular tumors		Endothelial cells, megakaryocytes and platelets, mast cells
Fascin	Reed–Sternberg cells in classic Hodgkin's lymphoma, follicular and interdigitating dendritic cell tumors	Adenocarcinoma of breast, colon, biliary tract, pancreas, lung, ovary, skin; papillary transitional cell carcinoma of bladder, anaplastic large cell lymphoma, diffuse large B-cell lymphoma	Interdigitating and follicular dendritic cells, endothelial cells
Fli-1	Ewing's sarcoma, vascular tumors	Lymphoblastic lymphoma	Endothelial cells, T-cells
Galectin-3	Papillary thyroid carcinoma	Parathyroid carcinoma, carcinomas of breast colon, liver and tongue	Endothelial cells, peripheral nerves, granulocytes
Glial fibrillary acidic protein (GFAP)	CNS tumors (astrocytoma, glioblastoma, oligodendroglioma, medulloblastoma, ependymoma), retinoblastoma, neurolemoma, neurothekoma, MPNST	Salivary gland tumors (myoepithelial adenoma/carcinoma, basal cell adenoma/carcinoma, pleomorphic adenoma)	Astrocytes, oligodendrocytes, subset of CNS ependymal cells, Schwann's cells, Kupffer cells, chondrocytes
Glycophorin C	AML (M6)		Erythroid cells

(continued)

Table 2.6 *(Continued)*

Antigen	Main diagnostic use	Possible expression in other tumors	Expression in normal cells
Glypican-3 (GPC3)	Hepatocellular carcinoma, hepatoblastoma	Yolk sac tumor, choriocarcinoma	
Granzyme B	T-cell and NK-lymphomas		Cytotoxic T- and NK-cells
Gross cystic disease fluid protein 15 (GCDFP-15)	Breast carcinoma	Salivary gland tumors, skin adnexal tumors, apocrine tumors	Apocrine cells
HBME	Mesothelioma, thyroid carcinoma		Mesothelial cells
Hep Par1	Hepatocellular carcinoma, hepatoblastoma	Mucosa with intestinal metaplasia, tumors with hepatoid differentiation	Hepatocytes
HER-2 (c-erb-2)	Breast carcinoma, detection of HER-2 overexpression for immunotherapy	Ovarian and endometrium carcinomas	Breast epithelium
HMB-45	melanoma, Spitz and cellular blue nevi, PEComa	Angiomyolipoma, sugar tumor of lung, pheochromocytoma,	Melanocytes/nevus cells, mononuclear cells
Human chorionic gonadotropin (hCG)	Trophoblasts in germ cell tumors		Trophoblasts
Human placental lactogen	Placental site trophoblastic tumor		Trophoblasts
Inhibin A	Adrenocortical tumors, granulosa cell tumor, sex-cord stromal tumors	Thecoma and fibrothecoma, choriocarcinoma and trophoblastic lesions	Sertoli cells, granulosa cells, theca interna, adrenal cortex, brain tissue
Langerin	Langerhans cell histiocytosis		Langerhans cells
MAGE-1	Melanoma	Hepatocellular carcinoma	Germ cells: spermatogonia and spermatocytes, placenta
Mart-1 (melan A)	Adrenal cortical tumors, melanoma sex-cord stromal tumors	Angiomyolipoma, osteosarcoma,	Adrenal cortex, nevus cells, brain tissue, granulosa ant theke cells, Leydig cells
Mesothelin	Mesothelioma, ovarian surface carcinomas	Some adenocarcinoma and squamous cell carcinoma types	Mesothelial cells, kidney, tracheal mucosa

Table 2.6 *(Continued)*

Antigen	Main diagnostic use	Possible expression in other tumors	Expression in normal cells
MDM2	Liposarcoma	Clear cell sarcoma, angiosarcoma, adrenal oncocytoma, Kaposi's sarcoma, rhabdomyosarcoma	
Microphthalmia transcription factor (MITF)	Nevus lesions, melanoma	Clear cell sarcoma, osteoclastoma, MPNST	Mast cells, melanocytes and nevus cells, Schwann cells, osteoclasts, smooth muscle
MUM-1	Plasma cell tumors	DLBCL	B-cells (centrocytes), plasma cells, activated T-cells
Myeloperoxidase (MPO)	AML (M4, M5)		Myeloid cells, monocytes
Myo D1	Rhabdomyosarcoma		Fetal muscle, myoblasts
Myf-3	Rhabdomyosarcoma		Fetal muscle, myoblasts
Myogenin (Myf-4)	Rhabdomyosarcoma	Wilms tumor	Fetal muscle
Myoglobin	Tumors with skeletal muscle differentiation/ rhabdomyosarcoma		Striated muscle
Myosin	Myogenic tumors		Skeletal muscle in sarcomeric form and smooth muscle in non-sarcomeric (heavy chain) form
Nestin	PNET, GIST, meningioma	Granular cell tumor, gliomas, melanoma	Neuroectodermal stem cells, fetal muscle
Neurofilament	Medulloblastoma, retinoblastoma, neuroblastoma, ganglioglioma	Merkel cell tumor, pancreatic endocrine neoplasms, carcinoid, small cell carcinoma, parathyroid tumors	Neuronal cells
Neuron-specific enolase (NSE) γ-subunit	Neuroectodermal and neuroendocrine tumors	Melanoma. Merkel cell tumor, meningioma, renal cell carcinoma	Neurons, neuroendocrine cells, megakaryocytes, T-lymphocytes, smooth and striated muscle
Oct-2	B-cell lymphomas, Reed Sternberg cells in classical Hodgkin's lymphoma		B-cells

(continued)

Table 2.6 *(Continued)*

Antigen	Main diagnostic use	Possible expression in other tumors	Expression in normal cells
Oct-4	Seminoma/IGCN, embryonal carcinoma	Ovarian dysgerminoma, CNS germinoma	Germ cells
PAX-5 (BSAP)	B-cell lymphoma/ leukemia, Reed–Sternberg cells in classic Hodgkin's lymphoma	Merkel cell carcinoma, small cell carcinoma	Pre-B- to mature B-cells
Placental alkaline phosphatase (PLAP)	Seminoma, embryonal carcinoma	Proximal GIT tumors, lung and ovarian carcinoma	Placental syncytiotrophoblasts
Podoplanin	Mesothelioma, adenomatoid tumor, lymphangioma	Vascular tumors, skin adnexal carcinomas	Mesothelial cells, endothelial cells of lymphatic vessels, granulosa cells, osteocytes, ependymal cells
Progesterone receptor (PR)	Breast and endometrium carcinoma		Breast and endometrial cells, endometrium stromal cells
Prostatic acid phosphatase (PAP)	Carcinoma of prostate		Prostate glandular epithelium
Prostate-specific antigen (PSA)	Carcinoma of prostate	Salivary duct carcinoma	Prostate secretory and ductal epithelium
Protein gene product 9.5	Malignant nerve sheet tumor, paraganglioma	Carcinoid tumors	Neurons and nerve fibers, neuroendocrine cells, renal tubular epithelium, spermatogonia, Leydig cells
Renal cell carcinoma marker (RCC, gp200)	Renal cell carcinoma (papillary carcinoma clear cell carcinoma)	Parathyroid adenoma, breast carcinoma, embryonal carcinoma	Renal proximal tubular brush border, epididymal tubular epithelium, breast parenchyma, thyroid follicles
S100	Melanomas, schwannoma, neuroendocrine tumors	Malignant peripheral nerve sheath tumors, chondrosarcoma and chondroblastoma clear cell sarcomas, myoepithelial tumors, granulosa cell tumor	Neural crest cells (glial cells, Schwann cells, melanocytes and nevus cells), chondrocytes, adipocytes, myoepithelial cells, macrophages, adrenal medulla and paraganglia, Langerhans cells, dendritic cells

Table 2.6 *(Continued)*

Antigen	Main diagnostic use	Possible expression in other tumors	Expression in normal cells
Synaptophysin	Neuroendocrine tumors: pituitary adenomas, medullary thyroid carcinoma, pheochromocytoma, islet cell tumors, small cell carcinoma, carcinoid and neuroendocrine carcinoma	Medulloblastoma, retinoblastoma, neurocytoma, ependymoma, neuroblastoma, adrenocortical tumors, Merkel cell tumors	Neuronal and neuroendocrine cells, carotid body cells, adrenal cortex and medulla
TdT	B- and T-ALL	AML	B- and T-cell precursors, cortical thymocytes
Thyroglobulin	Thyroid tumors		Thyroid follicle cells
Thyroid peroxidase	Thyroid tumors		Apical surface of thyroid follicle cells
TLE1 protein	Synovial sarcoma	Hemangiopericytoma, schwannoma, solitary fibrous tumor, fibroxanthoma, clear cell sarcoma, carcinosarcoma, chondrosarcoma	
Tryptase	Mast cell tumors		Mast cells
Thyroid transcription factor-1 (TTF-1)	Lung cancer (adenocarcinoma, small cell carcinoma), thyroid tumors	Small cell carcinoma of different locations	Type II pneumocytes and Clara cells of lung, thyroid follicular and parafollicular C cells, diencephalon
Villin	GIT adenocarcinoma	Lung adenocarcinoma, renal cell carcinoma	Brush border of GIT and biliary tract epithelium, renal proximal tubules
von Willebrand factor (vWF)	Vascular tumors	Osteosarcoma	Endothelial cells, megakaryocytes
VS38c (plasma cell marker)	Plasma cell tumors (myeloma, plasmacytoma), lymphoma with plasmacytic differentiation	Rare carcinomas, melanoma	Plasma cells and plasmablasts, melanocytes, epithelial cells (mucous glands, pancreatic epithelium, secretory breast cells, thyroid follicles), osteoblasts

(continued)

Table 2.6 *(Continued)*

Antigen	Main diagnostic use	Possible expression in other tumors	Expression in normal cells
Wilms tumor protein-1 (WT-1)	Nephroblastoma, mesothelioma, metanephric adenoma, ovarian serous carcinoma	AML, desmoplastic small round cell tumor, endometrial stromal sarcoma, uterine leiomyosarcoma, granulosa cell tumor, thecoma, rhabdoid tumor	Kidney, mesothelial cells, granulosa cells, Sertoli cells, fallopian tube endometrial stroma, spleen

2.5
Immunohistochemical Staining: Avidin–Biotin Horseradish Detection System

There are many methods used for the visualization of the antigen–antibody reaction in the immunohistochemistry but the avidin–biotin horseradish-based detection method (ABC) proved to be a simple, highly sensitive, robust and inexpensive system. The avidin–biotin horseradish detection method uses the advantage of the high affinity of avidin for biotin. Avidin is a 68-kDa glycoprotein having four potent binding sites for biotin and 1 mg of avidin is able to bind about 14.8 μg of biotin. The use of streptavidin instead of avidin helps to increase the sensitivity of the immunohistochemical reaction, decrease the background and increase the precision of the immune staining results. Biotin is a water-soluble vitamin (vitamin H), able to bind to several proteins without altering their properties. Biotinylated antibodies are used as secondary antibodies against the primary antibodies. The visualization of the antibody–antigen complex is based on the enzymatic conversion of a suitable chromogen into a colored precipitate by horseradish peroxidase. Horseradish peroxidase is a heme-containing protein isolated from horseradish roots composed of a mix of enzymes able to oxidize several substrates such as diaminobenzidine tetrahydrochloride (DAB).

Another simple detection method is the polymer-based detection system. This detection system does not depend on the avidin–biotin interaction for localization of the marker enzyme to the antigen but bases on an enzyme conjugated polymer backbone with attached secondary (link) antibody. This method has the advantage to join both secondary antibody step and the detection complex incubation step in a single step. Consequently, it is a more rapid method and the blocking of the endogenous biotin activity is not necessary.

The immunohistochemical reaction is usually performed on sections prepared from formalin-fixed paraffin-embedded tissue blocks. Several fixatives are used to conserve the tissue prior to paraffin embedding, but best results are obtained using formalin (formaldehyde water solution), which is the most common used fixative. Formalin reacts with the basic amino acids of the cellular protein forming methylene bridges (cross-linking) with minimal damage of cell structures without coagulation. For the optimal fixation of tissue the use of 4–10% buffered neutral formalin (pH 6.8–7.2) is essential and offers the best conditions for the immunohistochemical reaction (see below). Generally, if good H&E sections cannot be obtained from processed tissue, the immunohistochemistry will also be sub-optimal.

The rapid fixation of the tissue after excision is necessary to prevent the damage of antigens due to autolysis. Also, the fixation time must not exceed 24 h, as prolonged formalin-fixation can be deleterious to the immune reactivity of many antigens epitopes due to the chemical modification and cross-linking occurring within the antigen molecules.

Bone marrow trephine biopsies must be decalcified before paraffin embedding by the incubation in a saturated Na_2EDTA solution for 6–24 h. Heating by microwave can reduce this period.

2.5.1 Standard Staining Protocol (ABC Method)

Immunohistochemistry is a powerful and very useful tool in diagnostic pathology; consequently, the staining protocol must be optimized to ensure that the results are reproducible, reliable and accurate. A standard immunohistochemistry method includes the following steps:

1. Cutting optimal paraffin sections, mounting on adhesive slides and drying.
2. Deparaffinization and hydration.
3. Antigen retrieval if needed.
4. Blocking the endogenous peroxidase activity.
5. Elimination of non-specific background by incubation with non-immune serum.
6. Binding the epitope to the specific antibody.
7. Reaction with a biotinylated secondary (link) antibody.
8. Reaction with the detection complex.
9. Developing the reaction product with chromogen, counterstaining, dehydration, covering, and interpretation.

The following staining protocol offers an uncomplicated technique, which can be easily used in the routine histopathology performed on sections from formalin-fixed paraffin-embedded tissue blocks:

1. Cut thin (2–3 μm) paraffin sections, transfer on silane-coated slides (see below), dry overnight at 37 °C. Alternatively incubate 1–2 h at 55 °C or in a microwave oven for 3–4 min on high power.

2. Deparaffinize and hydrate to H_2O:
   - 30 min xylene;
   - 2 min absolute ethanol;
   - 2 min 96% ethanol;
   - 2 min 70% ethanol;
   - 2 min 40% ethanol;
   - H_2O; avoid drying of tissue sections.

3. Antigen unmasking (if necessary): Most antigens undergo different chemical modifications because of formalin-fixation, which can change the native three-dimensional protein conformation. Changing the normal three-dimensional structure of the epitope makes it more difficult or impossible for the antibody to bind to its target antigen. Two main methods are used to restore the original structure of antigens (unmasking) so as to be recognized by the primary antibody:

3.1. Heat-induced unmasking:
This method depends on the use of a high temperature, such as a microwave, pressure cooker or water bath, with various kinds of buffers. It seems that heat-induced unmasking breaks the molecular cross-links induced by formalin. The effective temperature ranges between 90 °C and 120 °C. The

heating time ranges between 20 min and 2 h but generally the higher the temperature the shorter the required time. The used buffers have different pH, usually ranging between 6 and 9. The citrate-based buffers (pH 6–7) are the most common used buffers, but other buffers such as EDTA or Tris-based buffers with higher pH (8.0–9.9) are recommended for some antibodies. Generally, the heat-induced unmasking method is suitable for more than 75% of the antigens. A further benefit of using a high temperature is the enhancement of the reactivity of any endogenous biotin present in the processed tissue, as the endogenous biotin may cause artifacts and a non-specific background.

For a simple and rapid heat unmasking, the following protocol based on microwave heating is recommended:

- Hydrate slides with distilled H_2O.
- Place the slides in a microwave jar or in a high-pressure cooker and load with suitable buffer.
- Heat by microwave for 5 min at 750 W, alternatively for 20–30 min in a suitable pressure cooker.
- Fill the jar again with the same buffer; wait for 5 min, keep microwave oven door open.
- Heat by microwave for another 5 min at 750 W.
- Fill the jar with the buffer; keep at room temperature for 15 min.

3.2. Enzymatic digestion:

The cleavage of the molecular cross-links between cellular proteins by the proteolytic enzyme allows the antigen to return to its normal conformation. For the enzymatic unmasking different proteolytic enzymes can be used such as pronase, protease type XXIV, pepsin, proteinase K, ficin or trypsin.

The following protocol is recommended for the majority of antibodies needing enzymatic digestion:

- Hydrate slides with distilled H_2O.
- Incubate in phosphate buffer saline (PBS) for 5 min.
- Add Pronase solution with 0.05% working concentration to cover the section (100–150 μl) and incubate for 10–15 min at 37 °C in humid chamber. Trypsin or protease with 0.1% working solution can be also used.–Wash in PBS for 5 min.

4. Blocking of endogenous peroxidase: 0.3% hydrogen peroxide solution gives good results. To prepare, add 1 ml of 30% H_2O_2 solution either to 100 ml H_2O or to 100 ml absolute methanol to give 0.3% final H_2O_2 concentration. The slides are incubated for 10 min at room temperature in a jar filled with the blocking solution.
5. Incubate in PBS for 1–2 min.
6. Blocking of endogenous biotin (optional if necessary): it is recommended to treat tissue types with highly endogenous biotin such as liver tissue, by biotin-blocking reagent. For this purpose, incubation in exogenous avidin such as

0.1% avidin solution or solution of egg white (two egg whites diluted with 200 ml distilled H_2O) is recommended. The exogenous avidin must be next inactivated by avidin-blocking reagent containing exogenous biotin (such as 0.01% biotin solution or skim milk). Incubate the sections for 15 min in each reagent in room temperature. High temperature can be also used to inactivate biotin.

7. Incubate in PBS for 5 min.
8. Add 100 μl of non-immune serum (normal serum such as horse or goat serum) and incubate for 15 min.
9. Add 50–100 μl (2–3 drops) of primary antibody with adequate concentration; incubate for 30–60 min at 25–37 °C. PBS or ready to use buffer can be used as diluents for the primary antibody. A low concentration of detergents such as Tween 20 increases the penetration of the antibody.
10. Wash with H_2O and incubate in PBS for 10 min.
11. Add 50–100 μl (2–3 drops) of biotinylated secondary antibody. Incubate for 15–30 min at 37 °C.
12. Wash with H_2O and incubate in PBS for 10 min.
13. Add 50–100 μl (2–3 drops) of the avidin/biotinylated horseradish peroxidase complex (ABC). Incubate for 15 min at 37 °C.
14. Wash with H_2O and incubate in PBS for 10 min.

15. Add 100 μl of one of the following pre-warmed (~25 °C) chromogen solutions and incubate for 8 min:

15.1. DAB: DAB (3,3′-diaminobenzidine tetra-hydrochloride, $C_{12}H_{14}N_4 \cdot 4HCl$) gives a brown hydrophobic end-product, resistant to alcohol and xylene. DAB chromogen seems to be the most economical, reliable and reproducible chromogen in the immunohistochemistry.
The use of metal salts such as copper sulfate increase the sensitivity of the reaction (see below).

15.2. AEC: AEC (3-amino-9-ethylcarbazole) gives a red-brown alcohol soluble end-product. AEC requires aqueous mounting medium.
Both chromogens are classified as carcinogens and contact with skin must be avoided.

16. Stain enhancement (optional step): this step is used to increase the stain intensity when staining is weak. Many salts are used for the stain enhancement, which give different stain colors. For this purpose, slides, after DAB staining, are incubated in one of the following salts for 4–5 min at 25 °C.

   - Copper sulfate: 5% copper sulfate solution. This salt gives dark brown stain.
   - Nickel ammonium sulfate: 0.05% nickel ammonium sulfate solution. This salt gives a black stain.
   - Nickel chloride: 0.05% nickel chloride solution. This salt gives a gray-black stain.

17. Wash with H_2O.
18. Staining with hematoxylin for 4–8 s or mount without counterstaining.
19. Wash with H_2O for 5 min.

20. Dehydrate with 40%–70%–96%–100% ethanol and xylene.
21. Mount using xylene-based mounting media.

2.5.2 Troubleshooting and Problems Occurring During Immunohistochemical Staining

2.5.2.1 Overstaining

1. High concentration of the primary antibody or long incubation time.
2. Too high an incubation temperature.
3. Long incubation time with either secondary antibody or detection system.

2.5.2.2 Non-Specific Background Staining

1. Inadequate washing between steps.
2. Tissue was allowed to dry before fixing or during staining.
3. Folds within the tissue section trapped reagents.
4. Tissue contains endogenous peroxidase.
5. Tissue contains endogenous biotin.
6. Antigen migrated in tissue (common in frozen sections).
7. Excessive tissue adhesive on slides.
8. Inadequate blocking with protein block.
9. Polyclonal antibodies that are not affinity purified may contain non-specific immunoglobulins.

2.5.2.3 Weak Staining

1. Low concentration of primary antibody or short incubation time.
2. Expired reagents.
3. Dilution of reagents due to inadequate removal of wash buffer between steps.
4. Counterstain or mounting media are incompatible, which may dissolve the chromogen reaction product.
5. Low incubation temperature or cold reagents.
6. Low target expression level or target damage including chemical modification due to fixation, over-fixation or paraffin embedding.
7. Excessive incubation with protein blocking reagents (normal serum or commercial reagents).

2.5.2.4 No Staining

1. Steps were inadvertently left out.
2. There is no antigen in the tissue.
3. Secondary antibody is not compatible with primary antibody (always use anti-mouse for mouse, anti-rabbit for rabbit, etc.).
4. Chromogenic substrate has been replaced with another that is not intended for use with the enzyme substrate.
5. Inactivation of one or more of stain components.

2.5.3
Preparing of Solutions and Slides for Immunohistochemical Staining

2.5.3.1 Buffered Neutral Formalin (10%)

- Formalin CH_2O (formaldehyde water solution >37%) 100 ml
- Sodium phosphate, monobasic monohydrate $NaH_2PO_4xH_2O$ 4.0 g
- Sodium phosphate, dibasic, anhydrous NaH_2PO 6.5 g
- Distilled water to 1 liter

2.5.3.2 Phosphate Buffer Saline 0.01 M

- Na_2HPO_4 1.48 g
- KH_2PO_4 0.43 g
- NaCl 7.2 g
- H_2O 1000 ml

Adjust pH to 7.2

To make washing buffer for immunohistochemistry, add 0.025% Tween 20.

2.5.3.3 Citrate Buffer, pH 6 (10 × Stock Solution)

- Sodium citrate tribasic $[HOC(COONa)(CH_2COONa)_2{\cdot}2H_2O]$ 29.41 g
- H_2O 1000 ml

Adjust pH to 6

Working solution: 1× (10mM)

Different modified citrate-based buffers are widely used and commercially available under different names such as Target Retrieval Solution, pH 6 (Dako) and Citra (BioGenex). Detergents could be added and slightly improve the efficiency of the buffer.

2.5.3.4 EDTA Buffer, pH 8 (1 mM Working Solution)

- EDTA 0.37 g
- H_2O 1000 ml

Adjust pH to 8 using NaOH

2.5.3.5 Tris/EDTA/Tween Buffer, pH 9 (10 × Stock Solution)

- Tris 14.4 g
- EDTA 1.44 g
- H_2O 550 ml

Adjust pH to 9 using 1M HCl

- Tween 20 0.3 ml

Add H_2O up to 600 ml

EDTA and Tris/EDTA-based high pH buffers are useful for antibodies with low expression levels or when the tissue is over-fixed, but the tissue sections suffer under the high pH conditions and a non-specific background appears. EDTA or Tris/EDTA-based buffers are also commercially available under different names such as "Target Retrieval Solution, pH 9" (Dako). Different observations reported non-specific nuclear staining using retrieval solutions containing Tris buffer, the addition of urea could solve this problem.

2.5.3.6 **EDTA Decalcification Solution**

- EDTA disodium salt dihydrate ($C_{10}H_{14}N_2O_8Na_2{\cdot}2H_2O$) 135 g
- 10% buffered formalin 900 ml

Adjust pH 7 with 40% NaOH

2.5.3.7 **DAB Stock Solution**

Dissolve 600 mg DAB ($C_{12}H_{14}N_4{\cdot}4HCl$) in 100 ml PBS. Filter the solution and deposit each 10 ml (1 ml) in separate tube, store at −20 °C.

Working solution: 10 ml (1 ml) DAB stock solution + 90 ml (9 ml) PBS + 10 μl (1 μl) 30% H_2O_2.

2.5.3.8 **Coated Slides for Immunohistochemical Staining**

In order to prevent detaching of tissue sections during immunohistochemical staining and unmasking, specially coated sides (with silane or poly-L-lysine) or electrostatically treated sides (commercially available) must be used. The following simple protocol provides suitable slides for immunohistochemical staining and *in situ* hybridization.

1. Wash slides in 70% ethanol
2. Incubate slides in 10% silane (3-aminopropyltriethoxysilane)-acetone solution for 1 min (20 ml stock silane solution mixed with 180 ml acetone)
3. Wash slides twice with acetone
4. Wash slides in distilled H_2O
5. Dry slides by incubation overnight in a 37 °C oven

3
Molecular Diagnosis of Tumors

3.1
Introduction

Each of the several hundreds of the known benign or malignant tumor types is the result of various multistep alterations of basic cellular functions due to non-lethal somatic changes at the genomic level, followed by clonal proliferation of a single precursor cell carrying these genetic changes. Consequently, the neoplastic transformation is a disease of cellular genome. The genetic changes occurring in neoplastic cells can be grouped in four categories:

1. Numerical aberrations of chromosomes or chromosomal segments: Numerical chromosomal aberrations, complete or segmental, are found in many malignant tumors, mainly in late, undifferentiated and anaplastic stages (G3–G4).
2. Structural aberrations of chromosomes, e.g. translocations: Structural chromosomal aberrations are common somatic mutations involved in carcinogenesis (Table 3.1). Often they occur as balanced or reciprocal translocations of chromosomal segments joining genes at their breakpoints together to form fusion-genes or upset their expression by joining genes to a different promoter region. Well established examples are synovial sarcoma associated with t(X; 18), promyelocytic leukemia associated with t(15; 17) and chronic myeloid leukemia associated with t(9; 22).
3. Gene amplifications or deletions: Gene amplification is associated with many tumors. An important example for gene amplification is the HER-2 neu (c-erbB) gene amplification associated with more than 30% of breast cancer. This gene encodes a tyrosine kinase receptor belonging to the family of growth factor receptors.
4. Mutations within the DNA sequences of genes: Mutations could be located anywhere within the genes. However, in some genes mutations are clustered around mutational hot spots.

Phenotypic and Genotypic Diagnosis of Malignancies. Muin S.A. Tuffaha
Copyright © 2008 WILEY-VCH Verlag GmbH & Co. KGaA, Weinheim
ISBN: 978-3-527-31881-0

Table 3.1 Most common recurrent translocations associated with solid tumors and hematological malignancies.

Solid tumor/hematological malignancy	Most common recurrent translocation(s)
Translocations associated with solid tissue tumors	
Alveolar rhabdomyosarcoma	t(2; 13)(q35; q14) t(1; 13)(p36; q14)
Alveolar soft part sarcoma	t(X; 17)(p11.2; q25.3)
Angiomatoid fibrous histiocytoma	t(12; 16)q13; p11)
Clear cell sarcoma	t(12; 22)(q13; q12)
Congenital fibrosarcoma/mesoblastic nephroma	t(12; 15)(p13; q26)
Dermatofibrosarcoma protuberans/giant cell fibroma	t(17; 22)(q22; q13)
Desmoplastic small cell tumor	t(11; 22)(p13; q12) t(21; 22)(q22; q12)
Endometrial stromal sarcoma	t(7; 17)(p15; q21)
Ewing's sarcoma/PNET	t(11; 22)(q24; q12) t(21; 22)(q22; q12) t(7; 22)(q22; q12) t(17; 22)(q12; q12) t(2; 22)(q33; q12)
Extraskeletal myxoid chondrosarcoma	t(9; 22)(q22; q12) t(9; 17)(q22; q11) t(9; 15)(q22; q21)
Inflammatory myofibroblastic tumor	t(2; 17)(p23; q23) t(1; 2)(q25; p23) t(19; 2)(p13.1; p23) t(2; 2)(p23; q13)
Lipoma	t(3; 12) (q27; q15) t(12; 13)(q15; q12) t(12; 15)(q15; q24)
Lipoblastoma	t(3; 8)(q12; q11.2) t(7; 8)
Low-grade fibromyxoid sarcoma	t(7; 16)(q34; p11) t(11; 16)(p11; p11)
Malignant hemangiopericytoma	t(12; 19)(q13; q13) t(7; 8)(q31; q13)
Mucoepidermoid carcinoma	t(11; 19)(q21; p13)
Myxoid/round cell liposarcoma	t(12; 16)(q13; p11) t(12; 22)(q13; p11)
Pleomorphic adenoma	t(3; 8)(p21; q12) t(3; 12)(p14; q15) t(9; 12)(p24; q15) t(5; 8)(p13; q12)

Table 3.1 *(Continued)*

Solid tumor/hematological malignancy	Most common recurrent translocation(s)
Papillary renal cell carcinoma	t(X; 1)(p11.2; q21.2) t(X; 17)(p11.2; q25)
Rhabdoid tumor	t(1; 22)(p36; q11.2)
Secretory breast carcinoma	t(12; 15)(p13; q25)
Solitary fibrous tumor	t(9; 22)(q31; p13) t(6; 17)(p11.2; q23) t(4; 15)(q13; q26)
Synovial sarcoma	t(X; 18)(q11.23; q11) t(X; 18)(p11.21; q11)
Thyroid carcinoma (papillary)	inv. (10)(q11.2; q21) t(10; 17)(q11.2; q23)
Thyroid carcinoma (follicular)	t(2; 3)(q13; p25)
Translocations in lymphoid and hematopoietic neoplasia	
Acute promyelocytic leukemia (M3)	t(15; 17)(q22; q12–21)
Acute myeloid leukemia (M2)	t(8; 21)(q22; q22)
Acute myeloid leukemia (M4 Eo)	inv. (16)(p13; q23) t(16; 16)(p13; q23)
Acute pre-B-lymphoblastic leukaemia	t(1; 14)(q21; q32)
Acute pre-T-lymphoblastic leukaemia	t(11; 14)(p15; q11)
Acute lymphoblastic leukaemia	t(12; 21)(p12; q22) t(9; 22)(q34; q11) t(1; 19)(q23; p13) t(4; 11)(q21; q23)
Anaplastic large cell lymphoma	t(2; 5)(p 23; q 35) t(1; 2)(q21; p23) t(2; 3)(p23; q21) inv. (2)(p23; q35) t(X; 2)(q11–12; p23)
Burkitt's lymphoma	t(8; 14)(q24; q32) t(8; 22)(q24; q11) t(2; 8)(p12; q24)
Chronic B-cell lymphocytic leukaemia	t(14; 19)(q32; q13) t(2; 18) t(18; 22)
Chronic myeloid leukemia	t(9; 22)(q34; q11)
Diffuse large B-cell lymphoma	t(14; 15)(q32; q11–13) t(8; 14)(q24;q32) t(8; 22)(q24;q11) t(3; 14)(p14.1;q32) t(3; 14)(q27;q32)

(continued)

Table 3.1 *(Continued)*

Solid tumor/hematological malignancy	Most common recurrent translocation(s)
Follicular lymphoma	t(14; 18)(q32; q21) t(3; 14)(q27;q32)
Lymphoplasmacytic lymphoma/immunocytoma	t(9; 14)(p13; q32)
Mantle cell lymphoma	t(11; 14)(q13; q32)
Marginal zone B-cell lymphoma of MALT type	t(11; 18)(q21; q21) t(1; 14)(p22; q32)
Plasmacytoma (multiple myeloma)	t(6; 14)(p25; q32) t(11;14)(q13; q32) t(4; 14)(p16; q32) t(14; 16)(q32; q23) t(16; 22)(q23; q11)

Neoplasia is a result of a complex multistep process and generally the genetic alterations associated with neoplastic evolution may include one or more of the following mechanisms (Table 3.2):

1. Inactivation of different tumor suppressor genes: mainly as a result of mutations appearing within a tumor suppressor-genes including missense, nonsense or splice site mutations as well as deletions and insertions. This mechanism has a recessive behavior as the inactivation of one allele does not have an effect on the cell, whereas the simultaneous loss of the second allele in the same cell (often lost by a larger deletion) leads to the complete "loss of function". This concept is known as the Knudson's two hit model (Knudson AG 1977) based on observations in retinoblastoma patients. Well documented tumor suppressor genes are the retinoblastoma, P53, WT1, and NF genes. Mutations in the retinoblastoma gene located on chromosome 13q14 are associated with retinoblastoma, Wilms tumor and hepatoblastoma. Mutations taking place in P53 gene located on chromosome 17p13.1 are associated with lung, colon and breast tumors.

2. Activation of proto-oncogenes stimulating the cell proliferation: Activating mutations in proto-oncogenes are usually missense mutations. Their location is limited to

Table 3.2 Mechanisms of neoplastic transformation.

Activation of proto-oncogenes	Inactivation of tumor suppressor genes	Changes in genes regulating apoptosis	Mutations of genes responsible for DNA repair	Epigenetic changes
↓	↓	↓	↓	↓
Loss of cell cycle control	Loss of cell cycle control	Cells resistant to programmed death	Loss of DNA repair	Activation of gene transcription

certain codons. Mutations activating proto-oncogenes convert these genes to oncogenes, which exert their "gain of function" in a dominant fashion. There are many classes of oncogenes. The following classes are the most important:

- Viral and viral like classes such as RAS oncogene.
- Growth factors such as platelet derivated growth factor (PDGF) and epidermal growth factor (EGF).
- Mutations in growth factor receptors, which lead to the activation of the cell cycle or the cell cycle functions without binding to growth factors.
- Activation of transcription factors, which stimulates DNA replication and cell proliferation.

3. Changes within genes regulating programmed cell death (apoptosis): a good example is the bcl-2 gene, which is activated as a result of the t(14; 18) translocation associated with follicular lymphoma.

4. Inactivation of mechanisms responsible for DNA-repair: This is a group of mutations affecting the ability to repair the DNA damage and include different syndromes such as HNPCC, Xeroderma pigmentosum, Ataxia telangectasia and Bloom syndrome.

5. Epigenetic changes: Epigenetic changes are a group of changes, in which the basic nucleic acid structures is not distorted. These changes play an important role in the malignant transformation and behavior of tumors; examples for such changes are the methylation of DNA and modification in chromatic structures. The methylation of DNA is a chemical modification of the DNA structure caused by adding a methyl group (CH_3) at the carbon 5 position of the cytosine ring, mostly in the CG sequence (called CpG islands). The hyper- or hypo-methylation of the promoter sequences leads to the activation or deregulation of the transcription of involved genes. The modification of chromatin structures is the results of the chemical modification of the chromatin-associated proteins such as the different chromatin histones that could result the activation or inactivation of different genes.

In most cases, neoplasia is an individual disease occurring sporadically. However, 5–15% of tumors (in certain types of cancer up to 25%; e.g. medullary thyroid carcinoma) is inherited due to a germ line mutation in a tumor suppressor gene or a proto-oncogene. It is noteworthy to mention that genes mutated in inherited cancer syndromes could also play a role in somatic carcinogenesis.

Usually a series of mutations accumulate in tumor cells, causing the progression from a normal somatic cell – probably via a benign tumor cell – to a disseminated neoplastic cell (Table 3.3). The loss of cell cycle control and the accelerated cell growth makes cellular genome more vulnerable to errors in replication, which results a higher incidence of chromosomal aberrations such as gene amplifications and gene mutations.

It is the aim of modern pathology to assess as many parameters as necessary to determine the phenotype and genotype of the tumor for a precise classification. This can be accomplished by conventional morphology, detection of mutations at the

Table 3.3 Stages of malignant transformation and tumor progression.

Accumulation of non-lethal genetic changes
Loss of growth control
Loss of contact inhibition
Invasion of surrounding tissue
Invasion of lymphatic and/or blood vessels
Adhesion to endothelium at distant sites
Invasion of distant tissue

genomic level, analysis of mRNAs expression and protein synthesis by means molecular analysis, ISH, CISH, immunochemistry or protein chemistry.

Proteins are usually detected by immunohistochemical procedures. Some proteins can be detected directly (e.g. by their enzymatic activity). However, the number of detected proteins depends on the availability of specific antibodies and often determined by the quality and the quantity of the surgical specimen. Immunohistochemical procedures discussed in the prior section have the advantage that the morphology of the tumor is still conserved and can be studied simultaneously with additional information regarding cellular antigens and their localization. In addition to immunohistochemistry, modern proteomics offers new approaches to detect proteins by western blotting, 2-dimensional gel electrophoresis or by protein arrays.

Related mRNAs encoding various proteins can be detected by different types of *in situ* hybridization. For this method, the number of specific mRNAs detectable per tumor sample is limited. However, the advantage of *in situ* hybridization is the same as in immunohistochemistry where the morphology of the tumor is still visible. Specific mRNA-species can be detected by northern blot, nuclease protection assay or reverse transcription (RT) combined with polymerase chain reaction (PCR). Using the modern real-time PCR protocols, reliable quantification of PCR targets is possible. A more complex approach is possible by using the micro-array technology, where hundreds or even more of mRNAs can be detected simultaneously in a semi-quantitative fashion.

3.1.1
Molecular Diagnosis of Tumors at the Genomic Level

Before the advent of molecular biology, classic morphological, histochemical and immunohistochemical methods served as tools to generate a treasure of empirical knowledge to classify tumors. The personal experience of the pathologist is still the most important factor for a competent interpretation; however, the future seems to show the increased importance of new technologies, which can assess many parameters either at the protein level or at the nucleic acid level or both. Evaluation of these data by automated computer programs gives a deeper insight into the cell biology of the individual tumor and serve as a basis for more refined classification, improved therapy methods and a better prognosis. At the same time, the precise characterization of the primary tumor gives a clue to detect disseminated tumor cells in the body allowing better therapy strategies.

Since the early 1990s, molecular techniques are more frequently used for tumor diagnosis, especially these methods based on polymerase chain reaction (PCR), reverse transcriptase PCR (RT-PCR) and gene sequencing. Molecular methods proved to be highly sensitive for detecting of minimal number of tumor cells compared with other conventional methods, which require a significant volume of tumor cells. PCR can be designed to detect ten cells in up ten million background cells targeting specific genes, allowing cancer detection at earlier stages when tumors are potentially more curable. This method is also effective to detect submicroscopic occult metastases as these metastases or isolated tumor cells represent the most important cause of treatment failure after putatively curative operations. Statistically about 30% of patients with no identifiable lymph node metastases die from disseminated tumor disease after putatively curative tumor resection.

Molecular diagnosis at the genomic level requires always the same series of steps including sample collection, DNA or RNA preparation, PCR (RT-PCR), Post PCR analysis and interpretation of results. The success of the analysis and the quality of obtained result depends on the selection of the optimal specimen, correct tissue sampling, transport and processing to gain intact nucleic acid representative for the tumor and suitable for further molecular and genetic analysis. As a source of DNA, fresh tissue, frozen tissue, archival formalin-fixed paraffin-embedded tissue, body fluids and sections from microscopic slides can be used. It requires some experience to select the right specimen from an excised tumor mass or from formalin-fixed paraffin-embedded tissue blocks or from a tissue section on a microscopic slide. Methods are also available to isolate nucleic acids from single cells dissected from a microscopic slide, e.g. micro-dissection by laser excision. The management of nucleic acid extraction is now highly standardized and different ready to use commercial kits are currently available.

Molecular methods developed for tumor diagnostics depend either on the detection of different types of non-random mutations or chromosomal rearrangements and translocations arising in tumor cells, or depend on the detection of RNA transcripts encoding factors of differentiation and specific antigens such as tumor markers, tissue-specific antigens and cytokeratins. Both methods have their limitations and are restricted to these tumors bearing specific gene transcripts or harboring specific non-random genetic abnormalities.

The new molecular technologies enable the miniaturization, parallelization and automation of mutation detection. Some of these technologies combine the PCR and the post-PCR step in homogeneous systems (e.g. real-time PCR). Fluorescence and electrochemical detection are utilized for the analysis of PCR-products. Another powerful technology is the MALDI-TOF (matrix-assisted-laser-desorption-ionization time of flight) technique. The PCR-products are analyzed by mass spectrometry, a precise physical measure to characterize nucleic acid fragments. Another option is the use of chip-technology, which allows the parallel detection of many target sequences in one reaction and has the potential to accelerate and simplify the detection of specific mutations.

The fundamental step in the analysis of nucleic acids is the amplification of the target sequence by the polymerase chain reaction (PCR). The amplified DNA is subsequently analyzed by post-PCR steps to detect pathological relevant mutations. If known mutations, like specific codon within an oncogene, have to be detected,

Table 3.4 Methods for scanning a gene for point mutations.

Method	Advantages	Disadvantages
Sequencing	Detects all changes; mutations fully characterized	Laborious; generates excessive information
Heteroduplex gel mobility	Very simple	Short sequences only (<200 bp); does not reveal position of change; limited sensitivity
Single-strand conformation polymorphism analysis (SSCP)	Simple	Short sequences only (<200 bp); does not reveal position of change
Denaturating gradient gel electrophoresis (DGGE)	High sensitivity	Choice of primers is critical; GC-clamped primers expensive; does not reveal position of change
Mismatch cleavage chemical enzymatic	High sensitivity; shows position of change	Experimentally difficult; toxic chemicals
Protein truncation test (PTT)	High sensitivity for chain terminating mutations; shows position of change	Chain-terminating mutations only; experimentally difficult; best with RNA; expensive

methods as described in Table 3.4 can be applied. For the detection of unknown mutations, a number of screening methods of PCR-products are described in the literature (Tables 3.4, 3.5). However, none of these methods has 100% sensitivity to detect mutations, especially if they are heterozygous or if tumor tissue is mixed with normal tissue.

Table 3.5 Methods for testing specified mutations.

Method	Comments
Restriction digestion of PCR-amplified DNA; check size of products on gel	Only when the mutation creates or abolishes a natural restriction site or one engineered by use of special PCR primers
Hybridize PCR-amplified DNA to allele-specific oligonucleotides (ASO) on dot-blot, slot-blot or southern blot	General method for point mutations
Multiplex allele-specific diagnostic assay (MASDA)	General method for multiplex detection of point mutations (combination of ASD and DNA-sequencing)
PCR using allele-specific primers (ARMS test)	General method for point mutations, primer design critical
Oligonucleotide ligation assay (OLA)	General method for point mutations
Allele discrimination with real-time PCR	General method for intragenic mutations (fast, high throughput)

In conclusion, the use of molecular methods for the detection of different genetic abnormalities provides a powerful tool for the diagnosis of many types of neoplasia and offers a genetic background for a modern and precise tumor classification. An example for the importance of the molecular and immunological methods is the modern classification of hematological and soft tissue tumors, where the genetic and phenotypic characterization makes it possible to classify or reclassify many tumors such as the new WHO lymphoma classification and to review other tumors of uncertain histogenetic origin such as the tumor group previously known as malignant fibrous histiocytoma. Furthermore, a PCR-based technique is an effective method to analyze the gene rearrangements for the detection of monoclonal cell proliferation, essential for the differentiation between reactive proliferation of lymphocytes and lymphoma.

In addition to the mentioned benefits, molecular methods based on RT-PCR are used to monitor the success of tumor therapy allowing the detection of minimal residual neoplastic cells after curative operations, chemotherapy or bone marrow transplantation. This approach allows the detection of minimal number of circulating and metastatic tumor cells in secondary sites such as surgical margins, lymph nodes, peripheral blood, bone marrow and other body fluids targeting mRNA of tissue-specific antigens like cytokeratins, specific hormones and antigens or tumor-specific genetic abnormalities. Many published data reveal that the molecular staging (based on RT-PCR) of lymph nodes and mapping of surgical margins may be a better predictor of recurrence of malignancy, as the so-called molecular relapse becomes detectable before clinical relapse.

Another useful aspect in oncology is the detection of tumor-associated viruses, which are important for tumor diagnosis, therapy and clinical follow-up. Using specific primers and probes complementary to virus genomes, provide a powerful and sensitive method to detect many viral genes and oncogenes associated with different types of malignancies such as HPV, EBV, HHV-8 and HCV.

Finally, it should be mentioned that the high sensitivity of molecular methods needs always a critical evaluation because of possible false-positive results accruing as a result of easy contamination by foreign DNA or RNA or by the presence of ectopic (illegitimate) transcription or pseudogenes.

3.2
Nucleic Acids and Techniques

3.2.1
Working with DNA

Genomic DNA Genomic DNA constitutes the total genetic information of any organism. Together with proteins, double-stranded DNA builds multiple linear chromosomes of different sizes. Most of the genomic DNA is located in the cell nucleus and in humans composed of $\sim 3–5 \times 10^{10}$ base pairs. Moreover, about 16 000 base pairs of additional genomic DNA are present in the mitochondria as a circular molecule in multiple copies.

Human genomic DNA contains approximately $30–40 \times 10^3$ genes in discrete regions that encode proteins or RNA. Each gene is composed of the coding DNA sequence as well as the associated regulatory elements that control the gene expression. Nuclear eukaryotic genes contain also introns, which are noncoding regions of hitherto unknown function. Coding DNA represents only 3% of the whole genomic DNA whereas the bulk of the DNA is noncoding, much of which composed of repetitive sequences. Somatic cells are usually diploid cells, having two sets of homologous chromosomes and hence two copies of each genetic locus, while germ cells are haploid and have only one copy of each chromosome.

Handling and Storage of Samples for Genomic DNA Extraction DNA is a relatively stable molecule. However, some precautions are recommended for handling and storage of specimen containing DNA or isolated DNA samples.

Genomic DNA consists of very large DNA molecules, which are fragile and can break easily. To ensure the integrity of genomic DNA, exposure to UV-light, heat, shearing, and aggressive chemicals (e.g. radicals) as well as excessive and rough pipetting and vortexing should be avoided. DNA is subject to acid hydrolysis when stored in water and therefore it should be stored in TE buffer. DNA contact with nucleases should be avoided, as these enzymes degrade DNA.

The quality of initial tissue or cells affects the quality and yield of the isolated DNA. The highest DNA yield and the best quality are achieved by purifying genomic DNA from freshly harvested tissues and cells. If samples cannot be processed immediately after harvesting, they should be stored under conditions that preserve DNA integrity. In general, genomic DNA yields decrease if samples are stored at 2–8 °C or at 20 °C without previous treatment. In addition, repeated freezing and thawing of samples should be avoided, as this leads to the reduction of genomic DNA size, and reduced the yields of pathogen DNA (e.g. viral DNA).

Blood samples must be treated with an anticoagulant before storage and transportation. For example, EDTA treated blood samples can be stored at 2–8 °C for a few days, but for many years if stored at −20 °C or −80 °C. The use of heparin is not recommended, as heparin affects as a PCR inhibitor.

Other body fluids such as plasma, serum, and urine can be stored at 2–8 °C for several hours. Freezing at −20 °C or −80 °C is recommended for long-term storage.

Stool is not an optimal source of DNA as it contains many compounds that can degrade DNA and inhibit downstream enzymatic reactions, e.g. *Taq*-polymerase.

Freshly obtained tissue from surgical samples and tissue culture can be immediately frozen and stored at −20 °C, at −80 °C or in liquid nitrogen. Lysed tissue samples are stable in suitable lysis buffer for several months at ambient temperature.

3.2.1.1 DNA Extraction

Disruption and Homogenization of Tissue for DNA Extraction The isolation and purification of nucleic acids is the first step for the majority of molecular techniques. Some sample sources contain substances that can cause problems during the DNA isolation and analysis and special considerations are required when working with such sample. The main steps for the isolation of nucleic acids include disruption of the tissue and cell lysis, inactivation of cellular enzymes, extraction and purification of nucleic acids from other tissue and cellular components.

The disruption of the tissue and lysis of the cell can be achieved by physical methods (mechanical disruption or hypotonic shock), chemical treatment (choatropic agents and detergents) or enzymatic digestion (protease, proteinase K). Fixation of the tissue (such as formalin-fixation) or the use of choatropic agents inactivates the cellular enzymes including nucleases. There are many methods for the extraction of nucleic acids that include precipitation (the salting-out method is popular), chromatography (such as silica or glass adsorption), centrifugation and electrophoresis.

Blood, different body fluids, fresh and fixed tissue are the most common DNA sources in molecular pathology and special precautions must be taken to handle these sources.

Blood Venous blood samples are routinely collected for different laboratory tests and are suitable for molecular analysis. Blood contains a number of enzyme inhibitors that can interfere with downstream DNA analysis. In addition, common used anticoagulants such as heparin and EDTA can interfere with downstream assays; as a consequence the DNA isolation from blood requires a standardized method to provide high-quality DNA without contaminants or enzyme inhibitors.

Since healthy human blood contains approximately 1000 times more erythrocytes than nucleated blood cells, which includes lymphocytes, monocytes, and granulocytes, removing of erythrocytes prior to DNA isolation can give higher DNA yields; this can be accomplished by several methods. One of the methods is the selective lysis of erythrocytes, which are more susceptible than leukocytes to hypo-osmotic shock and burst rapidly in the presence of a hypo-osmotic buffer. Alternatively, Ficoll® density-gradient centrifugation can be performed to recover mononuclear blood cells (lymphocytes and monocytes) and remove erythrocytes. This technique also removes granulocytes. A third method is the preparing of a leukocyte-enriched fraction from whole blood, known as buffy coat, by centrifuging of whole blood at 3300 *g* for 10 min at room temperature. After centrifugation, three different fractions are distinguishable: the upper clear layer consists of plasma, the intermediate layer is the buffy coat and the bottom layer contains concentrated erythrocytes.

Blood samples, including these treated to remove erythrocytes, can be efficiently lysed using lysis buffer and protease or proteinase K. Simultaneously with the human's genomic DNA, viral and bacterial DNA can also be isolated from blood samples.

Most other biological and body fluids can be treated in an identical way as blood samples for isolation of DNA.

Fresh or Frozen Tissue and Cells Human cell cultures and most human tissues can be efficiently lysed using lysis buffer and protease or proteinase K. Fresh or frozen tissue samples should be cut into small pieces to aid the lysis. Mechanical disruption using a homogenizer, mixer mill, or mortar and pestle prior to lysis can also reduce the lysis time. Skeletal muscle, heart, and skin tissue have an abundance of contractile proteins, connective tissue, and collagen, and special care should be taken to ensure complete digestion using protease or proteinase K.

Fixed Tissue For storage, human tissues are fixed in different fixatives. Generally we recommend the use of fixatives such as acetone, ethanol and neutral buffered formalin, however long-term storage of tissues in formalin results partial degradation, chemical modification or hydrolyzes of DNA. Archival formalin-fixed paraffin-embedded tissue blocks can be also used for nucleic acid extraction. Fixatives that cause cross-linking, such as osmic acid are not recommended if DNA has to be isolated from the tissue.

Fixatives should be removed prior to lysis. Formalin can be removed by washing the tissue in PBS. Paraffin should be removed from paraffin-embedded tissue sections by xylene or other organic solvents.

Genomic DNA yields depend on the size, type, age, and quality of the initial tissue. The DNA content of cell or tissue type also affects the extracted yield. A tissue composed of small cells has a higher cell density and therefore it is more likely to contain more DNA than a sample of the same size comprised of larger cells.

DNA Extraction from Formalin-Fixed Paraffin-Embedded Tissues Archival paraffin tissue blocks are a valuable nucleic acids source in histopathology. Nucleic acids, especially DNA extracted from sections obtained from formalin-fixed paraffin-embedded tissue can be used for different genetic analyses including PCR and the amplification of up to 1300-bp DNA segments has been reported. However, optimal size of targeted sequences should not exceed 500 bp. The quantity of the extracted DNA depends on the tissue type, the fixative used for tissue processing, the duration of fixation and the age of the paraffin blocks. Depending on the size of the biopsy, 3–5 sections (each 5–20 μm thick) are needed to extract sufficient DNA for further molecular analysis.

Deparaffinization

- *Xylene/ethanol method*
  1. Cut 3–5 tissue sections (each 10–15 μm thick) from the paraffin block. To avoid contamination, a new blade must be used for each tissue block.
  2. Add 1 ml xylene to each 1.5-ml microfuge tube containing one to three tissue sections (10–15 μm thick, paraffin-embedded), shake and vortex vigorously and allow to settle for 5 min.
  3. Centrifuge at 12 000 g for 15 min.

4. Remove the supernatant by pipetting. Do not remove any of the pellet.
5. Repeat steps 2–4.
6. Add 1 ml ethanol (96–100%) and shake, allow to settle for 5 min.
7. Centrifuge at 12 000 g for 15 min.
8. Carefully remove the ethanol (supernatant) by pipetting. Do not remove any of the pellet.
9. Repeat steps 6–8.
10. Dry the tissue in the microfuge tube by speed vacuum or incubate at 40 °C for 20 min.

Pellets may be stored at room temperature for several weeks before DNA extraction.

- *Microwave method*
  1. Add 100 µl 0.5% Tween 20 to a tube containing a tissue section (20 µm thick, paraffin-embedded).
  2. Heat in the microwave (650 W) for 45 s.
  3. Spin while warm at 1200 g for 15 min; place on ice.
  4. Remove solid paraffin by sterile pipette tip; digest.

Protein digestion

1. Add 50–100 µl of extraction buffer to the desiccated tissue produced by step 1 extraction buffer: 100 mM Tris-HCl, 1 mM EDTA and 0.5% Tween 20; pH 8.0
2. Add 2–5 µl to proteinase K solution 20 mg/ml H_2O. The optimal final working concentration of proteinase K is about 100 µg per 1 ml.
3. Disrupt the tissue pellet with a sterile pipette tip; incubate at 37 °C overnight or for 3 h at 55 °C.
4. Vortex for 30 s; boil for 7 min to inactivate proteinase K. Incubation in heat block at 96 °C for 7 min can also inactivate proteinase K.
5. Spin at 12 000 g for 5 min to pellet the residual tissue fragments.
6. Take the supernatant and store at −20 °C or precipitate. DNA included in the supernatant can be used directly for PCR.

Isopropanol Precipitation of Genomic DNA Ethanol precipitation is commonly used for concentrating, desalting and recovering of nucleic acids. Precipitation is mediated by high concentrations of salt for protein precipitation and the addition of either isopropanol or ethanol for DNA precipitation. Since less ethanol is required for isopropanol precipitation, this is the preferred method for precipitating DNA from large volumes. In addition, isopropanol precipitation can be performed at room temperature, which minimizes the co-precipitation of salt that interferes with downstream DNA applications. Do not use polycarbonate tubes for precipitation, as polycarbonate is not resistant to isopropanol. Use all solutions at room temperature to minimize the co-precipitation of salt.

- *Isopropanol precipitation*
  1. After digestion of the tissue add a highly concentrated salt solution and adjust the salt concentration if necessary, for example sodium acetate with (pH 5.2) to 0.3 M final concentration or ammonium acetate to 2.0–2.5 M final concentration.

The following protein precipitation solution is very efficient and composed of:

- 60 ml of 5 M potassium acetate;
- 11.5 ml of glacial acetic acid;
- 28.5 ml H_2O.

Add 35 µl of this precipitation solution to each 100 µl of the sample prepared in the previous step II. Mix vigorously for 20 s.

2. Pellet the precipitated protein complex by centrifugation at 13 00–16 000 g for 3 min. A protein pellet should be visible at the bottom of the tube.

3. Transfer the supernatant to a new tube and add 0.6–0.7 vol of room-temperature isopropanol, mix the solution by inversion until the white DNA strands be visible. If the expected amount of the DNA very little add a suitable DNA carrier e.g. glycogen, 0.5 µl of 20 mg/ml glycogen solution per 300 µl isopropanol..

4. Centrifuge the sample immediately at 10 000–15 000 g for 15–30 min at 4 °C. Centrifugation should be carried out at 4 °C to prevent overheating of the sample during centrifugation. When precipitating from small volumes, centrifugation may be carried out at room temperature.
   Alternatively, genomic DNA can be picked out by spooling the DNA using a glass rod after the addition of isopropanol. The spooled DNA should be transferred immediately to a microfuge tube containing an appropriate buffer and redissolved see step 10.

5. Carefully decant the supernatant without disturbing the pellet. Marking the outside of the tube before centrifugation makes it easier to locate the pellet. Pellets from isopropanol precipitation have a glassy appearance and may be more difficult to see than the fluffy salt-containing pellets resulting from ethanol precipitation. Care should be taken when removing the supernatant as pellets from isopropanol precipitation are more loosely attached to the side of the tube. Carefully tip the tube with the pellet on the upper side to avoid dislodging the pellet.

6. Wash the DNA pellet by adding 1–10 ml depending on the size of the preparation. of room-temperature 70% ethanol. This removes the co-precipitated salt and replaces the isopropanol by the more volatile ethanol, which also makes the DNA easier to redissolve.

7. Centrifuge at 10 000–15 000 g for 5–15 min at 4 °C. Centrifuge the tube in the same orientation as previously to recover the DNA into a compact pellet.

8. Carefully decant the supernatant without disturbing the pellet.

9. Air-dry the pellet for 5–20 min depending on the size of the pellet.. Do not over dry the pellet (e.g. by using a vacuum evaporator) since this makes the DNA, especially high-molecular-weight DNA, difficult to redissolve.

10. Redissolve the DNA in a suitable buffer. Choose an appropriate volume of buffer according to the expected DNA yield and the desired final DNA concentration.

Use a buffer with a pH of 7.5–8.0 such as TE buffer), as DNA does not dissolve easily in acidic buffers (if distilled water is used, check pH). Redissolve by rinsing the walls to recover the entire DNA, especially when glass tubes are used.

To avoid shearing the DNA, do not pipette or vortex. High-molecular-weight DNA, such as genomic DNA, should be redissolved very gently to avoid shearing. Dissolving at room temperature overnight or at 55 °C for 1–2 h with gentle agitation gives good results. Heating the DNA solution at 65 °C for 10 min helps to inactivate the DNases.

DNA Extraction by Silica Adsorption The silica DNA extraction method is an efficient method able to recover up to 90% of the sample DNA. The extracted DNA has various sizes ranging from 100 bp DNA fragments and up to high-molecular-weight DNA.

This method depends on the ability of the DNA to bind to the silica particles in the presence of a high salt concentration. This step must be followed by washing of the silica particles to remove salts, proteins, RNA and other contaminants that may be included in the original sample. Finally, the clean DNA is eluted by adding low salt solution (water or TE buffer). The silica DNA extraction method is faster and easier than the organic-based extraction method. Nevertheless, precautions should be taken during purification of large DNA fragments (20 kb) to avoid shearing.

1. Add at least 100 μl of 6 M guanidine-HCl or guanidine thioisocyanate solution to the tissue fragments (25–50 mg tissue) and mix. Pre-digestion of tissue by proteinase (proteinase K and 1% SDS solution for at least 1 h or overnight) improves the results. Incubate at 60 °C (or at room temperature) for 10–15 min to solubilize the proteins and DNA. If undigested tissue particles are visible, increase the incubation time up to 30 min, vortex, centrifuge at 11 000 g and transfer the clear supernatant to a new tube.
2. After digestion of the tissue add 20 μl of silica powder suspension (50% silica powder suspension in water) and incubate for 5 min at 57 °C. Mix occasionally to keep silica hanging in the suspension.
3. Centrifuge the suspension at 10 000 g for 15 s to form a pellet. Remove the supernatant.
4. Add 500 μl of ice cold wash buffer to the pellet. Resuspend the pellet in the wash buffer by vortexing.
   Wash buffer: 20 mM Tris-HCl pH 7.4; 1 mM EDTA; 50 mM NaCl; 50% ethanol.
5. Centrifuge for 15 s and discard the supernatant. Repeat the wash procedure two more times. During each wash, the pellet should be resuspended completely.
6. Centrifuge the tube and remove the remaining liquid by pipette tip. As ethanol may inhibit enzyme reactions such as PCR, the silica should be ethanol free. Add 10 μl of water to the silica suspension and heat to 70 °C or dry in vacuum for 10–15 min. Ethanol evaporates rapidly. The remaining water does not allow the silica to dry out.
7. Resuspend the silica suspension with bond DNA in 20–100 μl elution solution (TE buffer or deionized water) by vortexing. Incubate the tube at 57 °C for 5–10 min. Centrifuge for 30 s and carefully remove the supernatant containing the eluted

DNA. This solution includes the extracted DNA and can be used as template for PCR.

Columns with silica membrane are commercially available and are successfully used for the isolation of DNA and RNA.

3.2.1.2 Storage, Quantification and Determination of Quality of Genomic DNA

Quantification of DNA Spectrophotometry and fluorometry are commonly used to measure DNA concentration. Spectrophotometry can be used to measure microgram quantities of pure DNA samples (i.e., DNA that is not contaminated by proteins, phenol, agarose, or RNA). Fluorometry is more sensitive, allowing the measurement of nanogram quantities of DNA, furthermore the use of Hoechst 33258 dye or picogreen® allows more specific analysis of DNA.

Spectrophotometry DNA concentration can be determined by measuring of the absorbance at 260 nm (A_{260}) in a spectrophotometer using a quartz cuvette. For greatest accuracy, readings should be between 0.1 and 1.0. An absorbance of one optical density unit (1 OD) at 260 nm corresponds to 50 μg/ml of genomic DNA (see Table 3.6). This relation is valid only for measurements made at neutral pH; therefore, samples should be diluted in a low-salt buffer with neutral pH (e.g. Tris-HCl, pH 7.0). To measure multiple samples, the cuvettes must be matched.

Spectrophotometric measurements do not differentiate between DNA and RNA and contamination by RNA can lead to an overestimation of the DNA concentration. Phenol has an absorbance maximum of 270–275 nm, which is close to that of DNA. Phenol contamination mimics both higher yields and higher purity, caused by an upward shift in the A_{260} value.

Fluorometry Fluorometry allows specific and sensitive measurement of DNA concentration by the use of the fluorochrome Hoechst 33258, which shows increased emission at 458 nm when bond to DNA. This dye has little affinity for RNA, allowing accurate quantification of DNA samples that are contaminated with RNA.

DNA standards and samples are mixed with the Hoechst 33258 fluorochrome and measured in glass or acrylic cuvettes using a scanning fluorescence spectrophotometer or a dedicated filter fluorometer set at an excitation wavelength of 365 nm and an emission wavelength of 460 nm. The sample measurements are then compared with the standards to determine the DNA concentration.

Table 3.6 Determination of nucleic acid concentration.

Optical density	Nucleic acid	Approximate concentration (μg/ml H_2O)
1 OD unit at A_{260}	dsDNA	50
	ssDNA	37
	Oligonucleotides	30
	RNA	40

Purity of DNA The ratio of the readings at 260 nm and 280 nm (A_{260}/A_{280}) provides an estimation of the DNA purity with respect to contaminants that absorb UV light, such as proteins and RNA. The A_{260}/A_{280} ratio is significantly influenced by pH. Since water is not buffered, the pH and the calculated A_{260}/A_{280} ratio can significantly vary. Low pH readings with low A_{260}/A_{280} ratio reduce the sensitivity due to protein contamination. For accurate A_{260}/A_{280} values, we recommend to measure the absorbance in a slightly alkaline buffer (e.g. 10 mM Tris-HCl, pH 7.5) and be sure to zero the spectrophotometer with the appropriate buffer (blank).

Pure DNA has an A_{260}/A_{280} ratio of 1.8–1.9. An A_{260}/A_{280} ratio <1.8 is most probably a sign of contamination with proteins or aromatic substances such as phenol, while an A_{260}/A_{280} ratio >1.8 may be due to contamination with RNA (Pure RNA has A_{260}/A_{280} ratio of ~ 2.0). Absorption at A_{230} reflects the DNA contamination by other substances such as peptides, carbohydrates aromatic compounds and phenol. The A_{260}/A_{230} of DNA suitable for PCR should be >2.0. Scanning the absorbance with waves from 220 nm to 320 nm shows whether there are contaminants affecting the absorbance at 260 nm. Absorbance scanners should show a peak at 260 nm and an overall smooth shape.

RNA Contamination Depending on the method used for DNA isolation, RNA is co-purified with genomic DNA with different percentage if RNases are not added. RNA may inhibit some downstream applications, but it does not inhibit PCR. Spectrophotometric measurements do not differentiate between DNA and RNA, therefore RNA contamination can lead to an overestimation of DNA concentration.

Treatment with RNase A removes contaminating RNA and this can either be integrated into the purification procedure or performed after the DNA has been purified. Prior to use, ensure that the RNase A solution has been heat-treated to destroy any contaminating DNase. Alternatively, use DNase-free RNase purchased from a reliable supplier.

Advanced anion-exchange technology allows the isolation of high-molecular-weight genomic DNA that is free of RNA.

Integrity and Size of Genomic DNA The integrity and size of genomic DNA can be checked by pulse-field gel electrophoresis (PFGE) using an agarose gel. Genomic DNA isolated using anion-exchange chromatography should be sized up to 150 kb, with an average length of 50–100 kb. DNA of this size is suitable for Southern analysis, library construction, genome mapping and other demanding applications. Genomic DNA purified using silica-gel membrane chromatography should be sized up to 50 kb, with fragments of 20–30 kb dominating. DNA of this size is ideal for analysis by PCR. Different kits based on the silica gel membrane technology are now commercially available for the extraction and purification of DNA.

Storage of DNA For long-term storage, DNA should be dissolved in TE buffer (pH 7.0–8.0) and stored at $-20\,^{\circ}$C. DNA stored in water is subject to acid hydrolysis. Any contaminants in the DNA solution may lead to DNA degradation. Avoid repeated freeze-thawing as this leads to precipitates. Generally, it is recommended to store genomic DNA samples in aliquots at $-20\,^{\circ}$C.

Table 3.7 RNA content of a typical human cell.

Total RNA per cell	~10–30 pg
Proportion of total RNA in nucleus	~15%
DNA:RNA in nucleus	~2:1
mRNA molecules	2×10^5 to 1×10^6
Typical mRNA size	~1900 nt

3.2.2
Working with RNA

Ribonucleic Acid Ribonucleic acid (RNA) is a polymer consisting of nucleotides containing ribose and four different nucleotides: adenine, guanine, cytosine, and uracil. Three main classes of RNA molecules are transcribed from DNA by three different types of RNA polymerases: mRNA, tRNA and rRNA. Other RNA types are found in very small amounts including snRNAs, double-stranded RNA (dsRNA) and other non-coding RNAs such as the SRP RNAs. All RNA classes present in cells serve different functions.

Messenger RNA (mRNA) is synthesized by the transcription of DNA and transported to the cytoplasm where it serves as a template for the synthesis of proteins. Ribosomal RNA (rRNA) is a part of the ribosomes, which consist of rRNA and proteins. Eukaryotic ribosomes contain four different rRNA molecules: 18S, 5.8S, 28S, and 5S rRNA. rRNA makes up to 80% of the RNA molecules.

Transfer RNA (tRNA) molecules are small molecules composed of 74–95 nucleotides (nt). The function of tRNA is to deliver amino acids required for protein synthesis to the ribosomes. RNAs are also a part of riboproteins, involved in RNA processing.

Double-stranded RNA (dsRNA) is RNA with two complementary strands, similar to the DNA and found in Eukaryotic cells. The dsRNA forms also the genetic material of some viruses.

A typical mammalian cell contains 10–30 pg of total RNA. The majority of RNA molecules are tRNAs and rRNAs, which have a relatively long half-life. mRNA accounts for only 1–5% of the total cellular RNA. Some mRNA species have a half-life of minutes whereas others can have a half-life of many days. Some mRNAs comprise as much as 3% of the mRNA pool whereas others account less than 0.01%. Approximately 350 000–400 000 mRNA molecules are present in each single mammalian cell, made up of approximately 12 000 different transcripts with an average length of approximately 2 kb (Tables 3.7, 3.8).

The mRNA population signifies how genes are expressed under any given set of conditions. Analysis of RNA by hybridization technologies, such as *in situ* hybrid-

Table 3.8 RNA distribution in a typical human cell.

RNA species	Relative amount (%)
rRNA (28S, 18S, 5S)	8085
tRNAs, snRNAs, low MW species	1520
mRNAs	1–5

ization (ISH), northern blotting and microarray analysis, or by RT-PCR can provide a good image of an organism's gene-expression profile.

RNA is relatively an unstable molecule and significantly more susceptible to degradation than DNA, mainly because of the presence of ribonucleases (RNases), which represent different enzymes that break down RNA molecules. RNases are very stable, do not require cofactors, are effective in very small quantities and are difficult to inactivate. RNase contamination can come from human skin and dust particles, which can carry bacteria and molds. They may be present as well in cell lysates as in the surrounding environment, consequently the isolation and analysis of RNA requires highly specialized techniques.

Precautions for Handling RNA Contamination with RNases is one of the major problems facing the work with RNA. RNases are powerful and robust enzymes, difficult to inactivate and cannot be inactivated by autoclaving, thus it is essential to create and to maintain an RNase-free environment and to avoid inadvertently introducing of RNases into the RNA sample during or after the isolation procedure, which is critical for the successful work with RNA. RNases may be introduced during working procedures. In this case, the most common sources of RNase contamination are hands and dust or aerosol with microorganisms. Therefore, latex or vinyl gloves should be worn and frequently changed when handling reagents and RNA samples. Tubes containing samples or reagents should be kept closed wherever possible. Furthermore, it is also important to keep isolated RNA on ice.

Sterile, disposable polypropylene tubes are recommended for working with RNA. These tubes are generally RNase-free and do not require pretreatment to inactivate RNases. In order to destroy RNases, non-disposable plastic-ware can be rinsed with 0.1 M NaOH, 1 mM EDTA, followed by RNase-free water. Alternatively, chloroform-resistant plastic-ware can be rinsed with chloroform, which is sufficient to inactivate RNases.

Glassware used for RNA work should be cleaned with a detergent, thoroughly rinsed, and oven baked at 240 °C overnight or at 300 °C before use. Autoclaving alone does not fully inactivate many RNases. Alternatively, glassware can be treated with DEPC (diethylpyrocarbonate) as follows.

Fill glassware with 0.1% DEPC (0.1% in water), incubate overnight (12 h) at 37 °C, and then autoclave or heat to 100 °C for 15 min to eliminate residual DEPC.

Water and buffers should also be treated with 0.1% DEPC if possible as follows:

1. Add 0.1 ml DEPC to 100 ml of the solution to be treated. Shake vigorously to bring the DEPC into suspension.
2. Incubate for 12 h at 37 °C.
3. Autoclave for 15 min to hydrolyze any trace of DEPC.

As DEPC reacts with primary amines, it cannot be used directly to treat Tris buffers. When preparing Tris buffers, treat water with DEPC first, and then dissolve Tris to make the appropriate buffer.

Endogenous RNases are another source of RNase contamination released by the disruption of the cells during extraction. These RNases can be inactivated using various inhibitors of RNases. Inhibitors of RNases are used to suppress the activity of RNases preventing the degradation of RNA during the isolation, manipulation of RNA and cDNA synthesis. The most common used inhibitors are the protein inhibitors, which are broad-spectrum RNases inhibitors of different sources and commercially available from different manufacturers. As protein inhibitors form a non-covalent complex with the RNases, they are ineffective in the presence of choatropic agents (such as guanidine thiocyanate) and SDS.

3.2.2.1 RNA Extraction

Isolation of pure RNA is one of the most challenging techniques in molecular pathology. Most modern methods are based on the use of strong choatropic agents such as guanidinium thiocyanate, guanidinium hydrochloride and ammonium isothiocyanate, which are able to destroy the cell structures and to inactivate the cellular enzymes specially RNases. The most common used agent for this purpose is guanidine thiocyanate [guanidinium isothiocyanate, $NH_2C(:NH)NH_2.HSCN$], which combines the strong denaturing characteristics of guanidinium and choatropic action of thiocyanate due to the presence of patent cationic and anionic groups able to form strong hydrogen bonds. Commercially available ready to use solutions with choatropic agents can be also used for the extraction of RNA such as RNAzol and Trizol, which contain a high concentration of guanidine thiocyanate (>4 M) and urea salts. After the treatment by the choatropic agent, RNA can be precipitated by ultracentrifugation or ethanol.

RNA Extraction from Blood The following method is a cheap, simple and fast method to isolate RNA from blood, based on the strongly denaturing solution RNAzol, which inactivates RNase within a few seconds and stabilizes DNA.

1. Dissolve 0.2 ml freshly drawn EDTA blood under vigorous shaking or vortexing in 1.6 ml of RNAzol® in a 2.2-ml micro-tube (WAK Chemie Medical, Germany).
2. Add 0.2 ml chloroform to the tube and vortex for 15 s.
3. Keep 5 min on ice.
4. Centrifuge for 15 min at 4°C and 12 000 rpm.
5. Transfer the aqueous phase (supernatant) containing the RNA to a fresh micro-tube.
6. Add 1 vol isopropanol to the extracted RNA.
7. Keep at least 1 h on ice or in −20 °C freezer to precipitate the RNA.
8. Centrifuge for 15 min at 4 °C at 12 000 rpm.
9. Discard the supernatant and wash the pellet with 1 ml ice-cold 75% ethanol.
10. Centrifuge for 15 min at 4 °C at 12 000 rpm.
11. Discard the supernatant and air-dry the pellet of the RNA (or use a speed vac for desiccation).
12. Dissolve the RNA in DEPC-treated water or TE buffer.
13. Determine yield purity by photometric reading at 260 nm and 280 nm.
14. Store the RNA solution at −70 °C.

Various commercial RNA extraction kits are now available from several suppliers, which basically use the same principle of denaturation as described above. If RNA from blood or bone marrow required, the extraction system (kit) must guarantee the rapid denaturation by immediate collecting the blood or bone marrow after the puncture into a tube containing the denaturing solution. This causes the immediate denaturation of all enzymes including RNases and cell enzymes. The inactivation of the cell enzymes is an important factor to prevent the illegitimate transcription of different genes, which can be the source of false-positive results.

RNA Extraction from Formalin-Fixed Paraffin-Embedded Tissues Many protocols showed that formalin-fixed paraffin-embedded tissue from histopathology archives could be valuable source for RNA, suitable for extraction and expression analyses. In the tissue preserved in paraffin blocks after formalin-fixing and processing, RNA appears to be not completely degraded, and RNA fragments of 100–200 bases are still well preserved and can be successfully extracted.

- *Deparaffinization*

Cut 3–5 tissue sections from paraffin block. To avoid contamination always use a new blade for each tissue block. Deparaffinization is carried out as described in the previous section for DNA extraction.

- *Protein digestion*
  1. Add to each dried sample 100–300 μl of RNA digestion solution. 100 ml of RNA digestion solution contains:
     - 40 ml of 4 M guanidine thiocyanate.
     - 3 ml of 1 M Tris-HCl (pH 7.6).
     - 2.4 ml 30% sodium N-lauryl sarcosine.
     - 54.6 ml DEPC-treated H_2O.
     - 0.28 μl β-mercaptoethanol is finally added to each 100 μl extraction solution.
  2. Add proteinase K to a final concentration of 6 mg/ml (43 μl of 20 mg/ml proteinase K to 100 μl of digestion solution; incubate at 45 °C overnight with swirling.

- *RNA extraction*
  1. Add to each tube 1 vol of phenol-saturated water/chloroform in a 70:30 ratio.
  2. Mix by vortexing, put on ice for 15 min.
  3. Centrifuge at 12 000 g for 20 min.
  4. Take the upper aqueous phase avoiding the interface between the two phases and transfer to a new tube.

- *RNA precipitation*
  1. Add 5 μl of glycogen (1.0 mg/ml) and 1 vol of isopropanol to the aqueous phase and store at −20 °C for 48 h to obtain an efficient precipitation of RNA fragments.
  2. Centrifuge at 13 000 g for 20 min at 4 °C (RNA pallets do not adhere to the bottom of tube, keep the pellet insight).
  3. Wash the pellet with 100 μl of 75% ethanol; keep at −20 °C.
  4. Centrifuge for 5 min and air dry.
  5. Resuspend the RNA pellet in 25 μl of DEPC-treated water.
  6. Store at −80 °C.

- *Simultaneous extraction of DNA*

The remaining solution after the RNA extraction contains cell debris including cellular proteins and DNA. The DNA can be extracted for further analysis.

1. Equilibrate the bottom organic layer between the two phases (phenol-chloroform) from previous RNA extraction with 1 vol of Tris-HCl 50 mM (pH 8.0) and store at 4 °C overnight.
2. Centrifuge at 13 000 g for 10 min and remove the upper layer.
3. Add 5 μl of glycogen 1.0 mg/ml and 3 vol of ethanol to cause DNA precipitation, keep at −20 °C for 6–12 h. Isopropanol precipitation at room temperature is also possible (see section DNA extraction).
4. Centrifuge at 13 000 g for 15 min to receive the pellet, wash with 75% ethanol and air dry, resuspend in 25 μl of TE buffer.

Commercially available kits are successfully used to recover RNA and DNA from different sample types, including formalin-fixed paraffin-embedded tissue.

RNA Extraction from Fresh or Frozen Tissue using TRIzol-Like Reagent This method is suitable for fresh or frozen tissue. Formalin-fixed tissue can be also used, but with lower RNA yield.

Prepare TRIzol-like Reagent as follows:

- Phenol, pH 4.3, 380 ml, 38%
- Guanidine thiocyanate 118.6 g (1 M)
- Ammonium thiocyanate 76.12 g (1 M)
- Sodium acetate, 3 M, pH 5.0, 33.4 g (0.1 M)
- Glycerol, 50 ml
- Add H_2O to 1 liter

Other required reagents:

- 0.8 M sodium citrate/1.2 M NaCl
- Isopropanol
- Chloroform
- 75% ethanol
- DEPC-treated H_2O

1. Homogenize 1 g fresh, frozen or formalin-fixed tissue.
2. Add 10 ml TRIzol-like reagent (pre-heated to 60 °C) to tissue in a suitable tube. Vortex.
3. Incubate the sample at 60 °C for 5 min.
4. Centrifuge at 6000 g at 4 °C for 10 min. Transfer supernatant to 50 ml polypropylene centrifuge tube that has been dipped in chloroform to rinse out any residual RNases.
5. Add 2 ml chloroform. Vortex the sample until color shade is uniform, at least 5 s, and incubate at room temperature for 5 min.
6. Centrifuge at 6000 g for 15 min at 4 °C.

7. Collect the upper aqueous layer and transfer it to a new 50-ml centrifuge tube dipped in chloroform.
8. Add 0.5 vol isopropanol and sodium citrate/NaCl solution per 1 ml aqueous solution. Mix gently. Incubate at room temperature for 20 min.
9. Centrifuge at 6000 g for 10 min at 4 °C.
10. Wash the clear RNA pellet with 10 ml cold 75% ethanol. Centrifuge at 6000 g for 5 min at 4 °C.
11. Discard supernatant and air-dry the pellet for 10 min. Dissolve pellet in 300 μl molecular-biology-grade water. Resuspend by pipetting up and down a few times.
12. Add 2 μl suitable RNase inhibitor. Incubate at 55–60 °C for 10 min to resuspend.
13. Check concentration and purity. RNA extracted by this method has usually an A_{260}/A_{280} ratio of ~1.9–2.1. Store the RNA at −20 °C.

Alternative Method for RNA Extraction from Solid Tissue

1. Homogenize 100–200 mg fresh or frozen tissue with 1–2 ml lysis buffer. Lysis buffer consists of:
   - 4 M guanidium thiocyanate
   - 25 mM sodium citrate (pH 7)
   - 0.05% N-lauryl sarcosine
   - 0.1 M 2-mercaptophenol
2. Transfer the homogenized tissue to another tube and add 200 μl sodium acetate solution (4 M, pH 4.0), 2 ml phenol saturated water and 400 μl chloroform, shake for 15 s and keep on ice for 15 min.
3. Centrifuge at 6500 g for 20 min at 4 °C.
4. Transfer the upper aqueous phase to a new tube and add 1 vol ice-cold isopropanol and allow precipitation at −20 °C for 1–3 h.
5. Centrifuge at 6500 g for 20 min at 4 °C.
6. Dissolve the RNA pellet in 300 μl DNase buffer, add 30 μl RNase-free DNase (300 units) and incubate for 30 min at 37 °C.
7. Add 1 vol phenol/chloroform/isoamyl alcohol (25:24:1) and shake for 15 s.
8. Centrifuge at 3000 g for 10 min at 4 °C.
9. Transfer the aqueous phase to a new tube, add 1 vol chloroform and shake for 15 min.
10. Centrifuge at 3000 g for 10 min at 4 °C.
11. Transfer the aqueous phase to a new tube and precipitate with 0.1 vol 3 M sodium acetate (pH 7.4) and 2 vol ethanol for 1–3 h at −20 °C.
12. Centrifuge at 13 000 g for 30 min at 4 °C.
13. Wash the pellet with 80% ethanol.
14. Centrifuge at 13 000 g for 10 min at 4 °C.
15. Dissolve the pellet in sterile TE buffer.

RNA can also be extracted by adsorption chromatography. According to this method both DNA and RNA are reversibly bound to a matrix (usually columns with silica membrane) after the treatment by strong choatropic agents such as guanidinium isothiocyanate. The DNA is removed by the digestion with DNase while RNA is still bound to the matrix. Pure RNA is eluted using low salt elution buffer. The extracted RNA is larger than 200 bases and is suitable for different molecular biology reactions including cDNA syntheses.

The concentration of the RNA is measured by the A_{260} value (one A_{260} unit of ssRNA = 40 μg/ml H_2O). The purity of RNA is an A_{260}/A_{280} ratio between 1.9 and 2.1. An A_{260}/A_{280} ratio >1.9 may be due to the contamination by proteins or other aromatic substances such as phenol.

3.2.3 Polymerase Chain Reaction

3.2.3.1 Standard Polymerase Chain Reaction

Polymerase chain reaction (PCR) is an *in vitro* method for enzymatic synthesizing and amplification of defined DNA sequences out of a mixture of nucleic acids. PCR generates sufficient DNA copies for subsequent molecular analysis. The standard PCR is an end-point reaction, which produce adequate quantity of the PCR product suitable to be identified by size electrophoresis analysis, sequencing or by probe hybridization.

Reaction Components

1. *Template*:
   DNA of different origin containing target DNA sequences. The typical template used in molecular pathology is human genomic DNA from different tissue types in addition to blood, bone marrow and body fluids. RNA contents in the template sample must be low as large amount of RNA can chelate Mg^{++} necessary for polymerase activity. The used template must be free of polymerase inhibitors such as heparin, salts, detergents or proteases. The maximum amount of human genomic DNA should not exceed 500 ng but not less than 20 ng; 1–10 ng of bacterial or viral DNA are sufficient for PCR.

2. *Primers*:
   Primers (oligonucleotides) are short single-strand DNA sequences, usually synthesized on automated DNA synthesizer. Upstream (forward or sense) primer and downstream (reverse or antisense) primer are needed for each reaction. Optimal primers concentration ranges between 100 nM and 500 nM (0.1–0.5 μM) of each primer. The molar concentration of the primers can be calculated by the following formula:
   Micromolar (μM) concentration of the primer in the reaction = quantity of the primer (in picomoles)/volume of the PCR (in micromoles).
   A high primer concentration may cause the accumulation of non-specific reaction products. Primer purity must be more than 90%. For optimal primer design, different computer software programs are available. Ideal primers are 18–26 bases long, not complementary for each other, with no internal base-pairing sequences to build secondary structures, 40–60% G/C, and a balanced distribution of G/C- and A/T-rich domains. To increase the efficiency of annealing, primers should end with G, C

or GC, which builds a tight G–C bond. A close melting temperature (T_m) is recommended. The T_m can be adjusted by increasing of the primer length. Optimal melting temperature ranges over 55–65 °C. Melting temperature can be roughly calculated by the following formula: $T_m = [2\,°C \times (\text{number of A and T bases})]$ $[4\,°C \times (\text{number of G and C bases})]$.
If the sequence of target DNA is not exactly known or has a minimal deviation within the nucleotide sequence, degenerate primers are useful for optimal amplification. Degenerate primers are a mixture of two or more primers that differs in the base composition at the positions that correspond to the variable sequence within the target DNA. Degenerate primers are also useful when the primer design is based on protein sequence, as several different codons can code the same amino acid. To simplify writing of the sequence of degenerate primers the litters B, D, H, K, M, N, R, S, V, W and Y are used as the following: B = C + G + T; D = A + G + T; H = A + C + T; K = G + T; M = A + C; N = A + C + G + T; R = A + G; S = C + G; V = A + C + G; W = A + T; Y = C + T

3. *DNA polymerase*:
   DNA polymerases are different enzymes able to synthesize a new DNA strand on a template strand. DNA polymerases are important enzymes involved in the DNA repair and replication. The following thermostable polymerases are the most common used enzymes for PCR.

- *Taq* polymerase: DNA polymerase is a thermostable enzyme isolated from *Thermus aquarticus*, an enzyme with 5′–3′ polymerase and exonuclease activity. The half-life of the *Taq* polymerase at 95 °C is about 40 min and at 97.5 °C is only 2 min. *Taq* polymerase is the now the most common used PCR enzyme and amplifies DNA targets up to 3 kb. Optimal concentration of the enzyme is 1.25–2.5 units in 25–50 μl reaction volume. *Taq* polymerase has other modified thermally activated molecular forms to achieve a hot start PCR.
- *Pwo* polymerase: DNA polymerase isolated from pyrococcus woesei, an enzyme with 5′–3′ polymerase and 3′–5′ exonuclease activity. This DNA polymerase generates blunt-ended DNA fragments. It amplifies DNA targets up to 3 kb. Optimal concentration is 0.5–1.5 units/50 μl reaction volume with 2 mM $MgSO_4$.
- *Tth* polymerase: DNA polymerase isolated from thermus thermophilus, an enzyme with 5′–3′ polymerase and exonuclease activity, can use both Mg^{++} and Mn^{++} as cofactor but the efficient reverse transcriptase activity takes place in the presence of Mn^{++}. *Tth* polymerase amplifies up to 3 kb from DNA targets or up to 1 kb RNA template. Optimal concentration of this enzyme is 0.5–1.5 units/50 μl reaction volume.
- *Pfu* polymerase: DNA polymerase isolated from *Pyrococcus furiosus*. A thermostable DNA polymerase with high fidelity and low error rate.
- *Tfl* polymerase: DNA polymerase isolated from *Thermus flavis*. This enzyme is useful to amplify large DNA segments.
- *Tli* polymerase: DNA polymerase isolated from *Thermococcus litoralis*. A highly stable DNA polymerase with 3′–5′ activity and high fidelity.

- *KOD* DNA polymerase: a recombinant polymerase isolated from *Thermococcus kodakaraensis.*

4. *PCR buffer:*
   the universal PCR buffer includes only $MgCl_2$, KCl and Tris with a pH ranging between 8.3 and 9.0. A standard 10× PCR buffer composed of 15 mM $MgCl_2$, 500 mM KCl and 100 mM Tris-HCl. The PCR buffer can be individually modified by adding some other additives such as glycerol, DMSO, bovine serum albumin (BSA) and EDTA in a concentration ranging between 5% and 10% of the buffer. Detergents such as Triton X-100 (1–2%) or Tween (0.1%) can also be added. These additives may increase the efficiency of the amplification. To prepare the buffer special molecular-biology-grade nuclease-free distilled water must be used.

- $MgCl_2$: free Mg^{++} ions are co-factor for the Taq polymerase, essential for the enzyme activity. The optimal Mg^{++} concentration depends on the concentration of dNTPs and usually ranges over 1.5–5.0 mM, generally 1.5 mM $MgCl_2$ corresponds with 200 μM of each dNTP and any increase in the concentration of dNTPs requires more Mg^{++}, for example dNTPs in a concentration of 400 μM need 2 mM $MgCl_2$. In contrast, a high concentration of Mg^{++} can disturb the polymerase activity and increase a non-specific amplification, causing an unclear background.
- KCl: K^+ ions are an important component of the PCR buffer. The concentration of K^+ depends on the length of the amplified DNA template. Generally a DNA target of 100–1000 bp requires a K^+ concentration between 70 mM and 100 mM. A longer DNA target needs a lower K^+ concentration.

5. *Deoxynucleotides (dNTPs):*
   dNTPs (dATP, dCTP, dGTP, dTTP) are added to the reaction as a balanced solution of all dNTPs with a final concentration of 50–500 μM of each dNTP. The optimal concentrations range over 200–250 μg of each dNTP, which allow the synthesis of ~6.5 μg of PCR product. A high concentration of dNTPs needs an increased Mg^{++} concentration. The optimal concentration for *Taq* polymerase is 200 μM of each dNTP with 1.5 mM $MgCl_2$, which is usually sufficient to synthesize up to 6.5 μg of target DNA.

Reaction Volume The reaction volume ranges over 5–100 μl (5, 25, 50 or 100 μl). For small amounts of DNA, little reaction volume is recommended. A standard 25 μl PCR mixture includes the following components:

1. 1.0 μl DNA template (~100 ng/μl solution)
2. 2.5 μl buffer (standard 10× PCR buffer)
3. 0,4 μl primer mix (0.2 μl each primer from 100 μM primer solution)
4. 0.2 μl dNTPs mix (mix of 25 mM each dNTP)
5. 0.2 μl *Taq* polymerase (1.25–2.5 units)
6. 20.7 μl H_2O

If multiple samples are to be analyzed, a master mix can be assembled. A master mix is consisted of the total amount of all PCR components except DNA templates, which are added individually to each reaction.

Reaction Phases The PCR consist of three phases: denaturation, primer annealing and primer extension. All three phases make one reaction cycle, while the whole PCR includes several cycles, begins with an initial denaturation phase and ends by a final extension phase.

- *Initial denaturation*

This is the initial PCR phase necessary to separate the double-strand DNA to generate single-strand DNA template suitable for further amplification. The denaturation occurs at 94–95 °C for 2 min. This temperature is sufficient to break the hydrogen bonds between the base pairs of the dsDNA without breaking of the covalent bonds of the other macromolecules. A long denaturing time causes the degradation of polymerase molecules and decreases the polymerase activity. Hot start PCR improves the sensitivity and specificity of PCR and can be achieved manually by adding polymerase at 94 °C or by using modified thermally activated polymerase forms. Hot start prevents the non-specific primer annealing appearing at low temperatures, which may generate non-specific PCR products. Hot start PCR helps to optimize the multiplex PCR and it is also recommended to use if the targeted gene copies are too few.

- *Denaturation during cycles*

Each cycle begins with a denaturation phase. The denaturation temperature depends on the G + C content and is carried out at 94–95 °C for 20–45 s.

- *Primer annealing*

In this phase the primers complementary to the target DNA fragment are annealed to the denaturated single-strand DNA. The primer annealing time and temperature depends on primer sequences and melting temperature. Optimal annealing temperature ranges between 5–10 °C below the melting temperature of the primers. Highly specific primer annealing takes place when annealing temperature is more than 62 °C. The annealing duration is usually 30–60 s at 54–65 °C. Long annealing time and low annealing temperature can enhance unspecific amplifications.
Touchdown PCR is a technique to improve the DNA amplification by decreasing the annealing temperature during the PCR cycles. The most common way to achieve a touchdown PCR is to set the initial annealing temperature for the first two cycles ~3 °C above the calculated melting temperature and then to decrease the annealing temperature by one grade each two or three cycles.

- *Primer extension (elongation)*

This is last phase in a PCR cycle. During this phase the thermostable polymerase catalyze the synthesis of the targeted DNA using the dNTPs. The optimal extension temperature ranges between 65 °C and 72 °C. Extension time depends on the size of the amplified sequence. For each kilobase of amplified sequence, 45–60 s are needed.

- *Number of PCR cycles*

The optimal number of PCR cycles ranges between 25 and 35 cycles. To amplify a single copy of a DNA target, at least 25 cycles are needed to achieve an acceptable amount of reaction product. An increased number of cycles could enhance unspecific amplification.

- *Final extension*

The final extension is not an essential phase, but added as a last cycle to allow the completion of all amplified sequences, usually 10–20 min at 72 °C.

Confirmation of Specific Target Sequence For the precise interpretation and analysis of PCR results, it is recommended to compare the sequence of the generated PCR product with the sequence of the targeted gene. The following techniques are the most common used to achieve this purpose:

- Hybridization with specific probes.
- Nested PCR using nested (inner) primers.
- Direct sequencing of PCR product.
- Restriction fragment length polymorphism (RFLP) achieved by the cleavage of the PCR product using restriction enzymes.

3.2.3.2 Multiplex PCR

Multiplex PCR is used for the simultaneous amplification of two or more targets in one PCR reaction using unique primer sets to produce various sized amplicons specific to different DNA sequences. The PCR products can be analyzed using one electrophoreses gel. The careful optimization of reaction components and conditions is essential for a successful multiplex PCR. The following factors must be considered in the designing of multiplex PCR:

- *DNA template*

Up to eight DNA templates can be simultaneously amplified by multiplex PCR. Large target number can disturb the reaction. DNA targets must have relatively a close length and the difference between the target sizes must not exceed 300–bp.

- *Primer design and concentration*

Primers should have close melting temperatures, longer primers (28–35 bases) can precise the reaction specificity. High primer concentrations may inhibit the reaction. The yield of the PCR product is reduced corresponding to the number of the primers. Generally, up to eight targets (eight primer pairs) can be successfully amplified by one multiplex PCR and the quantity of the PCR products can be detected using ethidium bromide and agarose gel. The amplicon sizes must be enough different to form distinct bands on gel electrophoresis.

- *PCR buffer*

Increased $MgCl_2$ concentration is required for multiplex PCR to match the increased amount of dNTPs needed for the reaction. Slightly reduced KCl concentration may be also necessary. Commercially PCR optimization kits, which include different buffers, are available and often recommended for the optimization of multiplex PCR.

- *Annealing temperature*

The annealing temperatures for each of the primer set must be optimized. Decreasing the annealing temperature (54–58 °C) can be useful to optimize the reaction. Longer annealing and extension time may be also necessary according to the number of the amplified targets.

3.2.3.3 Real-Time PCR

Real-time PCR (quantitative PCR, qPCR), is a quantitative method to measure the concentration of a gene of interest present in a PCR mixture. A special designed thermal cycler, with integrated fluorimeters, computer and soft ware are used to detect the emissions of the amplified DNA labeled by fluorescent molecules during the PCR cycles. The intensity kinetics of the fluorescence in the special designed PCR tube correlates directly with the quantity and accumulation of the DNA present in the tube, especially in the early PCR cycles, when the amount of the PCR product corresponds to the amount of initial DNA target in the reaction mix. Generally, no product is detected during the first few PCR cycles as the fluorescent emission signal is below the detection threshold of the real-time cycler. In the late cycles when the amplicon concentrations move toward the nanogram/ml level, the efficiency of the amplification falls constantly because the accumulated DNA amplicons are re-associated to dsDNA during the annealing phase. This leads to a stage during which the accumulation of the PCR product is approximately linear with a constant level of synthesis in each cycle. Finally, a plateau curve is reached when the net synthesis is around zero. Real-time PCR cyclers are now designed to detect multiple dyes so that internal standards and multiplex PCR products can be monitored. Real-time PCR has many important applications and includes the following important aspects:

1. Estimation of the titer of pathogenic agents (different viruses or bacteria) in blood or body fluids to evaluate the success of the therapy;
2. Detection of minimal cancer disease and the concentration of tumor cells in blood or bone marrow to monitor the success of the tumor therapy;
3. Studying the gene expression levels;
4. Detection of single nucleotide polymorphisms (SNPs) in genomic sequences.

Different methods have been developed to measure the fluorescence of the amplified PCR product. The most popular methods are the SYBR Green I method and the FRET method in addition to the cycling probe technology method.

Detection of SYBR® Green I Fluorescence The SYBR Green I is a simple, inexpensive and sensitive method. SYBR Green I is an intercalating dye that binds to the minor groove of double stranded DNA (dsDNA) molecules, regardless its sequence. After binding the dye to the dsDNA, the intensity of the fluorescence emission increases more than 300-fold, which provide excellent sensitivity to quantify the dsDNA molecules in the reaction tube, which makes SYBR Green I ideal to monitor the amplification of any dsDNA sequence. Free dye molecules do not affect significantly the background fluorescence of the reaction mixture. To achieve real-time PCR, a small amount of SYBR Green I can be added to the PCR mixture before the first cycle. As the SYBR Green I dye becomes bond to the newly synthesized dsDNA amplicons during the PCR cycles, the increased emission intensity is measured by the real-time PCR cycler and the amount of fluorescence released during the PCR cycle is proportional to the amount of amplicon generated in each cycle. SYBR Green I binds also to non-specific dsDNA sequences, which may cause false-positive signals overestimating the target concentration. Nevertheless, in real-time PCR with single target and well designed primers, SYBR Green I works optimal and the non-specific background appears only in the very late cycles. To minimize the background noise the primers must be very specific and carefully designed with minimal mispriming possibility and the milting temperature must range between 60 °C and 65 °C. It is also important to chose a suitable target with a length ranging between 100 bp and 400 bp. Because SYBR Green I binds any amplicon this method is not suitable for multiplex reactions.

Fluorescence Resonance Energy Transfer Fluorescence resonance energy transfer (FRET) real-time PCR method depends on the introduction of fluorogenic-labeled probes affected by the 5′nuclease activity of Taq DNA polymerase. The FRET method includes the TaqMan®, molecular beacons scorpions and few other techniques, all of them are hybridization probes relying on fluorescence resonance energy transfer (FRET) to quantify PCR product via the generation of a fluorescent signal.

TaqMan® probes consist of a ssDNA molecule labeled with a reporter dye at the 5′ end and a quencher at the 3′ end. The quencher molecule (such as TAMRA® dye) inhibits the fluorescence emission of the reporter dye when located in its close proximity, resulting a non-fluorescent DNA fragment. During the PCR cycles, the labeled probes hybridize to their specific complementary sequence on the template DNA located between the forward and reverse primer positions. As soon as the DNA polymerase begins DNA synthesis from the forward primer during the extension phase, the probe interacts with the 5′ exonuclease activity of the DNA polymerase, which cleaves the probe's terminal 5′ nucleotide with attached reporter dye, releasing it into the reaction mixture. After the physical separation of the reporter dye from the quencher molecule, the reporter dye is able to emit strong fluorescence and the fluorescent signal increases in each cycle, proportional to the rate of probe cleavage and the amount of amplicon generated in each cycle corresponding to the number of PCR targets. TaqMan® probes are added to the PCR master mix in an excess amount, which provides a stable supply of uncleaved probes to the newly synthesized target molecules during each amplification cycle. Contrary to the SYBR Green I method, where SYBR Green I binds to any dsDNA molecule, TaqMan probes bind only to a specific target molecule. Different variants of TaqMan probes are designed and can be labeled with different, reporter dyes, which allows the amplification of two or more different sequences in the same PCR tube. Another variant is the MGB probes suitable for allelic discrimination assays especially when conventional probes exceed 30 nucleotides. The TaqMan MGB probes contain a quencher at the 3′ end and a minor groove binder at the 3′ end, which increases the melting temperature (T_m) of probes, allowing the use of shorter probes. The MGB probes show greater differences in T_m values between matched and mismatched probes, which provide more accurate allelic discrimination.

Molecular beacons consist of a sequence-specific region (loop region) with two inverted repeats and attached to fluorescent reporter and quenching dyes to each end of the probe. The fluorescence resonance energy transfer (FRET) occurs only when the quenching dye is directly adjacent to the fluorescent dye. Molecular beacons are designed to form a hairpin structure while free in solution, bringing the fluorescent dye and quencher in close proximity. When a molecular beacon hybridizes to a target, the fluorescent dye and quencher are separated, allowing reporter emission. In contrast to the TaqMan probes, molecular beacons are designed to remain intact during the amplification reaction, and must rebind to target in every cycle for signal measurement. Scorpions combine the labeled probe with the upstream primer allowing the scorpion to anneal to the template DNA strand. The annealing causes the separation of the quencher form the fluorescent dye allowing reporter emission.

Cycling Probe Technology Cycling probe technology (CPT) technology is a highly sensitive method initially developed by ID Biomedical and depends on the interaction between a labeled DNA–RNA probe and the RNase H activity. The probe is composed of two short ssDNA segments centrally attached to an RNA segment. The 5′ end of the probe is labeled by a fluorescent reporter molecule while the 3′ end is attached to a quencher. These probes have the same mechanism as in the FRET method when the quencher molecule inhibits the fluorescence emission of the reporter dye as long as the probe remains intact. When this probe hybridizes to its specific complementary sequence on the template DNA located between the forward and reverse primer positions, the RNase H enzyme activity cleaves the RNA region of the probe, causing the separation of the quencher and the free fluorescent reporter dye gives strong fluorescence emissions. The emitted fluorescence intensity correlates with the amount of amplified amplicon and the initial DNA target concentration can be calculated.

Final Quantitation of Real-Time PCR Results Two main approaches are commonly used to quantify the real-time RT-PCR results, the standard curve method and the comparative threshold method.

In the standard curve method, a reference curve is constructed from a nucleic acid of known concentration and this curve is used as a reference standard. The comparative threshold depends on measuring the cycle threshold (C_t). C_t is the time at which fluorescence intensity is greater than background fluorescence. According to this method the C_t value of each samples is compared with a control or calibrator.

3.2.4 Reverse Transcriptase Polymerase Chain Reaction

Reverse transcriptase polymerase chain reaction (RT-PCR) is a molecular reaction used for the amplification and detection of RNA (mRNA) after transcription in to cDNA.

The eukaryotic genes have a complex structure consisting of exons (coding DNA fragments) and introns (noncoding DNA fragments) and the transcribed mRNA must be processed prior to the translation into protein to remove the noncoding introns. This process takes place in the cell nucleus and called mRNA splicing, in which the introns are removed from the sequence and exon ends are joined together. As a result of splicing a continuous mRNA reading frame is formed, which encodes a functional protein. The spliced mRNA molecule is then exported into the cell cytoplasm, and translated into protein by cytoplasmic ribosomes. Consequently, RT-PCR is an essential method to study the gene expression at the mRNA level.

To analyze RNA by PCR an additional step has to be introduced to transcribe RNA into cDNA, which is a more stable single-strand DNA complementary to the targeted RNA. This can be achieved by the reverse transcription (RT) of the template RNA using a reverse transcriptase enzyme. RT and PCR can be carried out either sequentially in the same tube (one-step RT-PCR) or separately (two-step RT-PCR). One-step RT-PCR requires gene-specific primers for the reverse transcription reaction, whereas in two-step RT-PCR random primers can be used.

RT-PCR in a highly sensitive reaction and can be contaminated by DNA of different sources. To avoid amplification of contaminating genomic DNA, RNA samples can be treated with DNase I prior to reverse transcription. Alternatively primers for RT-PCR should be designed so, that one half of the primer hybridizes to the 3′ end of one exon and the other half to the 5′ end of the adjacent exon. Such primers do not anneal to genomic DNA. Another approach to distinguish between PCR products from genomic and cDNA is to position the primers in different exons, so that at least one intron is between them. Products amplified from cDNA (lacking introns) are smaller than these amplified from genomic DNA (containing introns).

3.2.4.1 **One-Step RT-PCR**

In one-step RT-PCR, both cDNA synthesis on the RNA temple and the cDNA amplification (PCR) take place in one tube and in the same buffer using sequence-specific primers.

Reaction components

1. RNA template: mRNA, total RNA.
2. Reverse transcriptase, DNA polymerase enzyme with buffer. The following enzymes are the most common used enzymes for one-step RT-PCR:

- *Tth* DNA polymerase is an enzyme with both reverse transcriptase and DNA polymerase activities, whereas the reverse transcriptase activity is manganese-dependent.
- *C. therm.* polymerase isolated from *Carboxydothermus hydrogenoformans* is a magnesium-dependent enzyme with both reverse transcriptase and DNA polymerase activities. *C. therm.* polymerase has a short half-life at 95 °C and *Taq* polymerase can be added to the reaction to improve the amplification.

3. Primers: site-specific primers selective for the mRNA template and PCR primers.
4. dNTPs (2 mM).

Reaction phases

1. Denaturation at 94 °C for 2 min.
2. Denaturation the cDNA during cycles at 94 °C for 30 s.
3. Primer annealing at 53 °C for 30 s.
4. Primer extension at 68 °C for 1 min; all three phases 2–4 are repeated for ten cycles.
5. Finally, 25–35 PCR cycles are used for amplification.

3.2.4.2 **Two-Step RT-PCR**

The first step in Two-step RT-PCR is the reverse transcription (RT) for the synthesis of cDNA on the RNA template using reverse transcriptase enzyme and oligo$(dT)_{12-18}$ or random hexanucleotides. The second step is the cDNA amplification by PCR.

RT Reaction components:

1. RNA template: total RNA or mRNA.
2. Reverse transcriptase with RNase inhibitor and buffer. The following viral reverse transcriptases are the most common used enzymes for the cDNA synthesis in two-step RT-PCR:

- AMV: Reverse transcriptase isolated from avian myeloblastosis virus. This enzyme catalyzes the polymerization of nucleotides and characterized by RNA-dependent DNA polymerase, DNA-dependent DNA polymerase activity besides the RNase H activity but lacks the 3′–5′ exonuclease activity. The RNase H activity can cause the degradation of the RNA strand of an RNA:DNA duplex, which is a disadvantage that can limit the complete synthesis of the total cDNA. The AMV Reverse Transcriptase is suitable for reverse transcription of fragments containing secondary structure due to its high optimum temperature (42 °C) for its activity.
- M-MulV: Reverse transcriptase isolated from *m*oloney *mu*rine *l*eukemia *v*irus. This enzyme is also characterized by RNA-dependent DNA polymerase, DNA-dependent DNA polymerase and weak RNase H activities but lacks the 3′–5′ exonuclease activity. M-MulV reverse transcriptase is able to synthesize full-length cDNA fragments from long mRNA. The temperature for the optimal enzyme activity is 37 °C. Because of the weak RNase H activities of the M-MulV reverse transcriptase, it is more suitable for the synthesis of long cDNA fragments.

3. Primers:

- Oligo$(dT)_{12-18}$ which hybridize the complementary endogenous poly A tail at the 3′ end of mammalian mRNA.
- Random hexanucleotides, which bind different complementary sites on the RNA template.
  Using the two above-mentioned primer systems, all cellular mRNA is transcribed to cDNA, which gives the advantage to amplify different targets in addition to positive controls simultaneously, also as a multiplex PCR.

- Site-specific primers selective for a specific mRNA template. Using site-specific primers only targeted RNA sequence is transcribed to cDNA.

4. dNTPs mix and $MgCl_2$.

Reaction phases

1. Incubation of reaction components at 22 °C for 10 min.
2. Synthesis of cDNA at 42 °C for 40–60 min.
3. Inactivation of the reverse transcriptase at 95 °C for 4 min.
4. RT followed by 25–35 cycles PCR using cDNA as template.

3.2.4.3 Notes for Performing RT-PCR

- All materials should be sterile and of molecular-biology-grade.
- All solutions should be prepared by autoclaved DNase- and RNase-free water.
- Use DEPC-treated water (diethyl pyrocarbonate, DEPC, a non-specific inhibitor of RNase activity). To treat add 0.1% of DEPC to the water; allow to stand overnight and then autoclave to stop DEPC activity.
- SDS sodium dodecyl sulfate (detergent) is unstable by autoclaving; incubate for 2 h at 65 °C to minimize the RNase activity.
- For accurate RT-PCR results, it is recommended to have a minimum of two independent positive results (two out of three).
- It is recommended to carry out the RT-PCR simultaneously with a negative control and an endogenous internal standard from a housekeeping gene (PBGD, GAPDH, β-actin, β-microglobulin) to confirm the presence of intact mRNA fragments.
- To maximize the success rate of RT-PCR targeting mRNA extracted from formalin-fixed paraffin-embedded tissue, it is important to target short sequences of RNA with a length ranging between 75 nt and 300 nt because of the high degradation rate of RNA after cell damage due to formalin-fixation.

3.2.5 Quality Control for DNA and RNA Extraction and Amplification

To control the integrity and the quality of the extracted DNA or RNA and the efficiency of the DNA amplification, it is important to carry out the PCR simultaneously with one or more primer sets designed to amplify housekeeping genes.

Housekeeping genes are constitutively expressed genes in the mammalian cells and can be used as an internal control during the gene amplification. Targeting gene sequences with different length (one of them with a similar length to the gene of interest), we can verify the presence and the integrity of DNA or RNA after extraction and estimate the grade of the nucleic acid degradation due to formalin-fixation or other negative factors.

Primers for the Amplification of Some Housekeeping Genes as Positive Internal Control

1. Beta actin gene:
   FP: 5′-TGA.CTT.TGT.CAC.AGC.CCA.AGA.TA -3′
   RP: 5′-AAT.CCA.AAT.GCG.GCA.TCT.TC -3′

PCR generates an 85-bp DNA fragment.

2. Beta actin gene:
   FP: 5′-CCACACTGTGCCCATCTACG -3′
   RP: 5′-AGGATCTTCATGAGGTAGTCAGTCAG -3′

PCR generates a 99-bp DNA fragment.

3. Porphobilinogen deaminase (PBGD) gene:
   FP: 5′-TGT.CTG.GTA.ACG.GCA.ATG.CGG.CTG.CAA.C-3′
   RP: 5′-TCA.ATG.TTG.CCA.CCA.CAC.TGT.CCG.TCT-3′

PCR generates a 127-bp DNA fragment.

4. Beta actin gene:
   FP: 5′-CAC.TGT.GTT.GGC.GTA.CAG.GT-3′
   RP: 5′-TCA.TCA.CCA.TTG.GCA.ATG.AG-3′

PCR generates a 154-bp DNA fragment.

5. Beta microglobulin gene:
   FP: 5′-CTT.GTC.TTT.CAG.CAA.GGA.CTG.G-3′
   RP: 5′-CCT.CCA.TGA.TGC.TGC.TTA.CAT.GTC-3′

PCR generates a 158-bp DNA fragment.

6. PGK (phosphoglycerate kinase) gene:
   FP: 5′-CAG.TTT.GGA.GCT.CCT.GGA.AG-3′
   RP: 5′-TGC.AAA.TCC.AGG.GTG.CAG.TG-3′

PCR generates a 247-bp DNA fragment.

7. GAPDH (glyceraldehyde -3-phosphate dehydrogenase) gene:
   FP: 5′-CGG.AGT.CAA.CGG.ATT.TGG.TCG.TAT-3′
   RP: 5′-AGC.CTT.CTC.CAT.GGT.GGT.GAA.GAC-3′

PCR generates a 306-bp DNA fragment.

8. Beta microglobulin gene:
   FP: 5′-CCT.GAA.TTG.CTA.TGT.GTC.TGG.GTT.TCA.TCC.A-3′
   RP: 5′-GGA.GCA.ACC.TGC.TCA.GAT.ACA.TCA.AAC.ATG.G-3′

PCR generates a 441-bp DNA fragment.

9. Beta actin gene:
   FP: 5′-CCACACTGTGCCCATCTACG -3′
   RP: 5′-GATCTTCATTGTGCTGGGTGCC -3′

PCR generates a 502-bp DNA fragment.

10. Beta actin gene:
    FP: 5′-CCACACTGTGCCCATCTACG -3′
    RP: 5′-CTGCTTGCTGATCCACATCTG -3′

PCR generates a 601-bp DNA fragment.

11. Hypoxanthine phosphoribosyl transferase (HPRT):
    FP: 5′-ACC.GGC.TTC.CTC.CTC.CTG.AGC.AGT.C-3′
    RP: 5′-AGG.ACT.CCA.GAT.GTT.TCC.AAA.CTC.AAC.TT-3′

PCR generates a 747-bp DNA fragment.

3.2.6 Electrophoresis and Detection of Mutations

Electrophoresis is a widely used method for the separation and analysis of macromolecules such as nucleic acids and proteins, using a gel matrix and an electrical field. During electrophoresis, positively charged molecules migrate toward the negative cathode and negatively charged molecules migrate toward the positive anode. Because of the negative charge presented on the phosphate groups of the DNA molecules, DNA fragments migrate toward the positive pole and the smaller DNA fragments migrate faster than the larger fragments through the gel matrix toward the positive anode. The size of the PCR product can be determined by comparing it with a DNA ladder, which contains DNA fragments of known size. After the separation of the nucleic acids the separated bands can be visualized using different dyes (such as ethidium bromide, SYBER Green or silver stain) or transferred onto a membrane for further analysis.

Agarose and polyacrylamide gels are the most common used matrixes for the separation of nucleic acids. Small size nucleic acids (less than 2 kb) are usually separated on agarose gels. For high resolution and for the separation of large DNA molecules polyacrylamide gel is recommended. Polyacrylamide gels are also used for the separation of proteins.

3.2.6.1 Polyacrylamide Gel

Polyacrylamide gels are composed of acrylamide and crosslinking molecule (bisacrylamide). The polymerization is initiated by the presence of free radicals (generated from ammonium persulfate or other reagents such as riboflavin) and accelerated by a catalyst such as TEMED. The concentration of polyacrylamide gel depends mainly on the size of the analyzed DNA fragment.

Table 3.9 shows suggested concentration optimal for good separation and high resolution of DNA fragments on polyacrylamide gel.

Components of a Non-Denaturing Polyacrylamide Gel A non-denaturing polyacrylamide gel is composed of the following components:

Table 3.9 Polyacrylamide concentration for optimal DNA separation.

Size of DNA fragment (bp)	Concentration of polyacrylamide (%)
10–100	20
25–150	15
40–200	12
60–400	8
80–500	5
1000–2000	3.5

1. Acrylamide prepared from the following components:
- 30 g acrylamide (C_3H_5NO, MW 71.08)
- 800 mg methylene-bisacrylamide ($C_7H_{10}N_2O_2$, MW 154.17)
- 100 ml H_2O
2. Ammonium persulfate (APS)
3. N,N,N′,N′-Tetra methyl ethylene diamine (TEMED)

Preparing a Non-Denaturing Polyacrylamide Gel Electrophoresis with an 8% non-denaturing polyacrylamide gel is commonly used and offers fine separation with high resolution of DNA fragments. Table 3.10 includes the components and the quantities to prepare an 8% non-denaturing polyacrylamide gel.

As a standard electrophoresis running buffer for polyacrylamide gels, 1.0–0.5% Tris-borate-EDTA (TBE) is used. To run electrophoresis, use low voltage ranging over 1–8 V/cm to prevent denaturation of small DNA fragments due to high temperature. The gel can be stained with ethidium bromide, SYBR gold or silver.

3.2.6.2 Agarose Gel

Agarose is a linear galactan hydrocolloid isolated from marine algae. Agarose colloid is a good matrix for the migration of biopolymers in electrical field. Agarose gel is the most common gel used for analyzing of DNA fragments ranging between 50 bp and 10 000 bp. The concentration of agarose depends mainly on the size of the analyzed DNA fragment.

Table 3.11 shows suggested agarose concentrations for optimal separation and best resolution of DNA fragments.

Components of Agarose Gel

1. Agarose
2. Electrophoresis running buffer: Tris-borate-EDTA, Tris-acetate-EDTA and Tris-phosphate are the most common used buffers for agarose gel.
   - Tris-borate-EDTA (TBE buffer) has more buffering capacity than TAE and provides a better resolution for the separation of small nucleic acid fragments (0.1–3.0 kb). A 5× stock TBE solution has the following components:
     –54 g Tris base
     –27.5 g boric acid

Table 3.10 Components of 8% non-denaturing polyacrylamide gel.

Reagent	100 ml gel	30 ml gel	10 ml gel
30% acrylamide (ml)	26.2	7.86	2.62
10× TBE (ml)	10	3	1
10% APS (μl)	700	210	70
TEMED (μl)	35.0	10.5	3.5
H_2O (ml)	62.7	18.81	6.27

Allow the gel 30–60 min to polymerize at room temperature.

Table 3.11 Agarose concentration for optimal DNA separation.

Size of DNA fragment (bp)	Concentration of agarose (% w/v)
50–200	2.0
200–300	1.5
300–500	1.2
500–800	1.0
800–1000	0.7
1000–5000	0.5

–20 ml of 0.5 M EDTA, pH 8
–H_2O to 1 L
Adjust pH to 8.0–8.3
Working solution 0.5×

- Tris-acetate-EDTA (TAE buffer) has less buffering capacity than TBE buffer and preferred for the separation of larger nucleic acid fragments, 10x stock solution has the following components:
  –48.4 g Tris base
  –11.4 ml acetic acid
  –20 ml of 0.5 M EDTA, pH 8.0
  –H_2O to 1 L

Adjust pH to 8.0
Working solution 1×

Preparing the Gel

1. Prepare the working solution of the running buffer (TBE or TAE buffer). Always use the same buffer for gel preparing and for electrophoresis.
2. Add agarose (amount depending on the needed concentration). Heat with microwave or water bath until the agarose is dissolved in the buffer.
3. Add suitable dye such as ethidium bromide (0.5 μg ethidium bromide/1 ml gel; 0.5 μl of 1% stock solution to each 1 ml TBE agarose gel); cool to 55–60 °C.
4. Pour the gel solution into a suitable electrophoresis tray, to give a 3–5 mm gel, insert the comb, leave the gel for 30 min.
5. Remove the comb, fill the electrophoresis tank and load PCR product into comb wells with a mixture of 16 μl of PCR product and 4 μl of load buffer, for good resolution, do not exceed 50 ng of PCR product.
6. Run electrophoresis (~10 V/cm gel length) for ~30 min.
7. Visualization of DNA fragments under UV light for gels with dye or by silver staining. Ethidium bromide gels allow the detection of 20 ng of DNA, SYBR Green I is more sensitive.
8. To estimate the size of analyzed DNA, use different molecular weight markers (ladders).
9. DNA can be preserved in agarose gel using 70% ethanol.

3.2.6.3 Principle of Denaturing Gel Analysis

In contrast to DNA, RNA has a high degree of secondary structure. Therefore, denaturing conditions as obtained in a formaldehyde or urea agarose gel are needed to facilitate migration in an electric field, which separates RNA molecules by size.

Detailed information on all types of analytical gels can be found in molecular biology manuals.

3.2.7 Detection of Mutations by Electrophoresis

DNA and RNA with different sequences and mutations show different electrophoretic migration behavior. There are simple electrophoretic methods, which can be used to detect mutations and minimal differences within the sequences of the PCR product. Popular methods are single-strand conformational polymorphism and heteroduplex analysis.

3.2.7.1 Single-Strand Conformational Polymorphism

Single-strand conformational polymorphism (SSCP) depends on the electrophoresis of single DNA strands. For the denaturation of DNA, formamide or high temperature are commonly used. On electrophoresis under non-denaturing conditions, different DNA heteroduplexes with different migration behavior are generated because of different intrastrand base pairing due to different sequences of the DNA products. Consequently, PCR products with mutated DNA show mobility different from the control DNA. These differences in electrophoretic mobility between nearly identical strands are the basis for this technique. Detailed SSCP information and protocols can be found in molecular biology manuals

3.2.7.2 Heteroduplex Analysis

Heteroduplex analysis depends on the electrophoresis of duplex DNA segments with same or different single strand sequences (homoduplexes and heteroduplexes). For the heteroduplex analysis, PCR product must be denatured by high temperature (94 °C for 5 min) and rapidly renatured at low temperature (4 °C for 30–60 min) to enforce the DNA duplex formation. On polyacrylamide gel electrophoresis under nondenaturing conditions, DNA duplexes with different strand sequences (heteroduplexes) show different migration behavior and speed, whereas DNA duplexes with the same strand sequences (homoduplexes) show rapid migration. This method is suitable for clonality analysis of lymphoid malignancies and effective to minimize false-positive or false-negative results due to improper primer annealing.

3.2.8 *In Situ* Hybridization

In situ hybridization (ISH) is another important molecular technique allowing the detection of specific nucleic acid sequences within cells or tissue structures using complementary oligonucleotide probes and suitable visualization system. This method is used to label both DNA and RNA sequences and to study the transcription activity of different cells by the detection of mRNA. A significant advantage of this technique is the preservation of tissue structures throughout the reaction, which provides optimal topological information of targeted genes. Two main *in situ* hybridization techniques are used:

1. Radioactive technique based on nucleic acid probes labeled by radioactive isotopes such as P_{32}, H_3, I_{131} and S_{35}.
2. Non-radioactive technique based on the use of non-radioactive-labeled nucleic acid probes.

The non-radioactive technique is now the most popular method practiced in histopathology. The reaction can be performed on sections prepared from frozen or formalin-fixed paraffin-embedded tissue, presenting a good correlation between tissue morphology and labeled targets. ISH can give useful answers for many diagnostic important questions. The following applications are only few examples for the benefits of ISH in surgical pathology:

- Detection of different infectious agents, mainly viruses in tissue sections, cell smears and body fluids such as human papillomavirus (HPV), human herpes-viruses including Epstein–Barr virus (EBV) and hepatitis C virus (HCV). ISH can detect latent viral genes and translation activity of infected cells as a criterion for latent infection.

- Assessment of the transcription type in the cells of interest, this is commonly used to detect kappa and lambda light chains as an important criterion for the diagnosis of lymphoid neoplasia, mainly plasma cell neoplasia.

- Detection of genetic diseases.

- Detection of tumor-associated genetic abnormalities including tumor-specific translocations such as t(9; 22) in CML, t(14; 18) in follicular lymphoma in addition to tumor-associated mutations.

- Quantitative estimation of gene copies, important to detect gene amplification. Gene amplification is a significant test to estimate the tumor sensitivity to specific anti-HER-2 antibodies.

Depending on the visualization system used for the detection of the probe-target hybrid, direct and indirect methods are used.

- Direct method in which the probe-target hybrid can be visualized after the hybridization. When fluorochrome-labeled probes are used, the visualization can be achieved by fluorescence microscopy. Various fluorochromes with different emission colors are now available such as AMA, FITC, fluorescein, rhodamine CY3 and Texas red. The use of different probes labeled by different fluorochromes with different emission colors allows the simultaneous detection of different nucleic acid sequences.

- Indirect method in which the probe-target hybrid detected by affinity cytochemistry when hapten-labeled probes are used. Different haptens such as, biotin, fluorescein and digoxigenin are used as reporter molecules and can be recognized by specific antibodies.

3.2.8.1 **ISH Technique**

1. *Pretreatment of tissue sections*
   This step includes many stages to prepare tissue sections for the hybridization. Initial preparing of tissue sections is the same as described for immunohistochemistry.
   - Inactivation of endogenous enzymes such as endogenous peroxidase DNase and RNase. Optimal fixation of tissue sections inactivates most of endogenous enzymes and preserve the morphology. Endogenous peroxidase can be successfully inactivated using hydrogen peroxidase (H_2O_2) as described for immunohistochemistry.
   - Digestion: cellular proteins must be digested in order to increase the target accessibility. Proteinase K and pepsin are the most common used proteases in ISH. After digestion, slides are dehydrated to 100% ethanol and dried at room temperature.
2. *Denaturation*
   The denaturation is a fundamental step in the ISH to separate the double strand DNA of both probe and target DNA and to prevent the formation of any secondary RNA structures such as hairpin loops. Heating the tissue sections to 80–95 °C for 8–10 min is sufficient to break the hydrogen bonds between the base pairs. At the end of the denaturation, the temperature must be rapidly decreased to the hybridization temperature of 37 °C. Incubation in high-pH solutions also causes the denaturation of nucleic acids. Both denaturation and hybridization steps are usually performed simultaneously with the probes and a suitable hybridization buffer. The universal hybridization buffer contains 50–80% formamide, sodium salts and dextran sulfate.
3. *Hybridization*
   In this phase, hydrogen bonds are created between the bases of the probe and the complementary sequences inside targeted nucleic acids. The stringency of the hybridization depends on many conditions such as the temperature of the hybridization. The hybridization is usually performed at 37 °C for 1–3 h or overnight, and increasing the temperature enhances the stringency. A low salt concentration and a high formamide concentration in the hybridization buffer also increase the specificity of hybridization.

4. *Post-hybridization wash*
   This step is necessary to remove excessive probe or non-specifically hybridized probes from the surface of tissue sections.

5. *Visualization of hybridization product*
   Direct or indirect visualization methods can be used to detect the labeled nucleic acid sequences.
   - Direct visualization is performed by fluorescence microscopy when fluorochrome-labeled probes or florescence-labeled antibodies are used (fluorescence *in situ* hybridization, FISH). The color of fluorescence depends on the type of used fluorochrome, blue (AMCA), green (fluorescein) and red (rhodamine, Texas red, CY3 and TRITC).
   - Indirect visualization when hapten-labeled probes are used. Haptens are detected by specific antibodies, which react with biotinylated secondary antibody. Streptavidin-horseradish peroxidase or streptavidin-alkaline phosphatase complexes are bound to the biotin molecule. Finally, a substrate/chromogen system (such as DAB/hydrogen peroxidase, AEC/hydrogen peroxidase, fast red/naphthol phosphate or BCIP/nitroblue tetrazolium chloride) is added and the conversion of the substrate/chromogen system generates a precipitate seen by light microscopy (chromogenic *in situ* hybridization, CISH). To finish, slides are counter-stained with eosin or hematoxylin and mounted with a permanent mounting medium for DAB- and BCIP/NBT-stained slides and with an aqueous medium for AEC- and fast red-stained slides.

In situ hybridization can be combined with immunohistochemistry. This double staining approach allows the simultaneous detection of proteins (antigens) and specific nucleic acid sequences, which presents a good correlation between the genotype and the phenotype of examined cells, in addition to the phenotypic characterization of cells harboring viral genome.

Because of the high temperature and aggressive enzymes used in ISH, which can destroy the protein molecules, immunohistochemistry should precede the ISH.

Spectral karyotyping (SKY) method is a modification if the FISH technique. SKY uses a mixture of many probes specific to each cellular chromosome, each with unique color. These probes are hybridized to the metaphase chromosomes and give a general picture of all cellular chromosomes and structural chromosomal abnormalities.

The DNA microarray (gene chips) is a further modification of the hybridization technique allowing the large-scale analysis of gene expression pattern of tumors. According to this method, cDNA fragments transcribed from tumor and normal RNA or oligonucleotides are spotted on microscope slides. For tumor gene profiling, fluorochrome (red for tumor RNA and green for normal RNA control) labeled cDNA prepared from tumor RNA is hybridized to the spotted cDNA/oligonucleotide probes. The hybridization signals are detected and analyzed using laser scanner and software. Colored spots indicate the expression of the targeted gene. Black spots (no signal) indicate that cDNA is not expressed.

3.3
Diagnosis of Tumors, Micrometastases and Circulating Tumor Cells by Molecular Detection of Specific Chromosomal Translocations

3.3.1
Introduction

The neoplastic transformation begins after a sequence of different acquired genetic mutations in the cellular genome, which include mutations within the gene sequences and different types of chromosomal abnormalities. Chromosomal translocations are one of the most important and interesting types of chromosomal abnormalities associated with neoplasia.

Chromosomal translocation is a rearrangement in which a part of a chromosome is detached by breakage in a specific breakpoint and becomes attached to another chromosome or inversion in the same gene. Chromosomal translocations generate hybrid genes as a result of the fusion of different genes from two chromosomes and may be found as genetic abnormalities in many tumors. Compared with other tumor-associated cytogenetic abnormalities such as point mutations, deletions or amplifications, tumor-associated translocations are specific for the related tumors and can be used as specific cytogenetic markers for precise diagnosis and monitoring of these tumor. These genotypic markers can be detected by several methods including reverse transcriptase polymerase chain reaction (RT-PCR), fluorescent *in situ* hybridization (FISH) and southern blot with rapid and accurate results.

The detection of these specific translocations and the generated fusion transcripts provide highly specific objective molecular markers for the diagnosis of many tumor identities. They are very helpful for the diagnosis of many hematopoietic malignancies; good examples are chronic myeloid leukemia, follicular lymphoma and mantle cell lymphoma, in addition to several sarcomas such as synovial sarcoma and the group of small round cell tumors of childhood, beside some carcinomas such as papillary and follicular thyroid carcinoma and mucoepidermoid carcinoma. Additionally, the detection of neoplastic cells bearing the specific fusion transcripts generated by these translocations in peripheral blood has proved to be a sensitive assay for the monitoring of hematogenous tumor spread as a risk factor for the development of lung or other systemic metastasis. Furthermore many translocations or translocation variants may have a significant prognostic value since many translocations or translocation variants may indicate an aggressive or less aggressive behavior of tumors bearing these translocations or translocation variants.

The detection of tumor-specific translocations can be used to monitor the response of these tumors to the specific therapy and also as a tool for the detection of minimal residual disease. Different studies demonstrate that the elimination of tumor cells in many hematopoietic malignancies below the level of one neoplastic cell in 10^5 normal cells predicts a stable remission, whereas a concentration of one neoplastic cell in less than 10^3 background cells is an indicator for poor prognosis, and considered as a molecular relapse. Note that tumor cell concentration of $1{:}10^3$ cannot be detected by conventional morphological methods.

As many tumor types including soft tissue tumors, hematopoietic malignancies and few carcinomas are associated with recurrent specific translocations, these chromosomal abnormalities seem to be of etiologic significance in the developing of these tumors. Many translocations found to provide an abnormal promoter such as the Ewing sarcoma gene (EWS) or the homologous FUS (TLS) gene to another gene encoding a protein with transcriptional activity or have tyrosine kinase activity or a suppressor of the normal cellular apoptotic death. Both EWS and FUS genes are members of the TET family of RNA-binding proteins, which also includes the TAFII68 (RBP56 or TAF2N) gene. The products of the TET gene family are the TET proteins having homologous structures. The TET proteins are ubiquitously expressed, and are supposed to act as adaptors between the transcription and the RNA processing.

However, many of the genes with transcriptional activity are involved in developmental or differentiation processes and the fusion with activated promoters enforces the expression of the gene product, while the protein-encoding sequence of these genes continued to be intact after the translocation. As a result, such translocations affect many biochemical pathways and may interrupt the differentiation program of the affected cells. The generated fusion gene acts as an oncogene with transforming ability.

The EWS gene (also called EWSR1 gene) is a common promoter involved in tumor-associated translocations. This gene is located on chromosome 22q12 and composed of 17 exons spans over 40 kb, where most breakpoints are clustered in a 7 kb region within intron 7 and 8. Table 3.12 shows some examples demonstrating the association of the EWS gene and the homologous FUS promoters with other potential oncogenes in different types of neoplasia. By now, it is known that the TET gene family members (mostly the EWS gene) are involved in more than 11 different fusions transcripts associated with different malignancies.

Although the generated fusion transcripts may vary in their exon composition according to the localization of the breakpoints, the encoded proteins have constant features that influence cellular metabolism, growth, differentiation and apoptosis, which can lead to an uncontrolled proliferation of cells bearing these genetic abnormalities. These mechanisms seem to play an important role in the initiation of many tumors.

Table 3.12 Examples demonstrating the association of the EWS gene and the homologous FUS promoters with other potential oncogenes in different types of neoplasia.

CHOP gene (12q13) + *FUS* gene (16p11)	Myxoid liposarcoma
CHOP gene (12q13) + *EWS* gene (22q12)	Myxoid liposarcoma
ERG gene (21q22) + *FUS* gene (16p11)	Acute myeloid leukemia
ERG gene (21q22) + *EWS* gene (22q12)	Acute myeloid leukemia
ERG gene (21q22) + *EWS* gene (22q12)	Ewing's sarcoma
EWS gene (22q12) + FLI-1 gene (11q24)	Ewing's sarcoma
EWS gene (22q12) + ETV1 gene (7 p22)	Ewing's sarcoma
EWS gene (22q12) + ATF1 gene (12 q13)	Clear cell sarcoma
EWS gene (22q12) + WT1 gene (11 p13)	Desmoplastic small round cell tumor
EWS gene (22q12) + ERG gene (21q22)	Desmoplastic small round cell tumor
EWS gene (22q12) + CHN gene (9q22)	Extraskeletal myxoid chondrosarcoma

In conclusion, the routine histopathologic approach in recognition and classification of tumors depends mainly on the resemblance of the tumor morphology to the adult mature tissue and the immunohistochemical profile. The molecular detection of tumor-specific transcripts and non-random mutations proved to be a further powerful and sensitive tool giving a new basis for diagnosis, modern classification and reclassification of many tumors specially tumors with controversial histogenesis or tumors with multilineage differentiation such as synovial sarcoma, malignant fibrous histiocytoma, Ewing's/PNET tumor family and many hematopoietic malignancies.

3.3.2 Molecular Diagnosis of Solid Tumors

3.3.2.1 Molecular Detection of Synovial Sarcoma

Synovial sarcoma is a highly malignant soft tissue tumor, arises in different locations mainly in the intra- and para-articular region, accounts about 10–15% of all soft tissue sarcomas and occurs most commonly in young adults between 15–40 years.

The histogenesis of synovial sarcoma remains uncertain but most likely to be derived from primitive multipotent stem cells, which are able to differentiate into mesenchymal and/or epithelial cells; consequently, synovial sarcoma has two major types of morphologic differentiation. The first and more common type is the biphasic type exhibiting both epithelioid and spindle cell sarcomatous differentiations. The second type exhibits only a monophasic morphology with sarcomatous (fibrous) or less common epithelioid differentiation. The third described and uncommon morphologic variant is the poorly differentiated or dedifferentiated type of synovial sarcoma characterized by an aggressive behavior and worse prognosis than the other two types. Synovial sarcoma has no specific immunophenotype and in classic cases shows positive nuclear stain with the antibody to the TLE1protein beside the coexpression of vimentin and different cytokeratins in addition to the epithelial membrane antigen (EMA), CD99 and bcl-2. The TLE1protein is encoded by the TLE1 gene (*T*ransducin-*L*ike *E*nhancer of split), one of four TLE genes (TEL 1–4) that encode human transcriptional repressors homologous to the *Drosophila* corepressor groucho.

The ambiguous histological differentiation and the controversial immunophenotype of synovial sarcoma make the microscopic diagnosis difficult, especially when the tumor occurs in atypical locations or when the tumor reveals a non-specific immunohistochemical phenotype, which may resemble the immunophenotypic profile of other malignancies such as different types of sarcoma (solitary fibrous tumor, malignant peripheral nerve sheet tumor, fibrosarcoma, epithelioid sarcoma or leiomyosarcoma), metaplastic carcinoma and mesothelioma. Since synovial sarcoma reveals a different clinical behavior than the other mentioned tumors, it is important to discriminate it from these malignancies and additional methods that are more specific may be necessary to confirm the diagnosis.

Synovial sarcoma is associated with few novel non-random and specific cytogenetic abnormalities that can be used as objective and precise markers for the diagnosis of this tumor. More than 90% of synovial sarcomas are associated with the specific and characteristic reciprocal chromosomal translocation t(X; 18)(p11; q11). This translocation involves the SSX gene (*s*ynovial *s*arcoma *X* chromosome) on chromosome Xp11 that encodes a transcriptional repressor and the SYT gene (*sy*novial sarcoma *t*ranslocation, also known as SS18 gene) located on 18p11 chromosome (senteny gene). The SYT gene is a ubiquitously expressed gene encoding a 387-nucleic-amino-acid protein, functioning as transcriptional activator.

To date, six highly homologous variants of the SSX gene have been identified, the SSX1, SSX2, SSX3, SSX4 SSX5 and SSX6 genes, all located on the X chromosome p11.2 and composed of six exons. The six SSX genes encode six closely related nuclear proteins, composed of 188 amino acids with a homology ranging between 88% and 95% and belonging to the cancer/testes family, sharing the feature transcribed in normal testicular tissue and in some types of human malignancies such as malignant melanoma. Studies demonstrate that all SSX proteins are able to suppress the transcription of reporter genes. Interestingly only SSX1 and SSX2 and in very rare cases SSX4 are able to fuse with the SYT gene to assemble a fusion transcript. Both SSX1 and SSX2 genes are highly homologous and encode homologous proteins composed of 188 amino acids.

In the vast majority of synovial sarcoma the t(X; 18)(p11; q11) translocation interrupts the SYT and the SSX genes in constant locations. In the SSX1, SSX2 and SSX4 genes, the breakpoints are located in codon 111 near the ornithine aminotransferase-like pseudogene clusters 1 and 2, between the exons 4 and 5. In the SYT gene, the breakpoint is located within codon 410. As a result, in the vast majority of cases exons 5 and 6 of the SSX gene fuse to the 3′ end of the SYT gene. Accordingly, the generated SYT-SSX fusion gene includes the domains of the SSX gene responsible for suppression of transcription and the domains of the SYT gene responsible for activation of transcription. The SYT-SSX1 and the SYT-SSX2 genes are the most common fusion transcripts associated with synovial sarcoma and generated by the following translocations:

- The *t(X; 18)(p11.23; q11)* translocation is the most common translocation found in about 65% of cases and generates the SYT-SSX1 fusion transcript with the following most common sequence (fusion point *G/AT* in italics):
  SYT...CCA.GCA.GAG.GCC.TTA.TGG.ATA.TGA.CCA.*G/AT*.CAT.GCC.CAA.GAA.
  GCC.AGC.AGA.GGA.CGA.AAA.TGA.TTC.GAA.GGG.AGT.GTC.AGA.AGC.
  ATC.TGG.CCC.ACA.AA. . .*SSX1*
- The *t(X; 18)(p11.21; q11)* translocation is found in about 35% of cases and generates the SYT-SSX2 fusion transcript with the following most common sequence (fusion point *G/AT* in italics):
  SYT. . .CCA.GCA.GAG.GCC.TTA.TGG.ATA.TGA.CCA.*G/AT*.CAT.GCC.CAA.
  GAA.GCC.AGC.AGA.GGA.AGG.AAA.TGA.TTC.GGA.GGA.AGT.GCC.AGA.
  AGC.ATC.TGG.CCC.ACA.AA. . .*SSX2*

The SYT-SSX fusion gene encodes a protein composed of the first 379 amino acids encoded by the SYT gene and the last 78 amino acids encoded by either SSX1 or SSX2 genes. The proteins encoded by the SYT gene and the SSX gene in addition to the SYT-SSX fusion gene are all localized in the nucleus of synovial sarcoma cells and can by detected by specific antibodies giving nuclear immunohistochemical stain.

Other rare sequences of the above-listed fusion transcripts due to different breakpoints or exon combinations were also described in different studies. Both fusion transcripts encode closely related nuclear proteins, able to cause transcriptional deregulation of some oncogenes. The SYT-SSX1 translocation is more common than SYT-SSX2 and detected in about 65% of synovial sarcomas where as SYT-SSX2 in about 35% of cases and SYT-SSX4 in less than 1%. These fusion transcripts are specific for synovial sarcoma and are detected in both epithelioid and spindle cell components of this tumor. The t(X; 18) translocation has not been detected in any other neoplasia or benign lesion. Nevertheless a few authors report the t(X; 18) translocation in association with some other soft tissue tumors such as malignant peripheral nerve sheath tumors, but these reports are very controversial and are rejected from the majority of researchers. Generally, tumors bearing the t(X; 18) translocation should be classified as synovial sarcoma.

The molecular detection of the transcripts generated by the above-described characteristic translocations are now used as a complementary assay for the precise diagnosis of synovial sarcoma especially for tumors with monophasic or poorly differentiated morphology or occurring in atypical locations. Furthermore, there is an important correlation found between the type of the fusion transcript, the differentiation pattern and the clinical behavior of synovial sarcoma. The SYT-SSX1 fusion transcript is associated in the majority of cases with biphasic morphology, while monophasic morphology can be associated with both SYT-SSX1 and SYT-SSX2 fusion transcripts. But the SYT-SSX2 fusion transcript is only associated with monophasic differentiated tumors and usually with a significantly higher proliferation rate (histologically estimated by the Ki-67 index) and worse prognosis. Interesting also is the male:female ratio, which is ~1:1 in cases associated with SYT-SSX1 and ~1:2 in cases associated with SYT-SSX2 fusion transcripts. Retrospective studies showed that tumors bearing the SYT-SSX1 fusion transcript have lower risk of early relapse and higher 5-year metastasis-free survival after adequate treatment compared with tumors bearing the SYT-SSX2 fusion transcript.

The amplification of the SYT-SSX transcripts can also be used to monitor the hematogenous spread of tumor cells using blood samples collected before surgical intervention. It can also be used for the detection of residual tumor cells in surgical margins after surgical resection. The detection of residual tumor cells in the surgical margins considered being an indication for early radiotherapy.

Other cytogenetic abnormalities were also described in association with synovial sarcoma. These abnormalities are non-specific for this tumor and mainly numerical chromosomal changes such as +2, +4, +7, +8, +9, +12, +15, −3, −11, −14, −17 and −22.

Method The method uses mRNA extraction, followed by RT-PCR. The mRNA needed for RT-PCR-based molecular analyses can be obtained from frozen or paraffin-embedded tissue, peripheral blood or bone marrow. For good results, short target sequences must be selected because of the degradation of mRNA extracted from formalin-fixed paraffin-embedded tissue. Amplification of a housekeeping gene is also recommended to determine the integrity of extracted mRNA.

- Primers for the cDNA amplification of the complete *SYT-SSX* fusion transcript:
  FP (SYT): 5′-AGA.CCA.ACA.CAG.CCT.GGA.CCA.C-3′
  RP (SSX): 5′-TCC.TCT.GCT.GGC.TTC.TTG-3′

PCR generates an 87-bp DNA fragment.

- Alternative primers for the cDNA amplification of the complete *SYT-SSX* fusion transcript:
  FP (SYT): 5′-CCA.GCA.GAG.GCC.TTA.TGG.ATA-3′
  RP (SSX): 5′-TTT.GTG.GGC.CAG.ATG.CTT.C-3′

PCR generates a 98-bp DNA fragment.

- Primers for the cDNA amplification of the *SYT-SSX1* fusion transcript:
  FP (SYT): 5′-AGA.CCA.ACA.CAG.CCT.GGA.CCA.C-3′
  RP (SSX): 5′-ACA.CTC.CCT.TCG.AAT.CAT.TTT.CG-3′

PCR generates a 110-bp DNA fragment.

- Alternative primers for the cDNA amplification of the *SYT-SSX1* fusion transcript:
  FP (SYT): 5′-CCA.GCA.GAG.GCC.TTA.TGG.ATA-3′
  RP (SSX): 5′-GTG.CAG.TTG.TTT.CCC.ATC.G-3′

PCR generates a 118-bp DNA fragment.

- Primers for the cDNA amplification of the *SYT-SSX2* fusion transcript:
  FP (SYT): 5′-AGA.CCA.ACA.CAG.CCT.GGA.CCA.C-3′
  RP (SSX): 5′-GCA.CTT.CCT.CCG.AAT.CAT.TTC-3′

PCR generates a 110-bp DNA fragment.

- Alternative primers for the amplification of the *SYT-SSX2* fusion transcript:
  FP (SYT): 5′-CCA.GCA.GAG.GCC.TTA.TGG.ATA-3′
  RP (SSX): 5′-GCA.CAG.CTC.TTT.CCC.ATC.A-3′

PCR generates a 118-bp DNA fragment.

- Alternative primers for the cDNA amplification of both *SYT-SSX1* and *SYT-SSX2* fusion transcripts:
  FP (SYT): 5′-AGA.CCA.ACA.CAG.CCT.GGA.CCA-3′
  RP (SSX): 5′-TGC.TAT.GCA.CCT.GAT.GAC.GA-3′
  RP (SSX1): 5′-GGT.GCA.GTT.GTT.TCC.CAT.CG-3′
  RP (SSX2): 5′-TCT.CGT.GAA.TCT.TCT.CAG.AGG-3′

PCR is carried out in three reactions. The first reaction using both SYT and SSX primers. This reaction generates a 401-bp DNA fragment. Second semi nested PCR using the SYTand SSX1 primers and generates a 151-bp DNA fragment in the case of the SYT-SSX1 fusion transcript. The Third semi nested PCR using the SYTand SSX2 primers and generates a 190-bp DNA fragment in the case of the SYT-SSX2 fusion transcript.

- Screening primers for the cDNA amplification of the *SYT-SSX1, SYT-SSX2* and *SYT-SSX4* fusion transcripts:

FP (SYT): 5′-CAG.CAG.AGG.CCT.TAT.GGA.TAT.GA-3′
RP (SSX): 5′-TCA.TTT.TGT.GGG.CCA.GAT.GC-3′

PCR generates a 101-bp DNA fragment.

- Primers for the cDNA amplification of the *SYT-SSX1, SYT-SSX2* and *SYT-SSX4* fusion transcripts:

  FP (SYT): 5′-AGA.CCA.ACA.CAG.CCT.GGA.CCA.C-3′
  RP (SSX1): 5′-ACA.CTC.CCT.TCG.AAT.CAT.TTT.CG -3′
  RP (SSX2): 5′-GCA.CTT.CCT.CCG.AAT.CAT.TTC-3′
  RP (SSX4): 5′-GCA.CTT.CCT.TCA.AAC.CAT.TTT.CT-3′

PCR generates a 108-bp DNA fragment.

- Primers for the cDNA amplification of the *SYT-SSX1, SYT-SSX2* and *SYT-SSX4* fusion transcripts:

  1. FP (SYT, exon 10): 5′-AGG.TCA.GCA.GTA.TGG.AGG.ATA.TAG.ACC-3′
  2. RP (SSX common, exon 5): 5′-GCT.GGC.TTC.TTG.GGC.ATG.AT-3′
  3. RP (SSX1, exon 5): 5′-GGC.CAG.ATG.CTT.CTG.ACA.CTC-3′
  4. RP (SSX2, exon 5): 5′-CAC.TTC.CTC.CGA.ATC.ATT.TCC.T-3′

The reaction is carried in two multiplex PCR. The first reaction for the SYT-SSX1 variant using the first three primers (1, 2 and 3). In the case of SYT-SSX1 translocation, PCR generates two products 102 and 146-bp DNA fragments. The second multiplex PCR for the SYT-SSX2 variant with the first two primers and the last primer (1, 2 and 4). In the case of SYT-SSX2 translocation, PCR generates two products 102 and 130-bp DNA fragments. If SYT-SSX4 is present, only the 102-bp fragment is generated in both reactions.

3.3.2.2 **Molecular Detection of Ewing's Sarcoma and Primitive Neuroectodermal Tumor**

Primitive neuroectodermal tumor (PNET) and Ewing's sarcoma in addition to Askin tumor of the thoracopulmonary region and esthesioneuroblastoma are highly malignant closely related neoplasms arising from primitive mesenchymal cells of the neural crest. All of them show an identical morphology, composed of closely packed small primitive cells occasionally with glycogen deposits in the cytoplasm and exhibit the similar immunophenotypic profile with expression of various neuroendocrine markers in addition to CD99 and c-myc oncoprotein. Using modern diagnostic tools, primitive neuroectodermal tumor accounts about 30% of all pediatric malignant soft tissue tumors. This tumor group is characterized by an aggressive clinical course with a high metastatic activity. About 30% of diagnosed cases have already distant metastases.

All the members of the PNET family including Ewing's sarcoma, Askin tumor of the thoracopulmonary region (primarily of chest wall) and esthesioneuroblastoma are characterized by a group of translocations involving the EWS gene and one of members of the ETS gene family encoding homologous transcriptional regulators. In rare cases the EWS gene is replaced by another homologous member of the TET family encoding RNA-binding proteins. These specific genetic abnormalities are found in more than 95% of studied cases.

- The EWS gene (*Ewing sarcoma* gene, also known as EWSR1 gene) is a member of the TET family of RNA-binding proteins, which are ubiquitously expressed proteins involved in transcription and processing of RNA. The EWS gene spans a 40-kb region on chromosome 22q12, contains an open reading frame of 1968 bp and is composed of 17 exons interrupted by 16 introns. This gene encodes a 656-amino-acid protein. As a result of alternative splicing two different transcripts are generated EWS and EWS-b that differs by the absence of the exons 8 and 9. The EWS protein includes an amino terminal domain, homologous to eukaryotic RNA polymerase II encoded by exons 1–7 and a carboxy terminal region, encoded by exons 11, 13 and 13 with a putative RNA-binding domain, functioning as a promoter. The breakpoints inside the EWS gene occurs usually inside the introns and are mainly clustered within a 7-kb region, frequently through intron 7, however other breakpoints are also reported through introns 8, 10 and 18. The first seven exons of the EWS gene are always a part of the fusion transcripts since they encode the transcription activator domain. The size of the generated transcripts varies and can range over 200–900 bp, depending on the location of the chromosome break site.

Other members of the TET family of RNA-binding proteins are the homologous FUS (TLS) and the TAF_{II} 68 (RBP56) genes. These genes are the fusion partners in other translocations associated with other malignancies.

- The ETS gene family (avian *e*rythroblastosis virus *t*ransforming *s*equence) is a gene family that includes many genes defined by the presence of a highly conserved 85-amino-acid domain termed the ETS domain that mediates a specific DNA-binding purine rich sequences characterized by invariant GGA/TA core element. The members of the ETS gene family are located on different chromosomes and include the following genes: EST1 and FL1 on chromosome 11q23–24, ERG and EST2 on chromosome 21q22, ETV1 on chromosome 7p22, E1AF on chromosome 17q21, FEV on chromosome 2q33, TEL on chromosome 12p13, ELK1 on chromosome Xp11 and SAP1 on chromosome 1q32.

The following genes of the ETS gene family are found to be the fusion partner of the EWS gene associated with Ewing's sarcoma and PNET family:

- FLI-1 gene (*f*riend *l*eukemia virus *i*ntegration site *1*) located on chromosome 11q24, a member of the ETS proto-oncogene family functioning as a transcriptional activator. Most breakpoints through the FLI-1 gene are clustered within a large, 40-kb region on chromosome 11q24 and exons 4–9 are frequently involved in this translocation, but the functionally important domain is encoded on exon 9, which is always a part of the fusion transcript. The t(11; 22)(q24; q12) translocation generates the fusion transcript EWS-FLI-1 encoding the EWS-FLI-1 fusion protein. This translocation is the most common and the most specific molecular marker for Ewing's sarcoma and PNET family that found in more than 90% of cases. According to the location of breakpoints and exon combinations, more than ten different variants of the EWS-FLI-1 fusion transcripts are described with a size ranging between 200 bp and 900 bp. Generally all variants of the EWS-FLI-1 are efficient transcriptional activators, usually 5- to 10-fold more active than the FLI-1 and able to promote malignant transformation *in vitro*. The product of the EWS-FLI-1 fusion transcript is able to deregulate the function of different genes involved in cell proliferation and differentiation such as MYC, EAT-2, MMP-3, FRINGE, ID2, CCND1, TGFBR2, CDKN1A, p19ARF and p57KIP. One study showed that the EWS-FLI-1fusion transcript is able to transform primitive progenitor cells *in vitro* to tumor cells with PNET phenotype including CD99 and NSE expression. Furthermore clinical observations assume a direct correlation between the expression levels of the fusion transcript and proliferation activity of tumors bearing this translocation. EWS-FLI-1 type 1 and type 2 are the most common variants of this translocation and account about 85% of all the EWS-FLI-1 fusions. In EWS-FLI-1 type 1, exon 7 of EWS gene is joined to exon 6 of the FLI-1 gene. Tumors bearing this translocation type are found to have the best therapy response and the best prognosis among this tumor family, and the detection of this translocation type considered to be a good prognostic factor, this type makes about 65% of the EWS-FLI-1 fusions. In EWS-FLI-1 type 2, exon 7 of EWS gene is joined to exon 5 of the FLI-1 gene; this translocation type shows a less favorite prognosis than type 1 and is found in about 20% of the EWS-FLI-1 fusion transcripts.

- ERG gene (*E*TS *r*elated gene) located on chromosome 21q22 and highly homologous to the FLI-1 gene encoding a transcriptional regulator (transforming protein ERG). The t(21;22)(q22;q12) translocation is the second common translocation type, associated with about 10% of this tumor family. Exons 6–9 of this gene can be involved in this translocation, but the functionally important domain is encoded by exon 9. This translocation generates the fusion transcript EWS-ERG functioning as a transcriptional activator. The amino acid sequences encoded by both EWS-ERG and EWS-FLI-1 have up to 98% identity, which suggest that both proteins regulate a common set of target genes. In very rare cases the EWS gene can be replaced by the homologous FUS gene, another member of the TET gene family.

The ERG gene is also described as a fusion partner of the TEL gene, a homologous gene to the EWS resulting the t (16; 21) (p11; q22) translocation reported in association with some types of acute myeloid leukemia.

- ETV1 gene (*E*TS *t*ranslocation *v*ariant *1*) located on chromosome 7p22, encoding a protein with a DNA-binding domain of the ETS family. The t(7; 22)(p22; q12) translocation is a rare translocation reported in association with the Ewing's sarcoma–PNET tumor family and generates the EWS-ETV1 fusion transcript.

- E1AF gene is another member of the ETS-DNA-binding protein family, located on chromosome 17q12, encoding the adenovirus E1A enhancer-binding protein. The E1AF protein is able to bind the adenovirus E1A enhancer, which plays an important role in the cell growth. The t(17; 22)(q12; q12) translocation is also described as a rare translocation associated with the Ewing's sarcoma tumor group and generates the EWS-E1AF fusion transcript.

- FEV gene consists of three exons, located on chromosome 2q33, a further member of the ETS family of transcription factors, possibly a repressor. The t(2; 22)(q33; q12) translocation is also one of the rare translocations associated with this tumor group and generates the EWS-FEV fusion transcript.

Recently rare cases of the primitive neuroectodermal tumors were described in association with translocations involving other members of TET family homologous to the EWS gene such as the FUS gene at 16p11 and one of the members of the ETS gene family.

The detection of the above-mentioned transcripts can be used as precise molecular markers complementary to the conventional histopathology and immunohistochemistry to distinguish the Ewing's sarcoma/PNET tumor family from other tumors with overlapped morphology. This method can also be used to monitor the progression of the disease and for the evaluation of treatment successes. Moreover, the presence of EWS-FLI-1 type 1 considered being a positive prognostic factor. Noteworthy is that the above-mentioned genetic abnormalities are not detected in neuroblastoma.

The above-mentioned translocations can be also used in combination with other molecular neuroendocrine markers expressed by this tumor group such as Chromogranin A and secretogranin II discussed in a later section.

Method The method used is mRNA extraction from fresh or fixed tissue in addition to FNA material, peripheral blood or bone marrow, followed by RT-PCR.

- Primers for the cDNA amplification of the *EWS-FLI-1* fusion transcript:
  FP (EWS, exon 9/10): 5′-CTG.GTG.GAC.CCA.TGG.ATG.AAG.GA-3′
  RP (FLI-1, exon 5/6): 5′-GAA.GGG.TCT.TCT.TTG.ACA.CTC.A-3′

PCR generates a 112-bp DNA fragment of the EWS-FLI-1 fusion transcripts.

- Primers for the cDNA amplification of both type 1 and 2 of the *EWS-FLI-1* fusion transcripts:
  FP (EWS, exon 7): 5′-TCC.TAC.AGC.CAA.GCT.CCA.AGT.CAA.TA-3′
  RP (FLI-1, exon 6): 5′-ATT.GCC.CCA.AGC.TCC.TCT.TCT.GAC-3′

PCR generates a 99-bp DNA fragment in type 1 and 165-bp DNA fragment in type 2 of the EWS-FLI-1 fusion transcripts.

- Alternative primers for the cDNA amplification of both type 1 and 2 of the *EWS-FLI-1* fusion transcripts:
  FP (EWS, exon 7): 5′-CTG.GAT.CCT.ACA.GCC.AAG.CTC.CAA.G-3′
  RP (FLI-1, exon 6): 5′-GTT.GAG.GCC.AGA.ATT.CAT.GTT.A-3′

PCR generates a 125-bp DNA fragment in type 1 and 191-bp DNA fragment in type 2 of the EWS-FLI-1 fusion transcripts.

- Primers for the cDNA amplification of type 1, 2 and 3 of the *EWS-FLI-1* fusion transcripts:
  FP (EWS, exon 7): 5′-TCC.TAC.AGC.CAA.GCT.CCA.AGT.C-3′
  RP (FLI-1, exon 9): 5′-ACT.CCC.CGT.TGG.TCC.CCT.CC-3′

PCR generates a 329-bp DNA fragment in type 1, a 395-bp DNA fragment in type 2 and 581-bp DNA fragment in type 3 of the EWS-FLI-1 fusion transcripts.

- Primers for the cDNA amplification of type 2 of the *EWS-FLI-1* fusion transcript:
  FP (EWS, exon 7): 5′-TAC.AGC.CAA.GCT.CCA.AGT.CA-3′
  RP (FLI-1, exon 5): 5′-TCG.GTG.TGG.GAG.GTT.GTA.TT-3′

PCR generates a 92-bp DNA fragment EWS-FLI-1 fusion transcripts.

- Primers for the cDNA amplification of type 3 of the *EWS-FLI-1* fusion transcript:
  FP (EWS, exon 10): 5′-CCA.TGG.ATG.AAG.GAC.CAG.AT-3′
  RP (FLI-1, exon 7): 5′-GGA.CTT.TTG.TTG.AGG.CCA.GA-3′

PCR generates a 100-bp DNA fragment EWS-FLI-1 fusion transcripts.

- Primers for the cDNA amplification of the both *EWS-FLI-1* and *EWS-ERG* fusion transcripts (nested PCR):
  –Outer primers:
  FP (EWS): 5′-GCA.GCA.GCC.TCC.CAC.TAG.TTA.CC-3′
  RP (FLI-1): 5′-CAG.GTG.ATA.CAG.CTG.GCG.TTG.G-3′
  RP (ERG): 5′-CAG.GTG.ATG.CAG.CTG.GAG.TTG.G-3′
  –Inner primers:
  FP (EWS): 5′-CTG.GAT.CCT.ACA.GCC.AAG.CTC.C-3′
  RP (FLI-1): 5′-CAG.GAG.GAA.TTG.CCA.CAG.CTG.G-3′
  RP (ERG): 5′-CAG.GAG.GAA.CTG.CCA.AAG.CTG.G-3′

PCR is carried out using the EWS foreword primer and both FLI-1 and ERG reverse primers as multiplex PCR. The size of generated DNA fragments ranges between 95 bp and 600 bp depending on the translocation type and involved exons. Using the above primers the EWS-ERG fusion generates short DNA fragments (<200 bp), whereas the EWS-FLI-1 fusion generates longer DNA fragments (250–600 bp). If the amplification reveals a positive result, two separated reactions with one EWS foreword primer and one reverse primer can be done to determine the type of the translocation.

- Primers for the cDNA amplification of the *EWS-ERG* fusion transcript
  FP (EWS, exon 9): 5′-TGGAGAGCGAGGTGGCTTCAATAA -3′
  RP (ERG, exon 6): 5′-CTCCTGGGGGGCTCATATGGTAAAT -3′

PCR generates a 90-bp DNA fragment.

- Alternative primers for the cDNA amplification of the *EWS-ERG* fusion transcript
  FP (EWS, exon 7): 5′-TCC.TAC.AGC.CAA.GCT.CCA.AGT.C-3′
  RP (ERG): 5′-GGT.TGA.GCA.GCT.TTC.GAC.TG-3′

PCR generates a 137-bp DNA fragment.

- Primers for the cDNA amplification of type 1e of the *EWS-ERG* fusion transcript
  FP (EWS, exon 7): 5′-TCC.TAC.AGC.CAA.GCT.CCA.AGT.CAA.TA-3′
  RP (ERG, exon 6): 5′-CTC.CTG.GGG.GGC.TCA.TAT.GGT.AAA.T-3′

PCR generates a 83-bp DNA fragment.

- Alternative primers for the cDNA amplification of the *EWS-ERG* fusion transcript (nested PCR):
  –Outer primers:
  FP (EWS, exon 7): 5′-TCC.TAC.AGC.CAA.GCT.CCA.AGT.C-3′
  RP (ERG): 5′-ACT.CCC.CGT.TGG.TGC.CTT.CC-3′

–Inner primers:
FP (EWS): 5′-CCA.ACA.GAG.CAG.CAG.CTA.C-3′
RP (ERG): 5′-CAG.GTG.ATG.CAG.CTG.GAG-3′
PCR generates a 642-bp DNA fragment.

- Primers for the cDNA amplification of the *EWS-FEV*, *EWS-ETV1* and *EWS-E1AF1* fusion transcripts:
  FP (EWS): 5′-GTC.AAC.CTC.AAT.CTA.GCA.CAG.GG-3′
  RP (FEV): 5′-GAA.ACT.GCC.ACA.GCT.GGA.TC-3′
  FP (ETV1): 5′-TAA.ATT.CCA.TGC.CTC.GAC.CAG-3′
  RP (E1AF): 5′-AAC.TCC.ATT.CCC.CGG.CC-3′

3.3.2.3 **Molecular Detection of Desmoplastic Small Round Cell Tumor**

Desmoplastic small round cell tumor is a highly malignant primitive tumor with an aggressive behavior and poor prognosis, affecting adolescents and young adults mainly males between 15 and 35 years of age. Desmoplastic small round cell tumor presents predominantly in the abdominal cavity including omentum, scrotum and ovary, but also reported to involve the pleura and the posterior mediastinum, which suggest that mesotheloblasts might be the origin of this tumor (mesotheloblastoma?). Morphologically, desmoplastic small round cell tumor composed of undifferentiated primitive small rounded cells arranged in well defined nests and masses separated by fibrous connective tissue stroma. The tumor cells are characterized by the coexpression of epithelial, mesenchymal and neural antigens. They react positively with the antibodies to different cytokeratins, EMA, desmin and vimentin, and in some cases with various neuroendocrine markers such as antibodies to NSE, S100 and leu 7, moreover most tumors reacts positively with the antibody against the Wilms tumor protein. This tumor must be distinguished from other lesions with similar histology or immunohistochemical profile such as PNET, embryonal rhabdomyosarcoma or neuroblastoma.

Few different specific cytogenetic abnormalities are found to be associated with the desmoplastic small round cell tumor, but the most common are the translocations involving the Wilms tumor suppressor gene (WT1). This gene is able to interact with other growth related genes resulting transcriptional suppression. The Wilms tumor suppressor gene is normally expressed in mesodermally derived tissues during early stages of renal embryogenesis and mutations appearing in this gene are associated with abnormal development of the urogenital tract including the WAGR syndrome counting Wilms tumor (nephroblastoma), aniridia, genitourinary abnormality and mental retardation.

The first and most common translocation results the in frame fusion of the Ewing's sarcoma gene EWS on chromosome 22q12 to the Wilms tumor suppressor gene (WT1). The WT1 gene is located on a ~50 kb region on chromosome 11p13, composed of ten exons and encodes a transcription factor with different isoforms due to alternative splicing containing four zinc-finger domains. The breakpoints inside the WT1 gene are mostly clustered within the intron between exons 7 and 8, however breakpoints within introns between exons 8 and 9 or between exons 9 and 10 are reported in rare cases. The breakpoints through the EWS gene are mainly clustered within the intron between exon 7 and 8 but other breakpoints between exons 8 and 9 or between exons 9 and 10 of this gene are also reported. Accordingly the t(11; 22)(p13; q12) translocation results an in frame junction between both genes generating the EWS-WT1 fusion transcript. The EWS-WT1 fusion transcript is a heterogeneous fusion transcript, in the majority of cases is composed of the first exons of the EWS gene and 8–10 exons of the WT1 gene. The gene product is a transcription factor with unique DNA-binding properties. This unique translocation converts the Wilms tumor suppressor gene to a potent transcriptional activator functioning as an oncogene.

The second but rare translocation associated with this tumor results the fusion of the above-mentioned EWS gene to the ERG gene on chromosome 21q22. The t(21; 22)(q22; q12) translocation generates the EWS-ERG fusion transcript, which is homologous to the fusion transcript found in Ewing's sarcoma.

The t(11;17)(p13; q11.2) translocation is a third translocation described in rare cases of desmoplastic small round cell tumor.

The detection of the above-described specific fusion transcripts for the desmoplastic small round cell tumor by RT-PCR can be used as molecular markers specific for this tumor, offering an accurate method to distinguish this tumor from other similar tumors, especially when the tumor occurs in an atypical location or shows an ambiguous immunophenotype.

Method The method used is mRNA extraction from frozen or paraffin-embedded tissue or FNA cells, followed by RT-PCR.

- Primers for the cDNA amplification of the *EWS-WT1* fusion transcript:
  FP (EWS, exon 7): 5′-TCC.TAC.AGC.CAA.GCT.CCA-3′
  RP (WT1, exon 8): 5′-CTT.CGT.TCA.CAG.TCC.TTG.AAG.T-3′

PCR generates a100-bp DNA fragment.

- Alternative primers for the cDNA amplification of the *EWS-WT1* fusion transcript:
  FP (EWS): 5′-CCA.ACA.GAG.CAG.CAG.CTA.CG-3′
  RP (WT1, exon 9): 5′-GAC.CAG.GAG.AAC.TTT.CGC.TGA.C-3

PCR generates a 168-bp DNA fragment in the case of EWS/exon 7 to WT1/exon 8 fusion. Other fragments of various sizes can be generated, depending on the location of breakpoints within the involved genes.

- Alternative primers for the cDNA amplification of the *EWS-WT1* fusion transcript:
  FP (EWS/exon 7): 5′-TCC.TAC.AGC.CAA.GCT.CCA.AGT.C-3′
  RP (WT1): 5′-GCC.ACC.GAC.AGC.TGA.AGG.GC-3

PCR generates a DNA fragment of 259–270 bp, depending on breakpoints within the involved genes.

- Alternative primers for the cDNA amplification of the *EWS-WT1* fusion transcript:
  FP (EWS, exon 7): 5′-TCC.TAC.AGC.CAA.GCT.CCA.AGT.C-3′
  RP (WT1, exon 9): 5′-GAC.CAG.GAG.AAC.TTT.CGC.TGA.C-3′

PCR generates approximately 197, 416 and 449-bp DNA fragments in translocations fusing exons 7, 9 or 10 of the EWS gene to exon 8 of WT1 gene respectively.

3.3.2.4 Molecular Detection of Clear Cell Sarcoma (Malignant Melanoma of Soft Tissue)

Clear cell sarcoma (malignant melanoma of soft tissue) was initially described by Enzinger as "clear cell sarcoma of tendons and aponeuroses". Clear cell sarcoma represents a highly malignant tumor of a neural crest origin with melanocytic differentiation and appears mainly in the adult life in the age between 20–40 years and presents in the deep subcutaneous soft tissue near to tendons and aponeurosis, mainly in the lower and upper extremities, but rare cases outside of soft tissue are also described. The tumor tends to metastasize to the regional lymph nodes and to the lung with a 5–year survival rate of ~50%. Morphologically appears the tumor as a well circumscribed not capsulated lobulated mass, composed of nests and fascicles of rounded or fusiform cells with clear cytoplasm, sometimes with marked atypia resembling malignant melanoma or epithelioid variant of malignant peripheral nerve sheath tumor. Clear cell sarcoma has an identical biochemical and immunohistochemical profile as malignant melanoma including melanin synthesis and positive immunohistochemical reactions with antibodies to S100, HMB45, leu 7, MITF (*mi*crophthalmia *t*ranscription *f*actor) and NSE but differs from malignant melanoma by specific cytogenetic abnormalities and clinicopathologic behavior.Clear cell sarcoma is characterized by a novel non-random balanced chromosomal translocation found in more than 90% of cases. The reciprocal t (12; 22)(q13; q12) translocation involves the EWS gene (*E*wing's *s*arcoma) on chromosome 22q12 and the ATF-1 gene (*a*ctivating *t*ranscription *f*actor *1*, also known as the TREB36 gene). The ATF-1 gene is located on chromosome 12q13 and encodes a DNA-binding nuclear protein, one of the members of a subgroup of bZIP transcription factors. This translocation generates different homologous EWS-ATF1 fusion transcripts in which the EWS encoded protein is lost and the ATF1 protein bZIP domain is conserved. The generated fusion transcripts are usually about 200-fold more active than the ATF-1 gene and activates the cAMP induction due to a partial deletion of the ATF1 kinase-inducible promoters. Depending on the location of breakpoints, four types of the EWS-ATF1 fusion transcript are described, whereas the types 1 and 2 are the most common. The first type is generated by the fusion of exon 8 of the EWS gene with exon 4 of the AFT-1 gene, while the second type results by the fusion of exon 10 of the EWS gene with exon 5 of the ATF-1 gene. The third and fourth types are less common and are generated by the fusion of exon 7 of the EWS gene with exon 5 and exon 7 of the AFT-1 gene respectively. Type 1 and 2 encode two transcription proteins differ by 40 amino acids but show identical properties and function as constitutive transcriptional activator of several cAMP-inducible promoters, which also interferes with the P53 function and causes the deregulation of the cell cycle.

In addition to this translocation, additional genetic abnormalities are also described in association with clear cell sarcoma such as the trisomy of the 2, 7, 8 and 22 chromosomes but they are not specific for this tumor. Furthermore, the translocations t(2; 22), t(1; 19)(q12; p13) and t/del(1)(p12–22) were also reported in rare cases of clear cell sarcoma.

As the t(12; 22)(q13; q12) translocation has not been reported in association of malignant melanoma or any other type of malignancies, it is suitable to be used as a specific molecular marker for the diagnosis and monitoring of clear cell sarcoma.

Method The method used is mRNA extraction from frozen or paraffin-embedded tissue or FNA cells, followed by RT-PCR.

- Primers for the cDNA amplification of the *EWS-ATF1* fusion transcript type 1:
  FP (EWS, exon 8): 5′-CAT.GAG.CAG.AGG.TGG.GCG-3′
  RP (ATF-1, exon 4): 5′-CCC.CGT.GTA.TCT.TCA.GAA.GAT.AAG.TC-3′

PCR generates an 81-bp DNA fragment.

- Primers for the cDNA amplification of the *EWS-ATF1* fusion transcript type 2:
  FP (EWS, exon 7): 5′-GCC.AAG.CTC.CAA.GTC.AAT.ATA.GC-3′
  RP (ATF-1, exon 5): 5′-CAA.CTG.TAA.GGC.TCC.ATT.TGG.G-3′

PCR generates an 83-bp DNA fragment.

- Primers for the cDNA amplification of both type 1 and type 2 of the *EWS-ATF1* fusion transcripts:
  FP (EWS): 5′-ATC.GTG.GAG.GCA.TGA.GCA-3′
  RP (ATF-1): 5′-ACT.CCA.TCT.GTG.CCT.GGA.CT-3′

PCR generates a 241-bp DNA fragment in the case of translocation type 1 and a 178-bp DNA fragment in the case of translocation type 2.

- Alternative primers for the cDNA amplification of both type 1 and type 2 of the *EWS-ATF1* fusion transcripts (nested PCR):
  –Outer primers:
  FP (EWS, exon 8): 5′-GAG.GCA.TGA.GCA.GAG.GTG.G-3′
  RP (ATF-1): 5′-GAA.GTC.CCT.GTA.CTC.CAT.CTG.TG-3′

PCR generates a 246-bp DNA fragment in the case of EWS-ATF1 type 1 and a 183-bp DNA fragment in the case of type 2.

  –Nested primers:
  FP (EWS, exon 8): 5′-GAG.GAG.GAC.GCG.GTG.GAA.TG-3′
  RP (ATF-1): 5′-CTG.TAA.GGC.TCC.ATT.TGG.GGC-3′

Nested PCR generates a 185-bp DNA fragment in the case of EWS-ATF1 type 1 and a 122-bp DNA fragment in the case of type 2.

- Primers for the cDNA amplification of type 3 of *EWS-ATF1* fusion transcript (nested PCR), carried out after the exclusion of type 1 and type 2:
  –Outer primers:
  FP (EWS, exon 7): 5′-TCC.TAC.AGC.CAA.GCT.CCA.AGT.C-3′
  RP (ATF-1): 5′-GAA.GTC.CCT.GTA.CTC.CAT.CTG.TG-3′

PCR generates a 124-bp DNA fragment in the case of EWS-ATF1 type 3.

  –Nested primers:
  FP (EWS, exon 7): 5′-TAT.AGC.CAA.CAG.AGC.AGC.AG-3′
  RP (ATF-1): 5′-CTG.TAA.GGC.TCC.ATT.TGG.GGC-3′

Nested PCR generates a 63-bp DNA fragment in the case of EWS-ATF1 type 3.

- Alternative primers for the cDNA amplification of the *EWS-ATF1* fusion transcript:
  FP (EWS): 5′-CCC.ACT.AGT.TAC.CCA.CCC.CA-3′
  RP (ATF-1): 5′-AAA.ACT.CCA.CTA.GGA.AAT.CCA.TTT-3′

PCR generates a 954-bp DNA fragment.

3.3.2.5 **Molecular Detection of Alveolar Rhabdomyosarcoma**

Rhabdomyosarcoma is a malignant soft tissue tumor derived from primitive mesenchymal cells with tendency to skeletal muscle differentiation. According to the international classification, rhabdomyosarcoma is classified into three major types based on morphological features, cytogenetic properties and clinical behavior:

1. Embryonal rhabdomyosarcoma (both spindle cell and botryoid types);
2. Alveolar rhabdomyosarcoma;
3. Pleomorphic or anaplastic rhabdomyosarcoma, a rare adult variant of rhabdomyosarcoma.

Both embryonal and alveolar rhabdomyosarcoma are associated with different cytogenetic abnormalities, which indicate the different molecular etiology of both tumor identities and explain the different clinical behavior.

Alveolar rhabdomyosarcoma is the second common type of rhabdomyosarcoma after the embryonal rhabdomyosarcoma and makes approximately 20% of all rhabdomyosarcoma. The tumor presents often at the age of 10–25 years and tends to appear in the muscle tissue of upper and lower extremities. Histologically, alveolar rhabdomyosarcoma is composed of aggregates of poorly differentiated oval to round proliferating cells with a mixture of solid and alveolar areas divided by fibrous septa resembling the morphology of skeletal muscle in 10–12 gestation weeks. This tumor shows the typical immunohistochemical profile of all types of rhabdomyosarcoma with the expression of various myogenic markers such as desmin, sr-actin and myoglobin in addition to the myogenic regulatory proteins such as MyoD1, myogenin, Myf-3, Myf-4 and Myf-5. Alveolar rhabdomyosarcoma is characterized by more aggressive clinical behavior and worse prognosis than embryonal rhabdomyosarcoma and needs therapy that is more aggressive. This tumor is also described to have an extensive bone marrow involvement, sometimes resembling leukemia. A further important feature of this tumor is the presence of characteristic genetic abnormalities, which include two novel translocations that have not been demonstrated in any other type of rhabdomyosarcoma or soft tissue tumor.

Alveolar rhabdomyosarcoma is a heterogeneous tumor and about 90% of this tumor are associated with the t(2; 13)(q35; q14) translocation, whereas the other 10% of this tumor are associated with another analogous genetic abnormality namely the t (1; 13)(p36; q14) translocation generating a homologous fusion transcript.

The first and most common t(2; 13)(q35; q14) translocation detected in about 75% of cases involves the FKHR (*fork head* in *r*habdomyosarcoma, also called

FOXO1A) gene and the PAX3 gene. The FKHR gene is a member of the fork head family of transcription factors, characterized by a conserved DNA binging domain termed the fork head domain, located on chromosome 13q14 and consists of 3 exons and encodes one of the fork head transcriptional regulators effecting as an oncoprotein. The PAX3 gene (*pa*ired bo*x* 3) is located on chromosome 2 q35, composed of ten exons with several alternatively spliced transcripts encoding a protein consists of a paired box domain, an octapeptide and paired type homeodomain. The gene product is a paired box transcriptional factor, mainly expressed in the early embryogenesis during mesenchymal and neurogenic differentiation.

The t(2; 13)(q35; q14) translocation disrupted the intron between exon 7 and 8 of the PAX3 gene and the intron between exon 1 and 2 of the FKHR gene. This in frame gene fusion generates the PAX3-FKHR fusion transcript containing the DNA-binding domain of the PAX3 genes and the transcriptional activation domain of the FKHR gene. The PAX3-FKHR fusion transcript encodes a 97-kDa, 836-amino-acid fusion protein, which is 10- to 100-fold more potent as a transcriptional activator than the PAX3 alone, and consequently effecting as an oncogene.

The second and less frequent genetic abnormality is the t(1; 13)(p36; q14) translocation described in about 10% of alveolar rhabdomyosarcoma and involves the same FKHR gene on chromosome 13 and the PAX7 gene (*pa*ired bo*x* 7) on chromosome 1p36, consists of eight exons and has a similar structure and functional organization as the PAX3 gene.

Clinical observations show that the two translocations are associated with different clinical presentations, different therapy response and different clinical behavior. Alveolar rhabdomyosarcomas bearing the first and more common t(2; 13)(q35; q14) translocation with the PAX3-FHHR fusion transcript show predilection for older patients, appear frequently in the chest, abdominal region and pelvis. These tumors are characterized by a high metastatic potential, frequent bone marrow involvement, more aggressive behavior and worse prognosis than alveolar rhabdomyosarcomas bearing the second t(1; 13)(p36; q14) translocation type with the PAX7-FHHR fusion transcript, which appear mostly in the extremities of younger patients (patients aged mainly two years or younger) and characterized by better prognosis.

In conclusion, the above-mentioned translocations are specific cytogenetic features for alveolar rhabdomyosarcoma, which reflect the different histogenetic features and clinical behavior of this tumor. Furthermore the molecular amplification of the generated specific fusion transcripts can be used as diagnostic tool to distinguish alveolar rhabdomyosarcoma from other types of rhabdomyosarcoma or from tumors with rhabdomyosarcomatous differentiation in addition to tumors with similar morphological features such as alveolar soft part sarcoma and malignant rhabdoid tumor. Moreover, the discrimination of the translocation type found to be a useful prognostic factor as the PAX7-FKHR fusion transcript indicates a longer disease-free survival and a better prognosis.

The amplification of the tumor-specific fusion transcripts can also be done simultaneously with the amplification of the mRNA encoding other myogenic markers discussed in later section as a multiplex PCR. This method can also be used for the screening of surgical margins after surgical resection of tumor, important for optimal management of tumor therapy.

Method The method used is mRNA extraction from frozen or paraffin-embedded tissue or FNA cells, followed by RT-PCR.

- Primers for the cDNA amplification of the *PAX3-FKHR* fusion transcript:
  FP (PAX3): 5′-CCG.ACA.GCA.GCT.CTG.CCT.AC-3′
  RP (FKHR): 5′-TGA.ACT.TGC.TGT.GTA.GGG.ACA.G-3′

PCR generates a 170-bp DNA fragment.

- Alternative primers for the cDNA amplification of the *PAX3-FKHR* fusion transcript (nested PCR):
  –Outer primers:
  FP (PAX3): 5′-AGC.TAT.ACA.GAC.AGC.TTT.GT-3′
  RP (FKHR): 5′-CTC.TGG.ATT.GAG.CAT.CCA.CC-3′
  –Nested primers:
  FP (PAX3): 5′-AGC.TAT.ACA.GAC.AGC.TTT.GT-3′
  RP (FKHR): 5′-TCC.AGT.TCC.TTC.ATT.CTG.CA-3′

Nested PCR generates a 158-bp final DNA fragment.

- Primers for the cDNA amplification of the *PAX7-FKHR* fusion transcript:
  FP (PAX7): 5′-CCT.CCA.ACC.(A/C)CA.TGA.ACC.C-3′
  RP (FKHR): 5′-GCA.CAC.GAA.TGA.ACT.TGC.TG-3′

PCR generates a 96-bp DNA fragment.

- Alternative primers for the cDNA amplification of the *PAX7-FKHR* fusion transcript (nested PCR):
  –Outer primers:
  FP (PAX7): 5′-GCT.TCT.CCA.GCT.ACT.CTG.AC-3′
  RP (FKHR): 5-CTC.TGG.ATT.GAG.CAT.CCA.CC-3′
  –Nested primers:
  FP (PAX7): 5′-GCT.TCT.CCA.GCT.ACT.CTG.AC-3′
  RP (FKHR): 5′-TCC.AGT.TCC.TTC.ATT.CTG.CA-3′

Nested PCR generates approximately 150-bp DNA fragment.

- Primers for the cDNA amplification of both *PAX3-FKHR* and *PAX7-FKHR* fusion transcripts in Multiplex PCR:
  FP (PAX3): 5′-TTG.GCA.ATG.GCC.TCT.CAC.C-3′
  FP (PAX7): 5′-CAA.CCA.CAT.GAA.CCC.GGT.C-3′
  RP (FKHR): 5′-ATC.CAC.CAA.GAA.CTT.TTT.CCA.G-3′

PCR generates a 106-bp DNA fragment in PAX3-FKHR and 89-bp DNA fragment in PAX7-FKHR fusion transcripts.

- Primers employed as an internal control for the amplification of the *FKHR* template:
  RP (FKHR): 5′-TGA.ACT.TGC.TGT.GTA.GGG.ACA.G-3′
  RP (FKHR): 5′-GCA.GAT.CTA.CGA.GTG.GAT.GG-3′

PCR generates a 117-bp DNA fragment.

3.3.2.6 Molecular Detection of Myxoid Liposarcoma

Liposarcoma is the most common sarcoma and makes about 20% of all soft tissue sarcomas. According to the WHO classification, liposarcoma is divided into four major types based on morphological features, cytogenetic properties and clinical behavior and includes the following entities:

1. Well differentiated liposarcoma (atypical lipoma or atypical lipomatous tumor)
2. Myxoid and round cell liposarcoma
3. Pleomorphic liposarcoma
4. Dedifferentiated liposarcoma

Myxoid liposarcoma is the second most common type of liposarcoma accounting for about 35–40% of all liposarcomas and represents about 10% of all adult sarcomas. This tumor appears usually after the age of 15 years with a peak incidence in the fourth and fifth decades. This sarcoma type tends to present in the extremities and trunk whereas two-thirds of the cases arise within the muscles of the thigh. Histologically, the tumor consists of proliferating lipoblasts with a myxoid matrix and plexiform capillary pattern. Myxoid liposarcoma shows two major differentiation types sharing morphologic, cytogenetic and clinical features. The first and more common type is the highly differentiated low-grade type with a typical myxoid pattern and low cellularity. The second type is the poorly differentiated high-grade round cell liposarcoma, characterized by more aggressive behavior and poor prognosis. The morphology of myxoid liposarcoma may resemble the architecture of other neoplastic lesions such as intramuscular myxoma, aggressive angiomyxoma and myxoid chondrosarcoma, which may have also a similar immunohistochemical profile. Other nonneoplastic lesions may also simulate myxoid liposarcoma such as fat atrophy with myxoid change after trauma.

Generally, all types of benign and malignant lipomatous tumors are associated with specific genetic alterations. Myxoid liposarcoma is found to be associated with two novel and specific translocations involving one of the homologous TLS or EWS genes (Both genes are members of the TET family of RNA-binding proteins) and the CHOP gene.

The t(12; 16)(q13; p11) translocation is the most common translocation, described in more than 90% of myxoid and round cell liposarcoma, mainly those arising within the thigh soft tissue. Myxoid liposarcoma arising elsewhere seems to be not related to this specific genetic abnormality. The t(12; 16)(q13; p11) translocation is caused by the fusion of the TLS gene and the CHOP gene.

The TLS gene (*t*ranslocation in *l*ipo*s*arcoma), also known as the FUS (fusion) gene is highly homologous to the EWS gene, located on the chromosome 16p11, composed of 15 exons and encodes a nuclear RNA-binding protein expressed in a housekeeping pattern. As a result of alternative splicing the gene product has two isoforms, one isoform includes exon 4a, composed of 526 amino acids, and a second isoform includes exon 4b, composed of 525 amino acids. The TLS gene products are involved in RNA splicing and able to promote the homologous DNA pairing. The TLS (FUS) gene is also involved in other translocations associated with various hematological malignancies including the t(16;21)(p11; q22) translocation associated with AML, additionally it may function as a regulator of BCR/ABL oncogenesis.

The CHOP gene (*C*CAT/enhancer-binding protein *ho*mologous *p*rotein), also known as GADD153 gene (*g*rowth *a*rrest and *D*NA *d*amage inducible gene *153*) is located on chromosome 12q13. It is a member of the leucine zipper (C/EBP) family of transcription factors, encoding a transcriptional factor involved in adipocytic differentiation, erythropoiesis, G1-S cell cycle progression and growth arrest. Note that the 12q13 chromosome region is the location of other genes, frequently involved in various lipomatous tumors such as MDM2, CDK4 and HMGI-C. The t(12; 16)(q13; p11) translocation encodes the fusion transcript TLS (FUS)-CHOP in which the RNA-binding domain of the TLS gene is replaced by the DNA-binding and leucine demerization domain of the CHOP gene. The generated fusion transcript functions as a potent transcriptional activator under the control of the FUS gene promoter. Depending on the chromosomal breakpoints and the involved exons, about nine different structural variants of the TLS (FUS)-CHOP fusion gene are described, whereas the next five types are the most common variants:

- Type 1, caused by the fusion of exon 7 of the TLS gene to exon 2 of the CHOP gene, found in about 20% of this translocation.
- Type 2, cased by the fusion of exon 5 of the TLS gene to exon 2 of the CHOP gene, found in about 65% of this translocation.
- Type 3, cased by the fusion of exon 8 of the TLS gene to exon 2 of the CHOP gene, found in about 10% of this translocation.
- Type 4, caused by the fusion of exon 5 of the TLS gene to exon 3 of the CHOP gene.
- Type 5, cased by the fusion of a part of exon 5 of the TLS gene to exon 2 of the CHOP gene.

Some studies show that there is no significant association between the fusion variant and the tumor behavior but other studies reported that the poorly differentiated round cell type of myxoid liposarcoma is mainly associated with the fusion transcript type 1.

The second but less common translocation associated with this neoplasia is caused by the fusion of the EWS gene on chromosome 22q12 and the CHOP gene. This translocation is described in 2–5% of all myxoid liposarcoma. The t(12; 22)(q13; q12) translocation generates the fusion transcript EWS-CHOP. Two structural variants of the EWS-CHOP fusion transcript are described. Type 1 caused by the fusion of exon 7 of the EWS gene to exon 2 of the CHOP gene and type 2 caused by the fusion of exon 10 of the EWS gene to exon 2 of the CHOP gene. A further rare genetic abnormality found in association with myxoid liposarcoma is the der(16) t(1; 16).

As the two above-described translocations are specific for myxoid and round cell liposarcoma, they can be used as sensitive molecular markers to distinguish both liposarcoma types from other neoplastic or nonneoplastic lesions with similar morphologic features. This method can be also used for the examination of surgical margins after surgical resection or for the detection of micrometastases. Additionally, the identification of different translocation variants may be useful as a prognostic factor, as type 1 of the TLS-CHOP fusion transcript may be associated with more aggressive tumor behavior and poor prognosis.

Method The method used is mRNA extraction from frozen or paraffin-embedded tissue or FNA cells, followed by RT-PCR.

- Primers for the cDNA amplification of the *TLS-CHOP* fusion transcript:
  FP (TLS): 5′-CAG.AGC.TCC.CAA.TCG.TCT.TAC.GG-3′
  RP (CHOP): 5′-GAG.AAA.GGC.AAT.GAC.TCA.GCT.GCC-3′

PCR generates a 653-bp DNA fragment in a type 1 TLS-CHOP fusion transcript, a 377-bp DNA fragment in type 2, a ~730-bp DNA fragment in type 3, a 329-bp DNA fragment in type 4 and a 284-bp DNA fragment in type 5.

- Alternative primers for the cDNA amplification of the *TLS-CHOP* fusion transcript (nested PCR):
  –Outer primers:
  FP (TLS): 5′-AGC.AAA.GCT.ATA.ATC.CCC.CTC.AG-3′
  RP (CHOP): 5′-GAA.GGA.GAA.AGG.CAA.TGA.CTC.A-3′
  –Nested primers:
  FP (TEL): 5′-GAC.AGC.AGA.ACC.AGT.ACA.ACA.GCA.G-3′
  RP (CHOP): 5′-GCT.TTC.AGG.TGT.GGT.GAT.GTA.TGA.AG-3′

PCR generates a 379-bp DNA fragment in a type 1 TLS-CHOP fusion transcript and a 103-bp DNA fragment in a type 2 TLS-CHOP fusion transcript.

- Primers for the cDNA amplification of the *EWS-CHOP* fusion transcript:
  FP: 5′-CCA.GCC.CAG.CCT.AGG.ATA.TGG.ACA-3′
  RP: 5′-CTG.GAC.AGT.GTC.CCG.AAG.GAG.AAA-3′

PCR generates a 441-bp DNA fragment in a type 1 TLS-CHOP fusion transcript and a 693-bp DNA fragment in a type 2.

- Alternative primers for the cDNA amplification of the *EWS-CHOP* fusion transcript (nested PCR):
  –Outer primers:
  FP (EWS): 5′-CCA.GCC.CAG.CCT.AGG.ATA.TTG.ACA-3′
  RP (CHOP): 5′-AGG.TGG.GTA.GTG.TGG.CCC.AAG.TG-3′
  –First nested primers:
  FP (EWS): 5′-AGC. TAC.CCC.ATG.CAG.CCA.GTC.AC-3′
  RP (CHOP): 5′-GCC.CAA.GTG.GGG.GAC.TGA.TGC.TC-3′

  PCR generates an 883-bp DNA fragment.
  –Second nested primers:
  FP (EWS): 5′-TGG.GCA.ACC.GAG.CAG.CTA.TG.AC-3′
  RP (CHOP): 5′-GTC.AGC.CAA.GCC.AGA.GAA.GCA.GGG-3′

  Second nested PCR generates a 423-bp DNA fragment.
  –Third nested primers:
  FP (EWS): 5′-TGG.GCA.ACC.GAG.CAG.CTA.TG.AC-3′
  RP (CHOP): 5′-GAG.AAA.GGC.AAT.GAC.TCA.GCT.GCC-3′

  Third nested PCR generates a 258-bp DNA fragment.

3.3.2.7 Molecular Detection of Extraskeletal Myxoid Chondrosarcoma

Extraskeletal myxoid chondrosarcoma is a rare low-grade malignant soft tissue tumor with chondroblastic differentiation, probably of neural-neuroendocrine origin. The tumor occurs mainly in adults, primarily in the deep tissue of proximal portions of the extremities and rarely in the trunk. Extraskeletal myxoid chondrosarcoma is characterized by protracted clinical course and low metastatic tendency. Histologically, extraskeletal myxoid chondrosarcoma is composed of small uniform ovoid cells with some resemblance to embryonal chondroblasts arranged in delicate chains and strands surrounded by myxoid stroma with a multilobular growth pattern. Most of the tumor cells reacts positively with the antibodies to vimentin, S100, NSE and leu 7, rarely EMA-positive but usually negative for all cytokeratins. The morphology and the immunohistochemical profile of extraskeletal myxoid chondrosarcoma may resemble the architecture of other neoplastic lesions such as intramuscular myxoma, aggressive angiomyxoma and myxoid liposarcoma.

A significant feature distinguishing extraskeletal myxoid chondrosarcoma from other tumors with similar morphology such as other types of chondrosarcoma or tumors with myxoid stroma is the association with various specific non-random cytogenetic abnormalities, namely the translocations t(9; 22)(q22; q12), t(9; 17)(q22; q11.2) and t(9;15)(q22; q21), which are not found in any other tumor even in chondrosarcoma of bony origin. In addition to these translocations, other genetic abnormalities such as trisomies of 1q, 7, 8, 12, 18, and 19 chromosomes were described in several cases of this tumor.

The most common cytogenetic abnormality is the t(9; 22)(q22; q12) translocation, found in about 80% of extraskeletal myxoid chondrosarcoma. This translocation involves the EWS gene (*Ew*ing *s*arcoma gene) on chromosome 22q12, a member of the TET family of RNA-binding proteins, described in previous sections and the CHN gene located on chromosome 9q22. The CHN gene (also known as NR4A3, TEC, CSMF, NOR1 or MINOR) encodes the orphan nuclear receptor, a member of the steroid/thyroid/retinoid receptor superfamily of nuclear hormone receptors and able to activate the c-fos promoter. This reciprocal translocation generates the transcriptionally active fusion transcript EWS-CHN encoding a fusion protein, which includes the whole sequence of the orphan nuclear receptor and the amino terminal domain of the EWS gene, functioning as a potent transcriptional activator, probably with oncogenic potential. This mechanism is an example emphasizing the role of activated nuclear receptors in the oncogenesis, another example is the activated retinoic acid receptor α (RAR-α) in the t(15;17) translocation associated with acute promyelocytic leukemia. According to the location of the breakpoints within the involved genes, different variants of the EWS-CHN fusion transcript are described, most common of them type 1 and 2. Type 1 is generated by the fusion of exon 12 of EWS gene with exon 2 of CHN gene, type 2 by the fusion of exon 7 of EWS gene with exon 2 of CHN gene and type 3 by the fusion of exon 11 of EWS gene with exon 2 of CHN gene.

The t(9; 17)(q22; q11.2) is the second translocation associated with about 15% of extraskeletal myxoid chondrosarcoma that involves the above-described CHN gene and the TAF2N gene (also named as RBP56 or hTAFii68), another member of the TET family of RNA-binding proteins similar to the EWS and FUS genes and located on chromosome 17q11.2. This translocation generates the TAF2N-CHN fusion transcript, also operating as a potent transcriptional activator.

A third but uncommon translocation associated with this tumor is the t(9; 15)(q22; q21) translocation, which involves the TCF12 gene located on chromosome 15q21 encoding a transcription factor and the above-described CHN gene. This translocation causes the fusion of the NH2 terminal domain of the TCF12 gene to the CHN gene. The TCF12-CHN fusion transcript functions also as a transcriptional activator.

Since extraskeletal myxoid chondrosarcoma has no specific immunohistochemical profile and its morphology resembles other mesenchymal tumors, the above-mentioned translocations are pathognomonic for this tumor and can be used as further specific molecular markers for the diagnosis of extraskeletal myxoid chondrosarcoma.

Method The method used is mRNA extraction from frozen or paraffin-embedded tissue or FNA cells, followed by RT-PCR.

- Primers for the cDNA amplification of type 1 of the *EWS-CHN* fusion transcript:
  FP (EWS, exon 12): 5′-GCG.ATG.CCA.CAG.TGT.CCT.ATG-3′
  RP (CHN, exon 3): 5′-ATA.TTG.GGC.TTG.GAC.GCA.GGG-3′

PCR generates 89-bp DNA fragment.

- Alternative primers for the cDNA amplification of type 1 of the *EWS-CHN* fusion transcript:
  FP (EWS, exon 12): 5′-AAG.GCG.ATG.CCA.CAG.TGT.C-3′
  RP (CHN): 5′-CCT.GGA.GGG.GAA.GGG.CTA.T-3′

PCR generates 109-bp DNA fragment.

- Primers for the cDNA amplification of type 2 of the *EWS-CHN* fusion transcript:
  FP (EWS, exon 7): 5′-CTC.CAA.GTC.AAT.ATA.AGC.CAA.C-3′
  RP (CHN, exon 2): 5′-GGA.CGT.CCG.GCG.AGG.CGA.AGC-3′

PCR generates a 146-bp DNA fragment.

- Alternative primers for the cDNA amplification of type2 of the *EWS-CHN* fusion transcript:
  FP (EWS, exon 7): 5′-TCC.TAC.AGC.CAA.GCT.CCA.AGT.C-3′
  RP (CHN): 5′-CCT.GGA.GGG.GAA.GGG.CTA.T-3′

PCR generates 275-bp DNA fragment.

- Primers for the cDNA amplification of type 3 of the *EWS-CHN* fusion transcript:
  FP (EWS, exon 11): 5′-TCT.GGC.AGA.CTT.CTT.TAA.GCA-3′
  RP (CHN, exon 2): 5′-GGA.CGT.CCG.GCG.AGG.CGA.

AGC-3′PCR generates a 133-bp DNA fragment.

- Primers for the cDNA amplification of the *TAF2N-CHN* fusion transcript:
  FP (TAF2N, exon 6): 5′-GAG.CAG.TCA.AAT.TAT.GAT.CAG.CAG.C-3′
  RP (CHN, exon 3): 5′-CCT.GGA.GGG.GAA.GGG.CTA.T-3′

PCR generates a 137-bp DNA fragment.

3.3.2.8 Molecular Detection of Dermatofibrosarcoma Protuberans and Giant Cell Fibroblastoma

Dermatofibrosarcoma protuberans and its pediatric counterpart giant cell fibroblastoma are two tumors sharing many histological, immunohistochemical and cytogenetic features.

Dermatofibrosarcoma protuberans is a slow infiltrative growing fibrohistiocytic tumor mainly affecting adults between the 3rd and 4th decade and appears in the deep dermis, composed of proliferating uniform small spindled cells arranged in a repetitive storiform pattern.

Giant cell fibroblastoma is a rare pediatric dermal or subcutaneous slowly growing tumor with a predilection for children under 10 years of age. Histologically the tumor consists of proliferating spindled cells mixed with multinucleated giant cells surrounded by solid fibromyxoid stroma dissected by vascular like spaces.

Dermatofibrosarcoma protuberans and giant cell fibroblastoma share the same immunohistochemical profile and show positive reaction with the specific antibodies to CD34, PDGF and vimentin. Both tumors reveal low or intermediate malignant behavior and tens to local recurrence after incomplete excision and characterized by very low metastatic tendency, which generally occurs after an inadequate surgical treatment.

Dermatofibrosarcoma protuberans and giant cell fibroblastoma are associated with the novel and specific unbalanced t(17; 22)(q22; q13) translocation generating the COL1A1-PDGFB fusion transcript, occasionally with a supernumerary ring chromosome r (17; 22) as a result of this translocation.

The t(17; 22)(q22; q13) translocation involves the genes COL1A1 and PDGFB. The COL1A1 gene is located on chromosome 17q21–22 and encodes the α1 chain of collagen type I. The second gene is the PDGFB gene located on chromosome 22q13 and encodes the B-chain of the platelet-derived growth factor ligand, effecting as mitogenic factor for cells of mesenchymal origin. The (17; 22)(q22; q13) translocation removes the upstream sequences from exon 2 of the PDGF gene and causes the in frame fusion of exon 2 of this gene to one of the variable regions of the COL1A1 gene generating the COL1A1-PDGFB-fusion transcript. Expression of the PDGF gene in this fusion transcript is controlled by the COL1A1 gene, causing the deregulation and overexpression of the platelet-derived growth factor B which is considered as a potential mitogen and a homologue of the transforming protein p28sis of simian sarcoma virus. The uncontrolled expression of the PDGFB seems to play an important role in the initiation of both tumors. The size of the COL1A1-PDGFB-fusion transcript depends on the localization of the breakpoints through the COL1A1 gene.

As the t (17; 22)(q22; q13) translocation is specific for both tumors and has not been described in association with any other benign or malignant lesions, the amplification of the COL1A1-PDGFB-fusion transcript can be used as a molecular marker to distinguish dermatofibrosarcoma protuberans and giant cell fibroblastoma from other tumors with similar morphology.

Method The method used is mRNA extraction from frozen or paraffin-embedded tissue or FNA cells, followed by RT-PCR. Because of different breakpoints within the

COL1A1 gene, various consensus forward primers (FP) to different exons of the COL1A1 gene have to be used for efficient detection of the t (17; 22)(q22; q13) translocation.

- Primers for the cDNA amplification of *COL1A1-PDGFB* fusion transcript:
  FP within the COL1A1 gene:
  –Exon 5: 5′-GCCGAGATGGCATCCCTGG-3′
  –Exon 8: 5′-TCAGGGTGCTCGAGGATTGC-3′
  –Exon 11: 5′-AAGGCTTCCAAGGTCCCCCTGG-3′
  –Exon 15: 5′-GGTGCTCGTGGAAATGATGG-3′
  –Exon 17: 5′-AAGGTCCCCAGGGTGTGCG-3′
  –Exon 20: 5′-AACCTGGTGCTCCTGGCAGC-3′
  –Exon 23: 5′-AAGCTGGTCGTCCCGGTGAAGC-3′
  –Exon 26: 5′-AAGGCTGGAGAGCGAGGTGTTC-3′
  –Exon 29: 5′-TGCTGGCAAAGATGGAGAGG-3′
  –Exon 32: 5′-TGAACGTGGTGCAGCTGGTCTTC-3′
  –Exon 35: 5′-TCCCACTGGAGCTCGTGG-3′
  –Exon 38: 5′-TGCTCCTGGAGCCAAAGGTGC-3′
  –Exon 40: 5′-TGCTGGCGAGAAAGGATCCCCTG-3′
  –Exon 43: 5′-TGGCAAGAGTGGTGATCGTGG-3′
  –Exon 46: 5′-TGGCTTCTCTGGCCTCCAGGG-3′
  –Exon 49: 5′-ACCTCAAGAGAAGGCTCACGATGG-3′
  –RP (Exon 2/PDGFB gene): 5′-ATCAAAGGAGCGGATCGAGTGGTC-3′

To simplify the detection of this translocation it is recommended to run out the reaction as multiplex PCR where each reaction includes four foreword primers in addition to the reverse primer. Size of generated DNA fragment depends on location of breakpoints and exon composition, ranging between 100 bp and 400 bp).

- Alternative primers for the cDNA amplification of *COL1A1-PDGFB* fusion transcript:
  FP within the COL1A1 gene:
  –Exon 5: 5′-AGATGGCATCCCTGGACAGC-3′
  –Exon 7: 5′-TTCGACGTTGGCCCTGTC-3′
  –Exon 8: 5′-TGGCGAGCCTGGAGCTTC-3′
  –Exon 9: 5′-CCCTGGAAAGAATGGAGATGAT-3′
  –Exon 10: 5′-AAGCTGGAAAACCTGGTCGTC-3′
  –Exon 11: 5′-CCTGGAATGAAGGGACACAGA-3′
  –Exon 12: 5′-ATGGTGCCAAGGGAGATGCT-3′
  –Exon 13: 5′-TGGCAGCCCTGGTGAAAA-3′
  –Exon 14: 5′-CTGCCTGGTGAGAGAGGTCG-3′
  –Exon 15: 5′-GAAATGATGGTGCTACTGGTGCT-3′
  –Exon 16: 5′-CCTGGTGCTGTTGGTGCTAA-3′
  –Exon 19: 5′-TGGCCTGCCTGGTGAGAG-3′
  –Exon 20: 5′-GGAGACACTGGTGCTAAGGGAGAG-3′
  –Exon 21: 5′CTGGAGAGGAAGGAAAGCGA-3′
  –Exon 22: 5′-GACCTGGTAGCCGTGGTTTC-3′

–Exon 23: 5′-TGAAGCTGGTCTGCCTGGT-3′
–Exon 24: 5′-TCCTGATGGCAAAACTGGC-3′
–Exon 26: 5′-AAGGCTGGAGAGCGAGGTGTT-3′
–Exon 27: 5′-ACCCCCTGGCCCTGCT-3′
–Exon 28: 5′-TGGCGAGAGAGGTGAACAAG-3′
–Exon 29: 5′-GGTCCTCCAGGTGAAGCAGG-3′
–Exon 30: 5′-TGGTGAGAGAGGTCGCCCT-3′
–Exon 31: 5′-GGTTTCCCTGGCGAGCGT-3′
–Exon 32: 5′-GGGAATGCCTGGTGAACGT-3′
–Exon 33: 5′-AAGATGGCGTCCGTGGTCT-3′
–Exon 34: 5′-GCCCTGCTGGTCCCACTG-3′
–Exon 35: 5′-GGAGACCGTGGTGAGCCTG-3′
–Exon 37: 5′-GGTAATGTTGGTGCTCCTGGAG-3′
–Exon 38: 5′-TTTCCCTGGTGCTGCTGG-3′
–Exon 39: 5′-CCTGGTGCTGATGGTCCTG-3′
–Exon 40: 5′-GGCTTCCCTGGTCTTCCTG-3′
–Exon 41: 5′-CCCTGGTGAATCTGGACGTGA-3′
–Exon 42: 5′-CCCCTGGACGAGACGGTT-3′
–Exon 44: 5′-TCCTGTCGGCCCTGTTGG-3′
–Exon 45: 5′-AGGGTCACCGTGGCTTCTC-3′
–Exon 46: 5′-CTGGTGAACAAGGTCCCTCTG-3′
–Exon 47: 5′-ACTGGTGATGCTGGTCCTGTT-3′
–Exon 48: 5′-CCACTCTGACTGGAAGAGTGGG-3′
–Exon 49: 5′-CTCAAGATGTGCCACTCTGACT-3′
–RP (Exon 2/PDGFB gene): 5′-GCGTTGGAGATCATCAAAGGA-3′

To simplify the detection of this translocation it is recommended to run out the reaction as multiplex PCR where each reaction includes four foreword primers in addition to the reverse primer. Size of generated DNA fragment depends on location of breakpoints and exon composition.

3.3.2.9 Molecular Detection of Inflammatory Myofibroblastic Tumor

Inflammatory myofibroblastic tumor (inflammatory pseudotumor, inflammatory fibrosarcoma, plasma cell granuloma) is a controversial mesenchymal tumor, presents mainly in children and young adults, nevertheless cases are reported between 15 and 76 years. The tumor occurs in soft tissues and in different visceral organs such as lung, bladder, pancreas and spleen. Inflammatory myofibroblastic tumor exhibits the clinical features of low-grade sarcoma, capable to local recurrence, but almost no metastatic potential has been reported. Histologically, the tumor is composed of proliferating myofibroblasts accompanied by a prominent infiltrate of chronic inflammatory cells, including plasma cells, lymphocytes and eosinophils. The myofibroblastic cells usually show a positive immunohistochemical stain with the antibodies to vimentin, actin, cyclin D1. The tumor is also positive for P80 (ALK, CD241) in up to 60% of cases, which is the most diagnostic marker. Another feature of the inflammatory myofibroblastic tumor is the association with different non-random cytogenetic abnormalities, which emphasize the neoplastic nature of this tumor identity and distinguish it from other highly malignant or reactive inflammatory lesions such as leiomyosarcoma, nodular fasciitis, GIST or desmoid tumor. Moreover, this tumor is frequently described in association with the HHV-8 infection.

Various cytogenetic studies found different chromosomal abnormalities associated with the inflammatory myofibroblastic tumor. These abnormalities involve the anaplastic lymphoma kinase gene (ALK) gene on chromosome 2p23 (discussed in the anaplastic large cell lymphoma section) and more than 5 other different genes located on different chromosomes such as TMP3 on 1p23, TMP4 on 19p13, CLTC on 17q23, CARS on 11p15, RANBP2 on 2q13 and SEC31L1 on 4q21. All these genes are permanently transcribed and the generated chromosomal rearrangements are able to cause an abnormal tyrosine kinase activity with similar mechanisms as described in anaplastic large cell lymphoma.

Tropomyosin 3 (TMP3) and tropomyosin 4 (TMP4) are two homologous genes expressed in mesenchymal cells. Both genes can be the fusion partner of the ALK gene generating the TMP3-ALK and TMP4-ALK fusion transcripts. Each fusion transcript encodes a 95-kDa oncoprotein characterized by constitutive tyrosine kinase activity. Tumors bearing these translocations show a positive cytoplasmic stain with ALK-specific antibodies.

Tropomyosin 3 (TMP3, also known as TRK or NEM1) gene located on chromosome 1q25 encoding a 284 amino acid protein with coiled coil structure involved in the calcium-dependent actin–myosin interaction. This gene is the fusion partner of the ALK gene in the t(1; 2)(q25; p23) translocation. This translocation seems to be the most common translocation associated with the inflammatory myofibroblastic tumor but it is also detected in other tumors such as the anaplastic large cell lymphoma discussed in a later section.

Tropomyosin 4 (TMP4, tropomyosin fibroblast, non-muscle type) gene is located on chromosome 19p13 encodes a 248 amino acid protein with coiled coil structure, an alpha tropomyosin, a member of the actin filament-binding proteins family. This gene is the ALK gene fusion partner in the t(19; 2)(p13; p23) translocation. Depending on the location of breakpoints and the involved exons two fusion variants are described.

The clathrin heavy chain gene (CLTC), located on chromosome 17q23 is a further fusion partner of the ALK gene. The CLTC gene encodes a 1675-amino-acid and 191-kDa protein that mediates the endocytosis of transmembrane receptors. The resulting t(2; 17)(p23; q23) translocation is one of the translocations described in association with the inflammatory myofibroblastic tumor and anaplastic large cell lymphoma and generates the transcriptionally active fusion transcript CLTC-ALK coupled with an abnormal tyrosine kinase activity. Tumors bearing this translocation show a positive granular cytoplasmic stain with ALK-specific antibodies.

The gene encoding the cysteinyl-tRNA synthetase enzyme (CARS) is a further fusion partner of the ALK gene in the t(2; 11; 2)(p23; p15; q31) translocation described in association with rare cases of the inflammatory myofibroblastic tumor. The generated CARS-ALK fusion gene encodes a 130-kDa fusion transcript exhibiting a cytoplasmic immunohistochemical stain with the ALK antibody.

The RANBP2 gene is another rare fusion partner of the ALK gene described in association with the inflammatory myofibroblastic tumor in the t(2; 2)(p23; q13) translocation generating the RANBP2-ALK fusion transcript. The RANBP2 gene (also known as NUP 358) is located on chromosome 2q13 and encodes the *ran-binding protein 2*, a 358-kDa protein located at the cytoplasmic side of the nuclear pore involved in nuclear transport. Tumors harboring this translocation show a nuclear membrane stain with the ALK antibody.

A sixth ALK gene fusion partner reported in association with the inflammatory myofibroblastic tumor is the SEC31L1 gene on chromosome 4q21 generating the t(2; 4)(p23; q21) translocation. The SEC31L1 gene encodes a protein involved in the intracellular transport from the endoplasmatic reticulum to the Golgi apparatus.

The detection of the above-mentioned chromosomal abnormalities can support the diagnosis of inflammatory myofibroblastic tumor, especially in atypical tumor locations or doubtful immunohistochemical profile. The association with HHV-8 virus is a further interesting aspect discussed in a later chapter.

Method The method used is mRNA extraction from frozen or paraffin-embedded tissue or FNA cells, followed by RT-PCR.

- Primers for the cDNA amplification of *TMP3-ALK* fusion transcripts (nested PCR):
  –Outer primers:
  FP (TMP): 5′-ACT.GAT.AAA.CCC.AAG.GAG.GCA.GAG.A-3′
  RP (ALK): 5′-AGG.TCT.TGC.CAG.CAA.AGC.AGT.AGT.T-3′
  –Inner primers:
  FP (TMP): 5′-AGA.GAC.CCG.TGC.TGA.GTT.TGC.TGA-3′
  RP (ALK): 5′-CGG.AGC.TTG.CTC.AGC.TTG.TAC.TC-3′

Reaction generates a 147-bp DNA fragment.

- Alternative primers for the cDNA amplification of *TMP3-ALK* fusion transcript:
  FP (TMP3): 5′-GTT.TGC.TGA.GCG.ATC.GGT.AGC.CAA.GC-3′
  RP (ALK): 5′-CGG.AGC.TTG.CTC.AGC.TTG.TAC.TC -3′

Reaction generates a 130-bp DNA fragment.

- Primers for the cDNA amplification of *TMP4-ALK* fusion transcript (type 1):
  FP (TMP4): 5′-CGA.GGC.TCC.CCC.GCC.TCG.TC-3′
  RP (ALK): 5′-TGC.AGC.TCC.ATC.TGC.ATG.GCT.TG -3′

Reaction generates a 609-bp DNA fragment.

- Primers for the cDNA amplification of *TMP4-ALK* fusion transcript (type 2):
  FP (TMP4): 5′-GCC.ATG.GCC.GGC.CTC.AAC.TCC-3′
  RP (ALK): 5′-TGC.AGC.TCC.ATC.TGC.ATG.GCT.TG-3′

Reaction generates a 716-bp DNA fragment.

- Primers for the cDNA amplification of *CLTC-ALK* fusion transcript (semi-nested PCR):
  FP (CLTC): 5′-TTA.GAT.GCT.TCA.GAA.TCA.CTG-3′
  RP (ALK): 5′-GCA.GTA.GTT.GGG.GTT.GTA.GTC-3′
  RP (ALK): 5′-CGA.GGT.GCG.GAG.CTT.GCT.CAG.C-3′

First reaction is carried out with the primers 1 and 2, the second reaction with the primers 1 and 3. Semi-nested PCR generates a 153-bp DNA fragment.

- Primers for the cDNA amplification of the *RANBP2-ALK* fusion transcript:
  FP (RANBP2, exon 17): 5′-CGT.TCT.ACA.CCG.TCT.CCT.ACC.AG-3′
  RP (ALK): 5′-CGA.GGT.GCG.GAG.CTT.GCT.CAG.C-3′

Reaction generates a 360-bp DNA fragment.

- Primers for the cDNA amplification of *SEC31L1-ALK* fusion transcripts (nested PCR):
  –Outer primers:
    FP (SEC31L1): 5′-AAT.GCT.GCT.GGT.CAG.CTT.CCC.ACA-3′
    RP (ALK): 5′-TCA.CCC.CAA.TGC.AGC.GAA.CAA.TG-3′

First PCR generates two 457 and 799 DNA fragments according to the fusion type.
  –Inner primers:
    FP (SEC31L1): 5′-TCC.AGC.ATG.GCG.GAC.CAG.GAG-3′
    RP (ALK): 5′-TCA.GGG.CTT.CCA.TGA.GGA.AAT.CCA.G-3′

Nested PCR generates a 447-bp DNA fragment.

3.3.2.10 Molecular Detection of Alveolar Soft Part Sarcoma

Alveolar soft part sarcoma is an uncommon enigmatic soft tissue tumor with controversial histogenesis, makes 0.5–1.8% of all soft tissue sarcomas and occurs mainly in adolescents and young adults, typically in the deep tissues of lower limb and limb girdle. In children, a large percentage of cases are reported in the head and neck region including the orbit and the tongue. The tumor is also described in other rare and unusual locations such as female genital tract, stomach, mediastinum and bone.

Alveolar soft part sarcoma is a slow-growing tumor, but it has a high metastatic activity mostly to brain and lung; and metastasis can be the first manifestation of this tumor. Alveolar soft part sarcoma reveals a characteristic morphology, composed of large round or polygonal uniform cells with epithelioid appearance, arranged in nests with alveolar appearance, surrounded by fibrovascular septa and delicate small sized vascular channels. The tumor cells are characterized by eosinophilic cytoplasm with PAS-positive diastase resistant crystalline inclusions. Recent analysis shows that these crystals are most probably composed of aggregates of monocarboxylate transporter protein. The tumor cells do not have a specific immunohistochemical profile but usually TFE3- and vimentin-positive, often desmin-, actin- and NSE-positive, while always cytokeratin-negative. Some studies suggested a myogenic origin of alveolar soft part sarcoma, which explain the positive immunohistochemical reaction with some antibodies to myogenic antigens, but this hypothesis was not definitely confirmed, as alveolar soft part sarcoma is negative for the myogenic differentiation markers such as myogenin and MyoD1. The tumor morphology and occasionally the immunophenotype can mimic other tumors such as metastatic renal cell carcinoma, alveolar rhabdomyosarcoma, granular cell tumor, malignant melanoma, adrenal cortical carcinoma or hepatocellular carcinomas and special studies may be necessary to discriminate the tumor.

An important and characteristic feature that distinct alveolar soft part sarcoma from other tumors are specific non-random cytogenetic abnormalities. The most common and specific genetic abnormality associated with this tumor involves the chromosomes X and 17 generating the t(X; 17)(p11.2; q25) translocation. Other chromosomal abnormalities such as trisomy of the chromosomes 8 and 12 in addition to structural abnormalities in the chromosomes 1, 8q and 16p are also reported in association with this tumor.

The t(X; 17)(p11.2; q25) translocation is the most common genetic abnormality associated with alveolar soft part sarcoma. In the majority of cases (~80%) it is a non-reciprocal translocation der(17)t(X; 17)(p11.2; q25) and only in about 20% of cases is it found as a balanced reciprocal translocation. It involves the ASPL (*a*lveolar *s*oft *p*art sarcoma) gene (also known as ASPSCP1 gene) located on chromosome 17q25 and the TEF3 gene on chromosome Xp11.2. The ASPL gene encodes an ubiquitously expressed protein which consists of 476 amino acids and has a high expression level in adult heart, skeletal muscle, pancreas and testis. The TEF3 gene encodes one of the members of the microphthalmia-TFE subfamily of basic-helix-loop-helix leucine zipper transcription factors (transcription factor for immunoglobulin heavy-chain enhancer 3). In the t(X; 17)(p11.2; q25) translocation the ASPL gene is connected in frame upstream of either the third or fourth exon of the TFE3 gene generating the ASPL-TEF3 fusion transcript in two different types, type 1 lacking exon 3 of the TEF3 gene, whereas this exon is included in type 2.

Noteworthy is that the TEF3 gene is also involved in other specific genetic abnormalities associated with a portion of papillary renal cell carcinoma, namely both reciprocal translocations t(X; 1)(p11.2; q21.2) and t(X; 17)(p11.2; q25), mainly described in pediatric renal cell carcinomas, which may indicate a genetic relationship between both tumor identities.

As the t(X; 17)(p11.2; q25) translocation was not detected in any other soft tissue tumor, the molecular detection of the specific fusion transcripts can be used as an additional diagnostic assay to confirm the diagnosis of alveolar soft part sarcoma, however we have to consider that similar genetic abnormalities may be associated primary and metastatic papillary renal cell carcinoma.

Method The method used is mRNA extraction from frozen or paraffin-embedded tissue or FNA, followed by RT-PCR.

- Primers for the cDNA amplification of *ASPL-TFE3* fusion transcript:
  1. FP (ASPL): 5′-AAAGAAGTCCAAGTCGGGCCA-3′
  2. RP (TEF3, type 1): 5′-CATCAAGCAGATTCCCTGACACA-3′

3. RP (TEF3, type 2): 5′-TGGACTCCAGGCTGATGATCTC-3′

The PCR is carried out with the forward primer and one of the reverse primers for fusion types 1 and 2. PCR with the primers 1 and 2 generates a 116-bp DNA fragment for type 1 fusion transcript. PCR with the primers 1 and 3 generates a 128-bp DNA fragment for type 2 fusion transcript.

- Alternative primers for the cDNA amplification of *ASPL-TFE3* fusion transcript:
  FP (ASPL): 5′-AAA.GAA.GTC.CAA.GTC.GGG.CCA-3′
  RP (TEF3): 5′-CGT.TTG.ATG.TTG.GGC.AGC.TCA-3′

Reaction generates a 195-bp DNA fragment in the case of type 1 fusion transcript and 310-bp DNA fragment in the case of type 2.

3.3.2.11 **Molecular Detection of Congenital Fibrosarcoma and Mesoblastic Nephroma**

Congenital (infantile) fibrosarcoma is a rare pediatric spindle cell tumor appears in newborn, infants and small children, usually before the age of two years and tends to present in the extremities and axial region.

Congenital mesoblastic nephroma is another rare pediatric tumor, which makes about 5% of all pediatric renal tumors, generally occurring in the first year of life (typically in the first 3–6 months). Congenital mesoblastic nephroma has 3 morphological variants: classic, cellular and mixed. The cellular variant seems to be the renal counterpart of congenital fibrosarcoma showing similar morphological and immunohistochemical features.

Histologically both tumors demonstrate the morphological criteria of malignant mesenchymal spindle cell tumors including nuclear atypia, marked mitotic activity and necrosis, but characterized by low metastatic tendency and good prognosis with survival rate of more than 90%, nevertheless they tend to recurrence after inadequate surgical treatment. Because of the benign clinical behavior of both congenital fibrosarcoma and mesoblastic nephroma, it is important to distinguish them from other malignant proliferating lesions with identical histological appearance and immunohistochemical profile but with malignant biological and clinical behavior such as adult fibrosarcoma, embryonal rhabdomyosarcoma and Wilms tumor, which have an aggressive behavior and need a more radical therapy. It is also important to distinguish these tumors from other benign lesions such as the richly cellular form of infantile fibromatosis, which has a benign course.

Congenital fibrosarcoma and the cellular variant of congenital mesoblastic nephroma are characterized by similar histogenetic abnormalities including trisomy or polysomy of the 8, 11, 17 and 20 chromosomes, where as trisomy 11 is found in almost all cases of these tumors. Furthermore both tumors are found to bear the novel chromosomal abnormality namely the t(12; 15)(p13; q26) translocation generating the ETV6-NTRK3 fusion transcript. Remarkable that this translocation was not detected in the classic variant of congenital mesoblastic nephroma, but reported in the mixed variant, which includes both cellular and classic components.

The t(12; 15)(p13; q26) translocation involves the ETV6 gene located on the chromosome 12p13 and the NTRK3 (TRKC) gene located on chromosome 15q26. The ETV6 gene (also known as TEL gene) encodes a member of the ETS family transcription factors, having a helix-loop-helix (HLH) dimerization domain. This gene is also involved in several translocations associated with some types of hematological malignancies. The NTRK3 (TRKC) gene encodes the neutrophin 3 receptor, primarily expressed in neural cells. The t(12; 15)(p13; q25) translocation generates a 710-bp fusion transcript ETV6-NTRK3 encoding the HLH dimerization domain coupled to the tyrosine kinase domain, which is able to interface and activate the Ras-MAP kinase mitochondric pathway and the phosphatidyl inositol-3-kinase signaling pathway.

Note that the t(12; 15)(p13; q26) translocation is also associated with the a rare type of breast cancer namely the secretory breast carcinoma, whereas the trisomy of chromosome 11 specific for both congenital fibrosarcoma and mesoblastic nephroma was not found in association with this type of breast carcinoma.

As the t (12; 15)(p13; q26) translocation was not reported in any other benign or malignant mesenchymal tumor, the molecular amplification of the ETV6-NTRK3 fusion transcript can help to differentiate congenital fibrosarcoma and congenital mesoblastic nephroma from other benign or malignant lesion with similar morphology and immunohistochemical profile.

Method The method used is mRNA extraction from frozen or paraffin-embedded tissue including FNA, followed by RT-PCR.

- Primers for the cDNA amplification of the *ETV6-NTRK3* fusion transcript:
  FP (ETV6): 5′-ACC.ACA.TCA.TGG.TCT.CTG.TCT.CCC-3′
  RP (NTRK3): 5′-CAG.TTC.TCG.CTT.CAG.CAC.GAT.G-3′

PCR generates a 110-bp DNA fragment.

- Alternative primers for the cDNA amplification of the *ETV6-NTRK3* fusion transcript:
  FP (ETV6): 5′-AGC.CCA.TCA.ACC.TCT.CTC.AT-3′
  RP (NTRK3): 5′-CTC.GGC.CAG.GAA.GAC.CTT.TC-3′

PCR generates a 188-bp DNA fragment.

3.3.2.12 Molecular Detection of Endometrial Stromal Tumors

Endometrial stromal tumors are a group of tumors arising from the endometrial stromal cells and include the benign endometrial stromal nodules, the low-grade endometrial stromal sarcoma and the high-grade poorly differentiated endometrial stromal sarcoma. The diagnosis of the tumor depends mainly on the histological appearance of the tumor, which is composed of a neoplastic proliferation of uniform elongated and oval cells arranged concentrically around small vessels resembling the endometrial stromal cells of the proliferative phase. The endometrial stromal tumors have a remarkable but not specific immunophenotype, the cells are usually positive for the steroid hormones and CD10, but negative for the cytokeratins, CD34, CD117 and the smooth muscle-associated antigens such as actin, h-caldesmon and calponin.

A significant feature characteristic for the endometrial stromal tumors, specially the benign endometrial stromal nodules and the low-grade endometrial stromal sarcoma is the association with a recurrent and non-random genetic abnormality, specifically the t(7; 17)(p15; q21) translocation found in about 50% of the endometrial stromal tumors. This translocation causes the fusion of two zinc finger genes named as JAZF1 and JJAZ1 located on chromosomes 7p15 and 17q21 respectively generating the JAZF1-JJAZ1 fusion transcript. The t(7; 17)(p15; q21) translocation has not been reported in any other uterine or exrauterine tumor or normal uterine tissue. Note that this translocation is not distinctive for the high-grade poorly differentiated endometrial stromal sarcoma. In a case report the t(10; 17)(q22; p13) translocation was described in association with the high-grade variant, which may signify that this variant has a different pathogenesis or histogenesis.

As the endometrial stromal tumors does not have a specific immunophenotype and the expression of CD10 and steroid hormones can also be noted in some other tumors such as hepatocellular carcinoma and few types of renal cell carcinoma, the detection of the JAZF1-JJAZ1 fusion transcript can be used as a further diagnostic feature to distinguish the endometrial stromal tumors from other tumors, specially if the examined tissue represents a metastatic tumor of unknown primary.

Method The method used is mRNA extraction from frozen or paraffin-embedded tissue or FNA cells, followed by RT-PCR.

- Primers for the cDNA amplification of the *JAZF1-JJAZ1* fusion transcript:
  FP (JAZF1): 5′-AAC.AAG.AAT.TAC.AGC.AGC.CAA -3′
  RP (JJAZ1): 5′-CGA.GTT.CGA.AGA.AAT.CTA.TAG.A -3′

PCR generates a 280-bp DNA fragment.

3.3.2.13 Molecular Detection of Papillary and Follicular Thyroid Carcinoma

Papillary thyroid carcinoma is one of the most common endocrine malignancies, derived from follicular epithelium of thyroid gland and detectable in about 10% of population at autopsy. About 30% of adult and up to 80% of pediatric papillary thyroid carcinoma are associated with different specific somatic rearrangements of the RET gene. The RET gene is a proto-oncogene and a member of the cadherin family, located on chromosome 10q11.2 and consists of 21 exons. The RET gene encodes the tyrosine kinase receptor of GDNF (glial cell *d*erived *n*eurotrophic *f*actor), composed of an extracellular domain, a transmembrane portion and an intracellular domain, which includes the region exhibiting protein tyrosine kinase activity. This gene is normally expressed in cells of neural crest derivation, adrenal medulla and in parafollicular C-cells of the thyroid gland. Different mutations within the RET gene are associated with type IIA and IIB of multiple endocrine neoplasia syndrome (MENS), pheochromocytoma and medullary thyroid carcinoma, whereas loss function of this gene is associated with Hirschsprung disease.

The RET gene rearrangement in papillary thyroid carcinoma is caused by an intrachromosomal inversion or by a translocation involving the tyrosine kinase receptor domain of the RET/PTC (*p*apillary *t*hyroid *c*arcinoma) gene resulting the fusion of this domain to other heterologous genes located in different chromosomes. The breakpoints within the RET gene are usually restricted to a region within intron 11 leaving the tyrosine kinase domain intact, but the genes fused to this domain are active genes in thyroid follicular cells. All of these genetic rearrangements cause the permanent stimulation of the tyrosine kinase activity and the activation of various sequences expressed in thyroid follicular epithelial cells, which seems to be oncogenic for the thyroid follicular cells. About ten different RET/PTC rearrangement types are reported in association with papillary thyroid carcinoma, where RET/PTC1, RET/PTC2 and RET/PTC3 are the most frequent rearrangements.

The RET/PTC1 is the most common rearrangement, making 60–70% of all RET/PTC rearrangements and caused by the paracentric intrachromosomal inversion of the long arm of chromosome 10 resulting the fusion of the RET gene on q11 to the H4 gene located on q21 of the same chromosome generating the inv. (10)(q11.2q21) inversion. The encoded oncoprotein causes the activation of the tyrosine kinase function. This rearrangement is mainly associated with the spontaneous classic and solid types of papillary thyroid carcinoma.

The second RET/PCT2 rearrangement is the t(10; 17)(q11.2; q23) reciprocal translocation caused by the fusion of the RET gene to the RIα gene located on chromosome 17q23, encoding a subunit of cyclin AMP-dependent protein kinase A. The RET/PCT2 rearrangement is one of the rare genetic abnormalities associated with papillary thyroid carcinoma.

RET/PTC3 is a rearrangement caused by the paracentric intrachromosomal inversion of the RET gene within 10q11.2 and its fusion to the ELE1 (RFG) gene. The RET/PTC3 rearrangement is the second common genetic abnormality in papillary thyroid carcinoma making 20–30% of all REC/PTC rearrangements and mainly associated with radiation-induced papillary thyroid carcinoma and a well known example of RET/PTC3-associated papillary thyroid carcinoma is the post-Chernobyl papillary thyroid carcinoma.

The following genetic abnormalities are further rare RET rearrangements, also reported in radiation-induced papillary thyroid carcinoma:

- RET/PCT4 rearrangement.
- RET/PCT5 rearrangement, caused by the fusion of the RET gene with the GOLGA5 gene on chromosome 14q.
- RET/PCT6 rearrangement, caused by the fusion of the RET gene with the HTIF1 gene on chromosome 7q32.
- RET/PCT7 rearrangement, caused by the fusion of the RET gene with the RFG7 gene on chromosome 1p13.
- RET/KTN1 rearrangement, caused by the fusion of the RET gene with the KTN1 gene on chromosome 14q22.1.
- RET/RFG8 rearrangement, caused by the fusion of the RET gene with the RFG8 gene on chromosome 18q21–22.
- RET/PCM-1 rearrangement, caused by the fusion of the RET with the PCM1 gene on chromosome 8p21–22.

- RET/ELKS rearrangement, caused by the fusion of the RET gene with the ELKS gene located on chromosome 12p13. This rearrangement is reported in association with sporadic papillary thyroid carcinoma.

The NTRK1 gene is a further gene involved in other different rearrangements associated with papillary thyroid carcinoma. The NTRK1 gene is a proto-oncogene encoding a nerve growth factor receptor, located on chromosome 1q22 and expressed in the nervous system in addition to neuroblastoma. A number of different rearrangements of the NTRK1 gene are reported in association with radiation-induced papillary thyroid carcinoma. The most common one is an intrachromosomal inversion, inv. (1q) involving the NTRK1 gene and the tropomyosin (TPM3) gene resulting the TPM3-NTRK1 fusion gene. The NTRK1 gene is also a part of other rearrangements involving different sequences of the TPR gene (T1, T2, T4) on the same 1q25 chromosome and the TFG oncogene on chromosome 3 t(1; 3). All mentioned NTRK1 rearrangements cause a constitutive activation of the NTRK1 gene and the abnormal activation of the receptor tyrosine kinase.

Despite few controversial studies reporting the detection of RET/PTC rearrangements in benign or non-papillary thyroid lesions such as Hashimoto thyroiditis, follicular and Hürthel cell adenoma, the RET rearrangements seems to be characteristic for papillary thyroid carcinoma and not reported in any other thyroid carcinoma. The reported positive results in other thyroid lesion are possibly due to the presence of microfocus of papillary microcarcinoma, latent carcinoma or atypical follicular cells in transformation.

Follicular thyroid carcinoma is the second common thyroid carcinoma and frequently associated with genetic abnormalities involving the chromosomes 2, 3, 7 and 8. Recently, the t(2; 3)(q13; p25) translocation was described in about 45% of the follicular thyroid carcinoma. The t(2; 3)(q13; p25) translocation involves the PAX8 gene located on chromosome 2q13, which encodes a paired domain thyroid transcription factor, essential for thyroid development and the PPARγ1 gene located on chromosome 3p25, encoding the peroxisome proliferator-activated receptor, highly expressed in adipose tissue and the main target of the thiazolidinediones antidiabetic agents. The t(2; 3)(q13; p25) translocation generates the PAX8-PPARγ fusion transcript. This translocation is also found in up to 40% of the follicular variant of papillary thyroid carcinoma and in a significant number of follicular adenoma but not in the conventional variant of papillary thyroid carcinoma or in Hürthel cell tumors. The fact that this translocation is found in both follicular adenoma and carcinoma makes this genetic abnormality not a reliable marker to differentiate between thyroid adenoma and carcinoma. However, the presence of the t(2; 3) translocation in follicular adenoma may be due to focal malignant transformation.

In conclusion, the molecular detection of the RET/PTC rearrangements and in certain cases the t(2; 3)(q13; p25) translocation can be useful to distinguish papillary and follicular thyroid carcinoma from other types of carcinomas exhibiting a similar morphology. It can be also used to follow-up or to detect micrometastases of thyroid carcinoma if these rearrangements present in the primary tumor.

Method The method used is mRNA extraction from frozen or paraffin-embedded tissue or FNA cells, followed by RT-PCR.

- Primers for the cDNA amplification of *RET/PTC1* fusion transcript (nested PCR):
  –Outer primers:
  FP (H4): 5′-GCT.GGA.GAC.CTA.CAA.ACT.GA-3′
  RP (RET): 5′-GTT.GCC.TTG.ACC.ACT.TTT.C-3′

PCR generates a 165-bp DNA fragment.
  –Inner primers:
  FP (H4): 5′-ACA.AAC.TGA.AGT.GCA.AGG.CA-3′
  RP (RET): 5′-GCC.TTG.ACC.ACT.ACT.TTT.CCA.AA-3′

PCR generates a 151-bp DNA fragment.

- Primers for the amplification of *RET/PTC2* fusion transcript (nested PCR):
  –Outer primers:
  FP (RIα): 5′-GCT.TTG.GAG.AAC.TTG.CTT.TGA.TTT-3′
  RP (RET): 5′-GTT.GCC.TTG.ACC.ACT.TTT.C-3′

PCR generates a 202-bp DNA fragment.
  –Inner primers:
  FP (RIα): 5′-CTT.TGA.TTT.ATG.GAA.CAC.CG-3′
  RP (RET): 5′-ACT.TTT.CCA.AAT.TCG.CCT.TC-3′

PCR generates a 172-bp DNA fragment.

- Primers for the cDNA amplification of *RET/PTC3* fusion transcript (nested PCR):
  –Outer primers:
  FP (ELE1): 5′-AAG.CAA.ACC.TGC.CAG.TGG-3′
  RP (RET): 5′-CTT.TCA.GCA.TCT.TCA.CGG-3′

PCR generates a 242-bp DNA fragment.
  –Inner primers:
  FP (ELE1): 5′-CCT.GCC.AGT.GGT.TAT.CAA.GC-3′
  RP (RET): 5′-GGC.CAC.CGT.GGT.GTA.CCC.TG-3′

PCR generates a 219-bp DNA fragment.

- Primers for the cDNA amplification of *PAX8-PPARγ* fusion transcript:
  FP (PAX8, exon 7): 5′-ACC.CTC.TCG.ACT.CAC.CAG.AC-3′
  FP (PAX8, exon 8): 5′-CCC.TTC.AAT.GCC.TTT.CCC.C-3′
  FP (PAX8, exon 9): 5′-CTA.TGC.CTC.CTC.TGC.CAT.C-3′
  RP (PPARγ, exon 1): 5′-AGA.ATG.GCA.TCT.CTG.TGT.CAA.C-3′

PCR is carried out using the PPARγ primer and one of the PAX8 primers or as a multiplex PCR. The size of the generated DNA fragment depends on the location of the breakpoints and the number of involved exons within the PAX8 gene, ranging between 68 bp and 362 bp. A 362-bp DNA fragment matches the presence of exons 7, 8, and 9; a 173-bp DNA fragment matches the presence of the exons 7 and 9 and a 68-bp DNA fragment matches the presence of exon 8.

- Alternative primers for the cDNA amplification of *PAX8-PPARγ* fusion transcript:
  FP (PAX8, exon 6): 5′-CGC.GGA.TCC.GCA.TTG.ACT.CAC.AGA.GCA-3′
  RP (PPARγ, exon 1): 5′-CCG.GAA.TTC.GAA.GTC.AAC.AGT.AGT.GAA-3′

The size of the generated DNA fragment depends on the location of the breakpoints and the number of involved exons within the PAX8 gene. A 296-bp DNA fragment matches the presence of the exons 1–6; a 417-bp DNA fragment matches the presence of exons 1–7; a 519-bp DNA fragment matches the presence of exons 1–7 and 9 and a 606-bp DNA fragment matches the presence of exons 1–8.

3.3.3 Molecular Diagnosis of Hematopoietic and Lymphoid Neoplasia

3.3.3.1 Molecular Detection of Follicular Lymphoma

Follicular lymphoma is one of the most common types of non-Hodgkin lymphoma, supposed to be the neoplastic equivalent of normal germinal center cells, composed of a mixture of centrocytes and centroblasts with a follicular and diffuse growth pattern. According to the proportion of centroblasts, the WHO classification grades the follicular lymphoma into the following three grades:

- Grade 1, histologically with 0–5 centroblasts/HPF and shows indolent clinical behavior.
- Grade 2, histologically with 6–15 centroblasts/HPF and shows aggressive behavior.
- Grade 3, histologically with >15 centroblasts/HPF and shows very aggressive behavior. Grade 3 includes grade 3a with diffuse distribution of centroblasts and grade 3b with solid sheets of centroblasts. In the late stages as a result of the accumulation of genetic abnormalities and tumor progression, follicular lymphoma might undergo the transformation into high-grade diffuse large B-cell lymphoma.

About 90% of the low-grade follicular lymphoma (grade 1 and 2) is associated with the characteristic reciprocal t(14; 18)(q32; q21) translocation but this translocation is found in only about 13% of follicular lymphoma grade 3b. Alternatively, instead of the t(14;18)(q32; q21) translocation, follicular lymphoma can be associated with other recurrent genetic abnormalities involving other genes as about 10% of low-grade follicular lymphoma shows breaks in the chromosome 3q27, the gene encoding the bcl-6, half of which are the t(3;14)(q27; q32), t(3;22)(q27; q11) and t(2; 3)(p11; q27) translocations. The t(14; 18)(q32; q21) negative follicular lymphoma are mainly of extranodal location, commonly exhibit the high-grade morphology with low bcl-2 expression levels and less aggressive behavior. Further genetic abnormalities such as 6q21 and 17p13/p53 deletions and trisomy of the chromosomes 3, 7, 8, 12 and 18 were also identified in 5–15% of follicular lymphoma.

The (14; 18) (q32; q21) translocation takes place at the early pre-B stage in bone marrow, when the VDJ recombination of the IgH gene occurs and considered to be the first fundamental genetic change in developing of follicular lymphoma. This translocation causes the disruption of the bcl-2 gene (*B-c*ell *l*eukemia/*l*ymphoma-*2* gene) located on chromosome 18q21 and its fusion to the 5′ end of one of the six

JH segments of the immunoglobulin heavy chain locus on chromosome 14q32 (Figure 3.1).

The bcl-2 gene consists of three exons interrupted by large introns. As a result of alternative splicing, the bcl-2 gene is transcribed into three mRNA variants, which are translated into two homologous (~26 kDa) integral cell and mitochondrial membrane proteins (bcl-2 α isoform consists of 239 amino acids and bcl-2 β isoform consists of 205 amino acids). The bcl-2 protein has some homology to the BHRF-1 Epstein–Barr viral protein functioning as apoptosis suppressor. Chromosomal breakpoints are concentrated on the chromosome segment 18q21 in two well defined noncoding regions, leaving the bcl-2 coding sequence uninterrupted. First most common region known as the major breakpoint region (MBR) spanning a wide region (about 150 bp) in the untranslated part of the third exon (Figure 3.1), where 50–75% of the breakpoints occur. The second but less common region is the minor cluster region (MCR), which is a ~500-bp segment located about 20 kb downstream from the MBR, involved in approximately 20–40% of cases (Figure 3.1). A third rare breakpoint region is the intermediate cluster region (ICR) situated between MBR and MCR. Other rare breakpoint regions in different locations of the bcl-2 gene are also reported. Note that the bcl-2 locus is also involved in the t(2;18)(p11;q21) and t(18;22)(q21;q11) translocations rarely associated with follicular lymphoma and with B-cell chronic lymphocytic leukemia. The breakpoints within the IgH gene are mainly clustered within the JH5 and JH6 segments.

Because the chromosomal breakpoints appear in noncoding sequences of the bcl-2 gene, the translocation results the deregulation and overexpression of the bcl-2 anti-apoptosis gene under the effect of the IgH enhancer sequence. The overexpression of the bcl-2 gene causes the suppression of the normal apoptotic programmed cell death of the cells carrying these genetic abnormalities; furthermore, this mechanism seems to be responsible to block the chemotherapy-induced apoptosis. B-cells

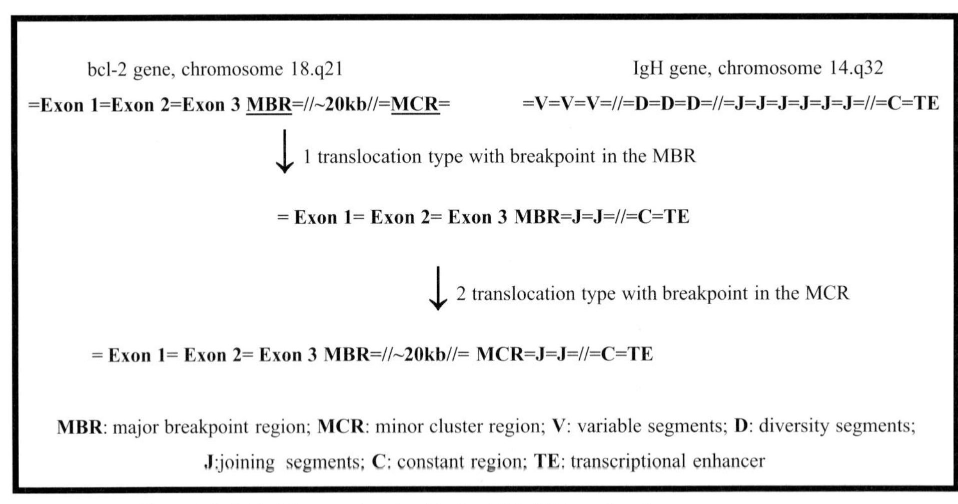

Figure 3.1 JH/bcl-2 genes involved in the t(14; 18) translocation.

carrying the bcl-2 translocation undergo blast transformation in response to antigen stimulation but the inability to switch off the activated bcl-2 gene appears to be an important factor in the initiation of this lymphoma. It is important to mention that B-lymphocytes bearing the t(14; 18) translocation are found at low frequencies in healthy individuals, which signify that this translocation is not the direct cause of lymphoma but needs further genetic factors for the compete malignant transformation.

The same t(14; 18)(q32; q21) translocation described in the follicular lymphomas has been also identified in 15–35% of diffuse large B-cell lymphomas, a part of which seems to be transformed from the follicular lymphoma after the accumulation of other genetic abnormalities, in these cases the presence of the t(14; 18) translocation indicates a poor prognosis. The overexpression of bcl-2 is also found in many other lymphoid, myeloid and solid neoplasms. It is important to mention that the expression or the overexpression of bcl-2 is not equivalent to the presence of the t(14; 18) translocation-specific follicular lymphoma as it is expressed in many normal cells and in different tumors. The bcl-2 is normally expressed in breast epithelium and many non-neoplastic lymphocytes including the small lymphocytes of mantle and marginal zones as well as thymus and many T-cell populations. Furthermore, bcl-2 is frequently expressed in many soft tissue tumors such as solitary fibrous tumor, synovial sarcoma and rhabdomyosarcoma in addition to breast carcinoma and many lymphoma types as CLL and marginal cell lymphoma.

The t(14; 18) translocation in follicular and diffuse large B-cell lymphoma can be detected by PCR-based molecular methods using consensus primers to both bcl-2 and JH regions. The PCR-based method allows the use of smaller amounts of DNA to detect this translocation, even low molecular DNA extracted from formalin-fixed paraffin-embedded tissue, bone marrow or fine needle aspiration. This method is efficient for the primary diagnosis of follicular lymphoma or to detect minimal residual tumor cells that are below the limit of detection using conventional diagnostic methods especially after chemotherapy or bone marrow transplantation. It is also suitable to detect tumor cells in autologous bone marrow prepared for transplantation. Long-term studies show that patients how achieved complete molecular remission after chemotherapy or stem cell transplantation have an improved outcome compared with patients with PCR detectable disease. However, it is important to point out that some studies demonstrated the presence of this translocation in circulating lymphocytes of healthy individuals, but in very rare cases.

As a result of the extensive somatic mutations occurring within the immunoglobulin loci of germinal center cells, the detection of follicular lymphoma by the immunoglobulin heavy chain gene rearrangement assay (described in a later chapter) is associated with frequent false-negative results, which makes the rearrangement method effective only in less than 70% of cases and the detection of the t(14; 18) translocation can provide further significant information for the diagnosis of follicular lymphoma.

Method The method used is DNA extraction from blood, different tissue types, bone marrow or FNA (lymph node aspiration), followed by PCR. Best results are achieved using primers directed to the MBR and JH regions combined with rearrangement analysis using primers directed to the conserved framework region 3 (FR3) and the joining region (JH) discussed in a further section. The use of this combination allows the detection of up to 80–90% of follicular lymphomas; and it can be also used for the detection of minimal residual neoplastic cells.

- Primers for the amplification of *t(14; 18)* DNA template (nested PCR) in the *MBR*:
  –Outer primers:
  FP (bcl-2, MBR): 5′-CAG.CCT.TGA.AAC.ATT.GAT.GG-3′
  RP (JH): 5′-ACC.TGA.GGA.GAC.GGT.GAC.C-3′
  –Nested primers:
  FP (bcl-2, MBR): 5′-TCT.ATG.GTG.GTT.TGA.CCT.TTA.G-3′
  RP (JH): 5′-ACC.AGG.GTC.CCT.TGG.CCC.CA-3′

The size of the generated DNA fragments varies depending on the location of the breakpoint, and each neoplastic clone generates a different DNA fragment (usually ranging between 100 bp and 150 bp).

- Alternative primers for the amplification of *t(14;18)* DNA template in the *MBR*:
  FP (bcl-2, MBR): 5′-GAG.TTG.CTT.TAC.GTG.GCC.TG-3′
  RP (JH): 5′-ACC.TGA.GGA.GAC.GGT.GAC.C-3′

The size of the generated DNA fragments varies depending on the location of the breakpoint and each neoplastic clone generates a different DNA fragment (usually ranging between 100 bp and 300 bp).

- Alternative primers for the amplification of *t(14;18)* DNA template in the *MBR* (semi-nested PCR):
  –FP (bcl-2, MBR3): 5′-TTT.GAC.CTT.TAG.AGA.GTT.GCT.TTA.CG-3′
  –FP (bcl-2, MBR4): 5′-GCA.ACA.GAG.AAC.CAT.CCC.T-3′
  –RP (JH 19): 5′-ACC.TGA.GGA.GAC.GGT.GAC.C-3′

PCR is carried out in two reactions: The first reaction using the primers 1 and 3, the second reaction using the primers 2 and 3. The size of the generated DNA fragments varies depending on the location of the breakpoint and each neoplastic clone generates a different DNA fragment (120–300 bp in size).

- *BIOMED-2* primers for the amplification of the *t(14;18)* DNA template. Multiplex PCR is carried out in three tubes (A, B, C) each contains one of the groups of the forward primers (MBR, 3′ MBR or MCR primers) and the reverse primer of the JH region.
  –*Forward primers:*
  1. Tube A (MBR primers):
     FR (MBR-1, exon 3): 5′-GAC.CAG.CAG.ATT.CAA.ATC.TAT.GG-3′
     FR (MBR-2, exon 3): 5′-ACT.CTG.TGG.CAT.TAT.TGC.ATT.ATA.T-3′
  2. Tube B (3′ MBR primers):
     FR (3′ MBR-1, exon 3): 5′-GCA.CCT.GCT.GGA.TAC.AAC.ACT.G-3′
     FR (3′ MBR-2, exon 3): 5′-AAA.CTA.GCA.GGG.TGT.GGT.GGC-3′
     FR (3′ MBR-3, exon 3): 5′-GGT.GAC.AGA.GCA.AAA.CAT.GAA.CA-3′
     FR (3′ MBR-4, exon 3) 5′-ACT.GGT.TGG.CGT.GGT.TTA.GAG.A-3′

3. Tube C (MCR primers):
   FR (5′ MCR, end of exon 3): 5′-CCT.TCT.GAA.AGA.AAC.GAA.AGC.A-3′
   FR (MCR-2): 5′-TAG.AGC.AAG.CGC.CCA.ATA.AAT.A-3′
   FR (MCR-2): 5′-TGA.ATG.CCA.TCT.CAA.ATC.CAA-3′

–*General reverse primer:*
  RP (JH primer) 5′-CCA.GTG.GCA.GAG.GAG.TCC.ATT.C-3′

PCR generates various-sized DNA fragments depending on the combination of genomic breakpoints (generally ranging between 200 bp and 600 bp).

- Primers for the amplification of *t(14; 18)* DNA template in the *MBR* and *MCR*:
  FP (bcl-2, MBR): 5′-GTG.GCC.TGT.TTC.AAC.ACA.GA-3′
  FP (bcl-2, MCR): 5′-GGA.CCT.TCC.TTG.GTG.TGT.TG-3′
  RP (JH): 5′-GTG.ACC.AGG.GTC.CCT.TGG.CCC.CAG-3′

The PCR is carried out in two reactions with one FP and RP or as multiplex reaction with all three primers. The size of the generated DNA fragments varies depending on the location of the breakpoint and each neoplastic clone generates a different DNA fragment (usually shorter than 500 bp in MBR and more than 600 bp in MCR).

- Primers for the amplification of *t(14; 18)* DNA template in the *MBR* and *MCR*:
  FP (bcl-2, MBR): 5′-CCA.AGT.CAT.GTG.CAT.TTC.CAC.GCT-3′
  FP (bcl-2, MCR): 5′-ACA.GCG.TGG.TTA.GGG.TTA.GGT.CGT.A-3′
  RP (JH): 5′-ACC.TGA.GGA.GAC.GGT.GAC.CAG.GGT-3′

The PCR is carried out in two reactions with one FP and RP or in one multiplex reaction with all three primers.

- Primers for the amplification of *t(14; 18)* DNA template in the *MCR*:
  FP (bcl-2, MCR): 5′-GAC.TCC.TTT.ACG.TGC.TGG.TAC.C-3′
  RP (JH): 5′-ACC.TGA.GGA.GAC.GGT.GAC.CAG.GGT-3′

The size of the generated DNA fragments varies depending on the location of the breakpoint and each neoplastic clone generates a different DNA fragment (ranging between 500 bp and 1000 bp).

3.3.3.2 **Molecular Detection of Mantle Cell Lymphoma**

Mantle cell lymphoma is a lymphoma arising from immature B-cells of the primary follicles and mantle zone of peripheral lymphoid organs, whereas the final malignant transformation seems to take place in the mature B-cell after antigen stimulation. Mantle cell lymphoma is described in the Kiel classification as centrocytic lymphoma and listed in the WHO classification as mantle cell lymphoma. This lymphoma type makes 5–10% of all lymphomas, characterized by nodular and/or diffuse proliferation of small to medium sized neoplastic lymphoid cells with coexpression of B-cell-associated antigens and CD5. Mantle cell lymphoma occurs mainly in older adults with a high male: female ratio and shows low sensitivity to chemotherapy with a median survival less than 5 years.

More than 80% of mantle cell lymphomas are associated with a characteristic nonrandom genetic abnormality, namely the t(11; 14)(q13; q32) translocation, which is the molecular counterpart of the bcl-1 rearrangement and cyclin D1 overexpression typical for this lymphoma. This translocation involves the joining region (JH) in the

5'- = **bcl-1**=/120 kb/=**CCND1**=mTC2=/90 kb/=mTC1=/20 kb/=MTC= **-3'**
Chromosome 11q13

MTC: Major translocation cluster; mTC1 and mTC2: Minor breakpoint clusters

Figure 3.2 Structure of the q13 region on chromosome 11 involved in the t(11; 14)(q13; q32).

immunoglobulin heavy chain gene (IgH) located on chromosome 14q32 and the bcl-1 locus on chromosome 11q13. The bcl-1 locus is a noncoding region, but this translocation causes the deregulation of the CCND1 gene (known also as PRAD1 gene) located about 120 kb upstream to the bcl-1 locus (Figure 3.2). The CCD1 gene encodes a 36-kDa cell cycle protein known as cyclin D1. The breakpoints within chromosome 11 are widely scattered through a large region, but about 40% of the chromosomal breakpoints are clustered to a restricted DNA segment within 300 bp upstream to the CNND1 gene, which is known as the major translocation cluster region (MTC). Less common breakpoints found in other regions known as minor breakpoint clusters. The breakpoints within the IgH gene are clustered in the J region juxtaposing the IgH-Eμ enhancer to the bcl-1 sequence.

The abnormal activation and deregulation of the CCND1 gene by the enhancer sequence of the Ig heavy chain gene causes an unregulated overexpression of cyclin D1, which is transcriptionally silent in normal lymphohemopoietic cells, and control the cell cycle progression from G_1 to S phase through the activation of cyclin-dependent kinases. This mechanism promotes cell proliferation and may contribute to tumor development.

This translocation has been also sporadically described in few other malignancies such as some high-grade B-cell lymphoma (some of which may be transformed from mantle cell lymphoma), chronic lymphocytic leukemia, hairy cell leukemia and multiple myeloma.

The molecular detection of the t(11; 14)(q13; q32) translocation can be used as a sensitive method for the diagnosis of mantle cell lymphoma. It can also be used in combination with the immunoglobulin heavy chain (IgH) gene rearrangement assay – found in about 100% of mantle cell lymphoma – as an accurate genetic marker for the diagnosis of mantle cell lymphoma and as an efficient assay to detect minimal residual disease in bone marrow and peripheral blood. The described PCR-based methods use primers designed for the detection of these translocations with breakpoints located within the major translocation cluster.

Method The method used is DNA extraction from peripheral blood, bone marrow or tissue including formalin-fixed paraffin-embedded tissue, followed by PCR. This method is able to detect up to 60% of the t(11; 14)(q13; q32) translocation.

- Primers for the amplification of *t(11; 14)* DNA template (semi-nested PCR):
  –FP (outer bcl-1 primer): 5′-GAT.GGG.CTT.CTC.TCA.CCT.ACT.A-3′

–FP (inner bcl-1 primer): 5′-GCT.TAG.ACT.GTG.ATT.AGC-3′
–RP (JH primer): 5′-ACC.TGA.GGA.GAC.GGT.GAC.C-3′

Semi-nested PCR is carried out in two reactions. The first reaction uses primers 1 and 3, which generates various-sized DNA fragments depending on the combination of genomic breakpoints (ranging between 370 bp and 500 bp). The second reaction is carried out using primers 2 and 3, and this also generates various-sized DNA fragments (approximately 250 bp in size).

- Alternative primers for the amplification of *t(11; 14)* DNA template (nested PCR)
  –Outer primers:
  FR (bcl-1): 5′-CTA.CTG.AAG.GAC.TTG.TGG.GTT.GCT-3′
  RP (JH): 5′-TGA.GGA.GAC.GGT.GAC.C-3′
  –Inner primers:
  FP (bcl-1): 5′-ATA.AGG.CTG.CTG.TAC.ACA.TCG.GTG-3′
  RP (JH): 5′-GTG.ACC.AGG.GT(A/G/C/T).CCT.TGG.CCC.CAG-3′

Nested PCR generates various-sized DNA fragments depending on the combination of genomic breakpoints (ranging between 200 bp and 300 bp).

- Alternative primers for the amplification of *t(11; 14)* DNA template:
  –FP (outer bcl-1 primer): 5′-CCT.CTC.TCC.AAA.TTC.CTG-3′
  –FP (inner bcl-1 primer): 5′-TCA.GGC.CTT.GAT.AGC.TCG-3′
  –RP (JH1): 5′-ACC.TGA.GGA.GAC.GGT.GAC.CAG.GGT-3′

Semi-nested PCR is carried out in two reactions. The first reaction uses primers 1 and 3, which generates various sizes of DNA fragments depending on the combination of genomic breakpoints, ranging between 470 bp and 600 bp. The second reaction uses primers 2 and 3, and this also generates various sizes of DNA fragment (approximately 420 bp in size).

- *BIOMED-2* primers for the amplification of the *t(11; 14)* DNA template. This primer set is able to detect up to 50% of this translocation. 60 °C annealing temperature and 35 PCR cycles are recommended.
  FR (bcl-1 primer): 5′-GGA.TAA.AGG.CGA.GGA.GCA.TAA-3′
  RP (JH primer): 5′-CCA.GTG.GCA.GAG.GAG.TCC.ATT.C-3′

PCR generates various sizes of DNA fragments depending on the combination of genomic breakpoints (generally ranging between 150 bp and 400 bp).

3.3.3.3 Molecular Detection of Anaplastic Large Cell Lymphoma

Anaplastic large cell lymphoma (ALCL) is a lymphoma listed in both Kiel and WHO classification, accounting for about 2% of all adult and about 13% of pediatric non-Hodgkin's lymphomas, characterized by an aggressive behavior and a bimodal age distribution, with one peak in children and young adults and a second in old age. Anaplastic large cell lymphoma is a lymphoma arising from the activated mature cytotoxic T-cells and may show extranodal involvement such as skin, bone marrow, bone, soft tissue, lung and liver. The new WHO lymphoma classification distinguishes three morphological variants of anaplastic large cell lymphoma:

- The common variant consists of large bizarre cells with chromatin poor horseshoe-shaped nuclei including multiple nucleoli. These cells are characterized by a cohesive pattern, some times resembling anaplastic carcinoma or malignant melanoma. This morphological type is seen in about 70% of cases.
- The lympho-histiocytic variant seen in about 10% of cases and characterized by abundant number of histiocytes.
- The small cell variant composed of a mixture of small and medium-sized cells with large cells resembling the above-described large cells.

The large neoplastic lymphoid cells of anaplastic large cell lymphoma are characterized by the expression of the CD30 (Ki-1) antigen, which is a transmembrane cytokine receptor, a member of the tumor necrosis factor family. Additionally a frequent expression of the epithelial membrane antigen (EMA) is observed. Anaplastic large cell lymphoma is divided into two main groups depending on the presence of different non-random translocations causing the overexpression of the anaplastic lymphoma kinase (ALK) gene:

1. Primary systemic ALK-positive anaplastic large cell lymphoma (known also as *ALK*oma)
2. Primary systemic ALK-negative anaplastic large cell lymphoma

In the WHO classification, the primary cutaneous anaplastic large cell lymphoma is a distinctive type of lymphoma of T or null phenotype involving skin in the absence of pre-existing lymphomas and is usually ALK-negative.

More than 40% of all variants of ALCL are associated with the reciprocal t(2; 5)(p23; q35) translocation, which is associated with more than 80% of pediatric and with one-third of adult ALCL. This translocation found to be the most common cytogenetic abnormality in non-cutaneous form of this lymphoma. Further, but less common chromosomal abnormalities described in association with ALCL are the translocations t(2; 3)(p23; q21), t(1;2)(q25; p23), inv.(2)(p23; q35), t(2; 13)(p23; q34), t(X;2)(q11–12; p23), t(2; 17)(p23; q11) and t(2;19)(p23; p13). All of these translocations involve the ALK gene on chromosome 2 and seems to have a similar pathophysiological mechanism.

The t(2; 5)(p23; q35) translocation involves the ALK gene (*a*naplastic *l*ymphoma *k*inase gene) and the NPM gene. The ALK gene is located on chromosome 2p23 and encodes a novel 200-kDa protein, a membrane-associated tyrosine kinase receptor (clustered as CD246), which is a member of the insulin receptor family, a single chain protein composed of 1621 amino acids and includes an intracellular domain characterized by tyrosine kinase activity, transmembrane and an extracellular domains. The ALK gene is normally expressed by nerve cells but not in hematopoietic or lymphoid cells. The NPM gene is a housekeeping gene located on chromosome 5q35 and encodes the nucleophosmin protein. Nucleophosmin is a nuclear phosphoprotein with a conserved 37-kDa non-ribosomal RNA-binding protein, involved in the shuttling of ribonucleoproteins between nucleus and cytoplasm and taking part in the late stages of pre-ribosomal particle assembly. The expression of NPM is regulated by a constantly active promoter during the cell cycle and reaches the highest level just before the entry into the S phase. The in-frame fusion of both genes in the t(2; 5) translocation generates the transcriptionally active fusion transcript NPM-ALK, which is present in ~75% of ALK-positive ALCL and encodes an 80-kDa chimeric tyrosine kinase, located in the cytoplasm and able to enter the nuclear region, which explains both nuclear and cytoplasmic immunohistochemical stain with the ALK-specific antibody in lymphomas harboring this translocation. The constant transcription of the ALK gene under the stimulation of the NPM gene promoter causes the overexpression of the NPM-ALK fusion protein (also known as P80) that found to have an oncogenic activity. The overexpression of the ALK leads to a permanent activation of the kinase catalytic domain, abnormal intracellular phosphorylation and the stimulation of neoplastic transformation. The overexpression of ALK is also noted in few other tumors such as the inflammatory myofibroblastic tumor, malignant peripheral nerve sheath tumor and neuroblastoma.

The t(1; 2)(q25; p23) translocation is the second common chromosomal abnormality described in 10–20% of ALCL. This translocation involves the above-mentioned ALK gene and the TPM3 gene located on chromosome 1q25. The TPM3 gene encodes a 33-kDa protein composed of 284 amino acids, a non-muscular tropomyosin expressed in mesenchymal cells and taking part in the calcium-dependent actin myosin interaction. This translocation generates the transcriptionally active TPM3-ALK fusion transcript with an activated ALK domain. The encoded transcript is a 104-kDa protein restricted to cytoplasm and reveals a cytoplasmic immunohistochemical stain with the ALK-specific antibody. The t(1; 2) (q25; p23) translocation is also the most common genetic abnormality associated with the inflammatory myofibroblastic tumor. Noteworthy is that the TPM3 gene is a fusion partner in other neoplasia types such as papillary thyroid carcinoma and inflammatory fibroblastic tumor.

The t(2; 3)(p23; q21) is a further, but rare chromosomal abnormality described in 2–5% of ALCL and involves the ALK and TFG genes. The TFG (*TRK-fused* gene) gene encoding a 44-kDa widely expressed protein, composed of 406 amino acids. This translocation generates two fusion transcripts encoding two homologous proteins, a short transcript (TFG-ALKs) encoding an 85-kDa protein and a long transcript (TFG-ALKl) encoding a 97-kDa protein, both are restricted to cytoplasm with cytoplasmic immunohistochemical stain pattern with the ALK-specific antibody.

The chromosomal inversion inv. (2)(p23; q35) is a fourth alternative chromosomal abnormality, associated with 2–5% of the ALCL, involving the ALK gene and the ATIC gene. This inversion causes the fusion of the ATIC gene (also known as hPurH), which encodes a bifunctional enzyme (5-aminoimidazole-4-carboxamide ribonucleotide formyltransferase/IMP cyclohydrolase) involved in the purine nucleotide biosynthesis with the gene segment encoding the intracellular portion of the ALK receptor tyrosine kinase. The generated transcriptionally active ATIC-ALK fusion transcript is functionally similar to the NPM-ALK transcript and encodes a 96-kDa fusion protein restricted to the cytoplasm, consequently with a diffuse cytoplasmic immunohistochemical stain pattern.

Recently other genes were described as an alternative fusion partners for the ALK gene such as the MSN and the CLTC genes.

The moesin (MSN) gene is located on chromosome Xq11–12 and encodes a 75-kDa cytoskeleton protein sited under the cell membrane, playing role in the cell adhesion and motility. The t(X; 2)(q11–12; p23) generates the MSN-ALK fusion transcript encoding a 125-kDa fusion protein including the tyrosine kinase domain. Lymphomas bearing this translocation show a membranous immunohistochemical stain pattern.

The CLTC gene located on 17q23 encodes a 191-kDa protein, a component of the vesicles matrix. The CTCTgene is the fusion partner of the ALK gene associated with ALCL and inflammatory fibroblastic tumor.

Additional rare translocations reported in association with ALCL are the t(2;17) (p23; p25) and t(2; 22)(p23; q11.2) translocations generating the ALO17-ALK and MYH9-ALK fusion proteins respectively. The MYH9-ALK fusion protein shows a cytoplasmic immunohistochemical stain pattern.

In conclusion, the atypical activation of ALK signaling pathways caused by different cytogenetic abnormalities appears to have an etiologic significance in the initiation of the malignant transformation of ALCL and few other tumors such as inflammatory myofibroblastic tumor.

Clinical observations show that ALK-positive anaplastic large cell lymphoma with the t(2; 5) (p23; q35) translocation tends to occur in young patients but simultaneously indicates a better prognosis and associated with an improved survival rate (median survival rate 165 months, with more than 70% having 5 years survival) compared with these anaplastic large cell lymphomas lacking the ALK overexpression (median survival rate about 35 months, with less than 40% having 5 years survival) and usually with a predilection to older patient population.

The molecular detection of the t(2; 5)(p23; q35) translocation and other related chromosomal abnormalities can be helpful and a sensitive method for the detection, precise classification and monitoring of lymphomas bearing these translocations in addition to the prognostic value indicating an improved rate of 5 years survival. Furthermore, about 90% of this lymphoma type is associated with TCR γ and β rearrangements, which make the rearrangement analysis an additional diagnostic tool for the diagnosis of this lymphoma type.

Method The method used is mRNA extraction from blood, bone marrow or tissue (lymph node), followed by RT-PCR. Initially, it is necessary to rule out the most common t(2; 5) translocation, which can be predicted by the positive nuclear and cytoplasmic immunohistochemical stain using the specific monoclonal antibody. If the t(2; 5) translocation is not detected, other primer panels can be used to exclude other ALCL-associated translocations.

- Primers for the cDNA amplification of *NPM-ALK* fusion transcript in t(2; 5), nested PCR:
  –Outer primers:
  FP (NPM): 5′-TCC.CTT.GGG.GGC.TTT.GAA.ATA.ACA.CC-3′
  RP (ALK): 5′-CGA.GGT.GCG.GAG.CTT.GCT.CAG.C-3′

First PCR generates a 177-bp DNA fragment.

–Nested primers:
FP: 5′-CCA.GTG.GTC.TTA.AGG.TTG.-3′
RP: 5′-TAC.TCA. GGG.CTC.TGC.AGC.-3′

Nested PCR generates a 123-bp DNA fragment.

- Primers for the cDNA amplification of *TMP3-ALK* fusion transcript in t(1; 2):
  FP (TMP3): 5′-GAG.CTG.GCA.GAG.TCC.CGT.TGC.CG-3′
  RP (ALK): 5′-GCA.GCT.CCT.GGT.GCT.TCC.GGC.GG-3′

Reaction generates a 244-bp DNA fragment.

- Primers for the cDNA amplification of *TGF-ALK* fusion transcript in t(2; 3):
  FP (TFG): 5′-AGC.TTG.GAA.CCA.CCT.GGA.GAA.CC-3′
  RP (ALK): 5′-GTC.GAG.GTG.CGG.AGC.TTG.CTC.AGC-3′

Reaction generates a 141-bp DNA fragment in the case of TFG-ALKS transcript and a 306-bp DNA fragment in the case of TFG-ALKL transcript.

- Alternative primers for the cDNA amplification of *TGF-ALK* fusion transcript in t (2; 3):
  FP (TFG): 5′-AGC.TTG.GAA.CCA.CCT.GGA.GAA.CC-3′
  RP (ALK): 5′-GTT.GGG.CCG.GCT.TTC.AGG.CTG.ATG.TTG.C-3′

Reaction generates a 795-bp DNA fragment.

- Primers for the cDNA amplification of *ATIC-ALK* fusion transcript in inv. (2):
  FP (ATIC): 5′-CAC.GCT.CGA.GTG.ACA.GTG.GTG.TGT.GAA.CC-3′
  RP (ALK): 5′-GCA.GCT.CCT.GGT.GCT.TCC.GGC.GG-3′

Reaction generates a 213-bp DNA fragment.

- Alternative primers for the cDNA amplification of *ATIC-ALK* fusion transcript in inv. (2) semi nested PCR:
  –FP (ATIC): 5′-CGC.TCG.CCC.TGA.ACC.CAG.TG-3′
  –FP (ATIC): 5′-GTG.TCC.ACG.GAG.ATG.CAG.AG-3′
  –RP (ALK): 5′-CGA.GGT.GCG.GAG.CTT.GCT.CAG.C-3′

Reaction is carried out in two phases: the first with the primers 1 and 3, second with the primers 2 and 3. Semi nested PCR generates a 315-bp DNA fragment.

- Primers for the cDNA amplification of *MSN-ALK* fusion transcript in t(1; 2):
  FP (MSN): 5′-CAG.CTG.GAG.ATG.GCC.CGA.CA-3′
  RP (ALK): 5′-CTT.GGG.TCG.TTG.GGC.ATT.C-3′

Reaction generates a 298-bp DNA fragment.

3.3.3.4 **Molecular Detection of Chronic Myeloid Leukemia**

Chronic myeloid leukemia (CML) is a chronic myeloproliferative clonal disease of multipotential hematopoietic stem cells, mainly involving the myeloid cell lineage and usually associated with the reciprocal chromosomal translocation t(9; 22)(q34; q11). Chronic myeloid leukemia begins with an initial relatively benign chronic phase, progressing to a final blast crisis and characterized by marked leukocytosis due to large number of mature and immature neoplastic cells.

The t(9; 22)(q34; q11) translocation with its cytogenetic appearance as Philadelphia chromosome was the first demonstrated chromosome rearrangement in neoplasia. This translocation is the hallmark of the chronic myeloid leukemia; nevertheless it is also found in association with other hematological malignancies such as acute lymphoblastic leukemia, acute myeloid leukemia, essential thrombocythemia and multiple myeloma.

The t(9; 22)(q34; q11) translocation causes the head-to-tail fusion of the ABL and the BCR genes. The ABL (*Ab*elson *l*eukemia virus protein) gene is a proto-oncogene located on chromosome 9q34, encoding a cytoplasmic and nuclear 145-kDa protein, which is an ubiquitously expressed tyrosine kinase regulator with two isoforms resulting from the different splicing mechanisms of the first exon. ABL plays a key role in the cellular response to genotoxic stress and is believed to be the human homologue of the v-abl oncogene of the murine leukemia virus.

The BCR gene located on chromosome 22q11 consists of 25 exons, also encoding a ubiquitously expressed 160-kDa protein. The breakpoint position within the BCR gene determines the development of either chronic myeloid leukemia or acute lymphoblastic leukemia. In chronic myeloid leukemia the t(9; 22)(q34; q11) translocation disrupt the ABL gene on chromosome 9q34 within a region larger than 300 kb, mainly located upstream exon 1b, specifically between exon 1b and 1a, resulting the fusion of exon a2 to the truncated BCR gene. In the majority of CML cases, the breakpoint in the BCR gene appears within the major breakpoint cluster region (M-BCR) spanning in a 5.8-kb area between exon 13 and 14 of the hole BCR gene (exon 2 and 3 inside the BCR-ABL fusion gene) resulting the b2a2 (also know as e13a2) rearrangement, or between exon 14 and 15 (exons 3 and 4 inside the BCR-ABL fusion gene) resulting the b3a2 (also known as e14a2) rearrangement. Although the breakpoints occur within different exons of both ABL and BCR genes, both BCR-ABL fusion transcripts (b2a2 and b3a2) generated after RNA splicing encode the same cytoplasmic transcription p210 protein characterized by high tyrosine kinase activity. The BCR-ABL fusion is expressed in all CML associated with this translocation, whereas the ABL-BCR gene on the 9q+ is expressed in only 70% of cases. An interesting but rare fusion transcript is the e19a2 transcript associated with the neutrophilic CML generated by breakpoints within the μ-BCR region. Other rare transcripts are also reported in CML such as b2a3, b3a3, e1a2 and e6a2, which are also able to cause the abnormal activation of the tyrosine kinase gene.

The t(9; 22)(q34; q11) translocation is a common translocation in many hematological malignancies and found in more than 95% of cases of chronic myeloid leukemia (CML), in up to 30% of adult acute lymphoblastic leukemia (ALL) and in about 5% ALL in children. About 70% of Ph1-positive ALL reveal the same chromosomal rearrangement as in CML, while the other 30% differ by other breakpoints on the BCR gene occurring within the minor breakpoint cluster region (m-BCR) upstream of the M-BCR in the first intron (e1a2 rearrangement), generating a different transcription protein (p190) discussed in the next section.

It is worth mentioning that the BCR-ABL translocation is only a co-factor and not the direct cause of CML as this translocation may be detected at very low frequency in the blood of healthy individuals, which indicates that the BCR-ABL translocation is not the only genetic abnormality required for a malignant transformation. Furthermore, malignant cells harboring the t(9; 22) translocation have an increased susceptibility to additional genetic mutations that induce tumor progression. All in all, CML seems to be the result of a cascade of different molecular mechanisms.

Chronic myeloid leukemia lacking the t(9; 22) translocation is a known variant of CML. About 5% of CML is associated with other cytogenetic abnormalities or seems to have a kind of link with a BCR-ABL gene located on one morphologically normal-looking chromosome (either 22 or 9). The translocation t(5; 12)(q33; p13) and t(8; 13)(p11; q12) are described in association with t(9; 22)-negative CML. Both translocations generate fusion genes with abnormal receptor tyrosine kinase activity. The platelet-derived growth factor receptor b (PDGFR-b) is involved in the t(5; 12) translocation; and the fibroblast growth receptor 1 (FGFR-1) is involved in the t(8; 13) translocation. The translocations t(8; 22)(p11; q11), t(4; 22)(q12; q11), t(5; 7)(q33; q11), t(5; 17)(q33; p13), t(5; 10)(q33; q21), t(9; 12)(q34; p13), t(9; 12)(p24; p13), t(9; 22)(p24; q11) and t(12; 14)(p12; q11–13) are also described in association with the t(9; 22)-negative CML.

The presence of the BCR-ABL fusion transcript enhances the tyrosine phosphokinase activity, causing the deregulation of cell cycle, activation of mitotic activity, alteration of adhesion cell properties and inhibition of aptoptosis. Since the presence of this translocation means a constitutively activated tyrosine kinase, neoplastic cells bearing this translocation can be the targets for specific tyrosine kinase receptor inhibitors, such as imatinib mesylate (also known as STI 571), which binds to the TK domain and shows a marked antileukemic activity in BCR-ABL-positive leukemia types, inducing a cytogenic remission.

The amplification of BCR-ABL mRNA can be used for the diagnosis and monitoring of CML in addition to other hematological malignancies bearing this translocation. It is also a useful assay for the detection of minimal residual leukemia cells after bone marrow transplantation or specific therapy (so-called molecular relapse). Clinical observations demonstrate that a molecular relapse detected within the first six months after bone marrow transplantation is associated with a high risk of disease relapse (in about 45% of cases), whereas a molecular relapse after a long period is less evident.

The molecular detection of the t(9; 22)(q34; q11) translocation must include the screening of both types of the BCR-ABL fusion transcript (b2a2 and b3a2). To avoid false-negative results, it is also recommended to use an internal control to confirm the presence and the integrity of mRNA (BCR-BCR internal control). This reaction can be done as multiplex PCR. Nested PCR is also useful to confirm the sequence of the mRNA target.

Method The method used is mRNA extraction from blood, bone marrow or tissue, followed by RT-PCR. The amplification of the BCR-mRNA as an internal control is recommended.

- Primers for the amplification of the *BCR-ABL* cDNA, *b2a2* rearrangement:
  FP (BCR, b2a2): 5′-ATC.CGT.GGA.GCT.GCA.GAT.G-3′
  RP (ABL, 2a2): 5′-CGC.TGA.AGG.GCT.TCT.TCC.TT-3′

PCR generates a 96-bp DNA fragment

- Primers for the amplification of the *BCR-ABL* cDNA, *b3a2* rearrangement:
  FP (BCR, b3a2): 5′-GGG.CTC.TAT.GGG.TTT.CTG.AAT.G-3′
  RP (ABL, 3a2): 5′-CGC.TGA.AGG.GCT.TTT.GAA.CT-3′

PCR generates a 74-bp DNA fragment.

- Primers for the amplification of the *BCR* mRNA (internal control):
  FP: 5′-CCT.TCG.ACG.TCA.ATA.ACA.AGG.AT-3′
  RP: 5′-CCT.GCG.ATG.GCG.TTC.AC-3′

PCR generates a 67-bp DNA fragment

- Primers for the amplification of the *BCR-ABL* cDNA template (nested PCR):
  –Outer primers:
  FP (BCR): 5′-GCT.TCT.CCC.TGA.CAT.CCG.TG-3′
  RP (ABL): 5′-CGA.GCG.GCT.TCA.CTC.AGA.CC-3′
  –Inner primers:
  FP (BCR): 5′-GGA.GCT.GCA.GAT.GCT.GAC.CCA.C-3′
  RP (ABL): 5′-CTG.AGG.CTC.AAA.GTC.AGA.TG-3′

Nested PCR generates a 119-bp DNA fragment in the case of the b2a2 transcript and a 194-bp DNA fragment in the case of b3a2 transcript.

- Alternative primers for the amplification of the *BCR-ABL* cDNA template (multiplex-nested PCR):
  –Outer primers:
  1. FP: (BCR -b2): 5′-ACA.GAA.TTC.GCT.GAC.CAT.CAA.TAA.G-3′
  2. RP: (BCR): 5′-ATA.GGA.TCC.TTT.GCA.ACC.GGG.TCT.GAA-3′
  3. RP: (ABL, a3): 5′-TGT.TGA.CTG.GCG.TGA.TGT.AGT.TGC.TTG.G-3′
  4. FP: (ABL, 1a): 5′-CTG.CAA.ATC.CAA.GAA.GGG.GCT.G-3′

Primers 1 and 3 are used to amplify the BCR-ABL fusion transcript. In the b2a2 translocation, PCR generates a 310-bp DNA fragment. In the b3a2 translocation, the generated DNA fragment is 385 bp in size.
Primers 1 and 2 are used for internal control to amplify an 808-bp DNA fragment of the BCR gene.
Primers 3 and 4 are used for internal control to amplify a 331-bp DNA fragment of the ABL gene.

  –Inner primers for *b2a2* and *b3a2* cDNA templates:
  1. FP (BCR): 5′-CTG.ACC.ATC.AAT.AAG.GAA.G-3′
  2. RP (ABL): 5′-TTC.ACA.CCA.TTC.CCC-3′

In the b2a2 translocation, nested PCR generates a DNA fragment ~243 bp in size. In the b3a2 translocation, the generated DNA fragment is ~318 bp in size.

- Alternative primers for the amplification of *BCR-ABL* cDNA template (nested PCR):
  –Outer primers:
  FP (BCR): 5′-GAG.CGT.GCA.GAG.TGG.AGG.GAG-3′
  RP (ABL): 5′-CCA.TTT.TTG.GTT.TGG.GCT.TCA.CAC.CAC.TCC-3′
  –Nested primers:
  FP (BCR): 5′-GAA.GAA.GTG.TTT.CAG.AAG.CTT.CTC.C-3′
  RP (ABL): 5′-TGT.GAT.TAT.AGC.CTA.AGA.CCC.GGA.GCT.TTT.C-3′

PCR generates a 320-bp DNA fragment in the case of the b2a2 transcript and a 395-bp DNA fragment in the case of b3a2 transcript.

3.3.3.5 Molecular Detection of Acute Lymphoblastic Leukemia

Acute lymphoblastic leukemia/lymphoma (ALL) is a high-grade malignant neoplasia derived from early lymphocyte precursors (pre and pre-pre lymphoblasts) and considered to be the most common malignancy of childhood. Depending on the morphologic and cytogenetic features of neoplastic lymphoblasts and the clinical behavior of the disease, acute lymphoblastic leukemia is divided into three subtypes according to the FAB classification and includes L1, L2 and L3.

L1: typical childhood leukemia, derived from pre-pre-lymphoblasts and can be of T- or B-cell lineage.
L2: leukemia common in adults, derived from pre-lymphoblasts and can be of T- or B-cell lineage.
L3: leukemia with Burkitt's morphology arises from more mature lymphoblasts and only of B-cell origin.

All three ALL subtypes are frequently associated with different translocations and genetic abnormalities reflected by different clinical behavior and different therapeutic response. Up to 50% of the 3 ALL types are associated with well defined and recurrent translocations. The following translocations are the most common recurrent genetic abnormalities described in association with ALL: t(9; 22), t(1; 19), t (17; 19), t(4; 11), t(7; 10), t(15; 17), t(12; 21), t(8; 14), t(1; 14), t(2; 8) and t(8;22) in addition to the IgH and TCR rearrangements. Many of genes involved in these translocations are transcription factors taking part in the regulation of the hematopoiesis. The related translocations are able to cause an unregulated overexpression of the involved genes or generate fusion transcripts with abnormal functions.

Further important feature of ALL is the IgH or/and TCR gene rearrangements found in nearly all ALL cases. IgH gene rearrangement is found in about 95% of B-ALL and in up to 20% of T-ALL, whereas Igκ is found only in B-ALL. TCR-γ gene rearrangement is found in about 95% of T-ALL and in up to 60% of B-ALL. TCR-β gene rearrangement is found in about 90% of T-ALL and in up to 30% of B-ALL.
TCR-δ gene rearrangement is found in about 70% of T-ALL and in up to 55% of B-ALL.

- *t(9; 22)(q34; q11) translocation* is a common translocation found in about 20% of adult ALL (L2) and in about 5% of pediatric ALL (L1). This translocation is characterized by poor prognosis and considered to be an indication for bone marrow transplantation. About 70% of Ph1-positive ALL harbor the same t(9; 22)(q34; q11) translocation found in association with the chronic myeloid leukemia. This translocation disrupts the ABL gene located on chromosome 9q34 within a region larger than 300 kb, between exon 1b and 1a, resulting the fusion of exon a2 to the truncated BCR gene. As in most cases of chronic myeloid leukemia, breakpoints through the BCR gene appear within the major breakpoint cluster region (M-BCR) spanning in a 5.8-kb area between exon 13 and 14 of the hole BCR gene (exon 2 and 3 inside the BCR-ABL fusion gene) resulting the b2a2 rearrangement, or between exon 14 and 15 (exons 3 and 4 inside the BCR-ABL fusion gene) resulting in the b3a2 rearrangement. Both transcriptions generate the fusion transcript gene BCR-ABL that encodes the p210 transcription protein characterized by a high tyrosine kinase activity. In the other 30% of Ph1-positive ALL, breakpoints through the BCR gene occurs within the minor breakpoint cluster region (m-BCR) upstream of the M-BCR in the first intron (e1a2 rearrangement) generating a different transcription protein known as p190. Other rare translocation type was also reported with breakpoints in the μ-BCR region between the exons e19 and e20 (e19a2 rearrangement) encoding the p230 transcription protein.

- *t(1; 19)(q23; p13) translocation* is one of the common translocations reported in about 7% of all ALL and in about 25% of pre-B ALL (L2), also associated with poor prognosis. This translocation involves the PBX1 and E2A genes. The PBX1 gene (*pre-B*-cell leukemia factor *1*) is located on chromosome 1q23, which interacts with the PAX6 gene and enhance its transcriptional activity. The E2A gene (also known as TCF3 gene) is located on chromosome 19p13 encoding the two helix-loop-helix proteins E12 and E47 functioning as Ig kappa chain enhancer-binding factors. Depending on the location of breakpoints, this translocation generates three types of the fusion transcript E2A-PBX1 (I, Ia and II), all of them operate as transcriptional activators with transforming potential.
- *t(17; 19)(q22; p13) translocation* is a rare translocation also involving the E2A gene on chromosome 19 and the HLF gene (*h*epatic *l*eukemia *f*actor) located on chromosome 17q22, mainly activated in hepatocytes and renal cells. The generated E2A-HLF fusion transcription factor seems to be a transcriptional activator with transforming potential.
- *t(4; 11)(q21; q23) translocation* is also one of the common translocations reported in about 8% of ALL and in about 75% of infantile ALL, characterized by poor prognosis and bad clinical outcome. This translocation involves the mixed-lineage leukemia (MLL) gene located on chromosome 11q23 and the AF4 gene located on chromosome 4q21, generating the fusion transcript MLL-AF4. The MLL gene is also involved in more than 40 translocations reported in association with many human tumors including ALL and AML. Blasts bearing the t(4;11)(q21; q23) translocation are arrested at an early progenitor stage.
- *t(12; 21)(p13; q2) translocation* is one of the most common translocations in pediatric ALL reported in about 25% of this leukemia type and can be detected in the initial leukemia development stages. ALL bearing this translocation are usually associated with a favorable prognosis and a good response to the standard chemotherapy. The t (12; 21) translocation involves the TEL (ETV6) gene located on chromosome 12p13 encoding a member of the ETS family transcription factors and the AML1 gene (also known as RUNX1 gene) located on chromosome 21q22, encoding a DNA-binding subunit. This translocation generates the transcriptionally active TEL-AML1 fusion transcript under the regulation of the TEL promoter. Few types of the fusion transcript TEL-AML1 are described depending on the location of the breakpoints. Type I TEL-AML1 is the most common fusion transcript, generated by the fusion of exon

5 of the TEL gene to the AML1 gene. Type II is a less common type, generated by the fusion of exon 3 of the TEL gene to the AML1 gene.

- *t(8; 14)(q24; q32) translocation* is also one of the most common translocations associated with the pediatric ALL, especially L3. The translocation between the c-myc proto-oncogene located on chromosome 8q24 and involved in the progression of cell cycle from S phase into G2 phase and the immunoglobulin heavy chain gene on chromosome 14q32, leads to the deregulation of the c-myc oncogene. The generated fusion transcript is difficult to detect by conventional PCR-based methods because of high variability in the breakpoint locations.
- *t(1; 14)(p32; q11) translocation* is a rare translocation, mainly reported in a small number of T-ALL and associated with a poor prognosis. This translocation involves the tal-1 gene (also known as *SCL* or *TCL-5*), located on chromosome 1p32 and the TCR α/δ gene on chromosome 14q11. The tal-1 gene encodes a 42-kDa nuclear protein essential for mammalian hematopoiesis and normal yolk sac angiogenesis. The tal-1 gene is also involved in the 90-kb deletion del.(1), associated with up to 30% of pediatric ALL. In this deletion, the breakpoints occur in two locations, tal-1^{d1} found in about 85% and tal-1^{d2} found in about 15% of this deletion. As a result of this deletion the tal-1 gene is activated by the sil gene promoter.

The detection of the different ALL-specific translocations is one of the sensitive methods for the diagnosis and monitoring of ALL. Beside clonal analysis, discussed in a later section, the detection of the specific translocations is an important method for the detection of minimal residual cancer cells after therapy or bone marrow transplantation as the detection of minimal residual disease considered to be an independent prognostic factor. It is also an important prognostic factor and a therapy guide as some translocations such as the t(9; 22) translocation are associated with poor prognosis and an indication for bone marrow transplantation, whereas the t(12; 21) translocation is usually associated with a good prognosis.

Method The method used is mRNA extraction from blood, bone marrow, body fluids or tissue (lymph node), followed by RT-PCR.

- Primers for the cDNA amplification of the *BCR-ABL* fusion transcript (multiplex PCR):
  1. FP (M-BCR-b2): 5′-ATC.CGT.GGA.GCT.GCA.GAT.G-3′
  2. FP (M-BCR-b3): 5′-GAG.TCT.CCG.GGG.CTC.TAT.GG-3′
  3. FP (m-BCR-e1): 5′-AGA.TCT.GGC.CCA.ACG.ATG.GCG.A-3′
  4. RP (ABL-a2): 5′-TCA.GAT.GCT.ACT.GGC.CGC.TGA-3′
  5. FP (ABL-a2): 5′-ACC.CAG.TGA.AAA.ATG.ACC.CCA.A-3′
  6. RP (ABL-a3): 5′-TGT.GAT.TAT.AGC.CTA.AGA.CC-3′

- PCR with M-BCR-b2 and ABL-a2 primers (primers 1 and 4) generates a 111-bp DNA fragment (b2a2 transcript).
- PCR with M-BCR-b3 and ABL-a2 primers (primers 2 and 4) generates a 98-bp DNA fragment (b3a2 transcript).

- PCR with m-BCR-e1 and ABL-a2 primers (primers 3 and 4) generates a 73-bp DNA fragment (e1a2 transcript).
- PCR with ABL-a2 and ABL-a3 primers (primers 5 and 6) used as an internal control.
- Primers for the cDNA amplification of the *BCR-ABL* fusion transcript (multiplex PCR):
  1. FP (BCR-e1): 5′-ACC.GCA.TGT.TCC.GGG.ACA.AAA.G-3′
  2. FP (BCR-b2): 5′-ACA.GAA.TTC.GCT.GAC.CAT.CAA.TAA.G-3′
  3. RP (BCR-rev): 5′-ATA.GGA.TCC.TTT.GCA.ACC.GGG.TCT.GAA-3′
  4. RP (ABL-a3): 5′-TGT.TGA.CTG.GCG.TGA.TGT.AGT.TGC.TTG.G-3′
  5. FP (ABL-1a): 5′-CTG.CAA.ATC.CAA.GAA.GGG.GCT.G-3′
- PCR with BCR (b2) and ABL (a3) primers (primers 2 and 5) generates a 310-bp DNA fragment in the case of the b2a2 transcript and a 385-bp DNA fragment in the case of the b3a2 transcript.
- PCR with BCR (e1) and ABL (a3) primers (primers 1 and 4) generates a 481-bp DNA fragment in the case of the e1a2 transcript.
- PCR with BCR (b2) and BCR (rev) primers (primers 2 and 3) used as an internal control and generates an 808-bp DNA fragment of the BCR gene.
- PCR with ABL (a3) and ABL (1a) primers (primers 4 and 5) used as an internal control and generates a 331-bp DNA fragment of the ABL gene.

Nested primers for the *e1a2* transcript:

FP (BCR): 5′-ATG.GAG.ACG.CAG.AAG-3′
RP (ABL): 5′-GGC.CAC.AAA.ATC.ATA.C-3′

Nested PCR generates a 155-bp DNA fragment in the case of e1a2 transcript.

- Primers for the cDNA amplification of the *MLL-AF4* fusion transcript generated by the *t(4;11)* translocation:
  FP: 5′-AAA.GCA.GCC.TCC.ACC.ACC-3′
  RP: 5′-GGT.TAC.AGA.ACT.GAC.ATG.CTG-3′

PCR generates an approximately 388-bp DNA fragment depending on the location of translocation breakpoints.

- Primers for the cDNA amplification of the *E2A-PBX1* fusion transcript generated by the *t(1; 19)* translocation:
  FP: 5′-GCG.CTG.GCC.TCA.GGT.TT-3′
  RP: 5′-TAA.CTC.CTC.TTT.GGC.TTC.CTC-3′

PCR generates ~625, 652 and 325-bp DNA fragments in type I, type Ia and type II translocation respectively.

- Primers for the cDNA amplification of the *TEL (ETV6)-AML1* fusion transcript generated by the *t(12; 21)* translocation:
  1. FP: 5′-TGC.ATA.GGG.AAG.GGA.AG-3′
  2. FP: 5′-TTC.TTT.TTT.CAC.CAT.TCT.TCC-3′
  3. RP: 5′-GCA.GAG.GAA.GTT.GGG.GCT-3′

PCR generates a DNA fragment ranging between 321 bp and 359 bp in length in a type I translocation using primers 1 and 3 and a DNA fragment between 281 bp and 320 bp in a type II translocation using primers 2 and 3.

- Alternative primers for the cDNA amplification of the *TEL (ETV6)-AML1* fusion transcript generated by the *t(12; 21)* translocation (nested PCR):
  –Outer primers:
  FP: 5′-CGT.GGA.TTT.CAA.ACA.GTC.CA-3′
  RP: 5′-ATT.GCC.AGC.CAT.CAC.AGT.GAC-3′

PCR generates a DNA fragment ranging between 210 bp and 260 bp in length depending on the location of breakpoints.
  –Nested primers:
  FP: 5′-AAC.CTC.TCT.CAT.CGG.GAA.GA-3′
  RP: 5′-CAG.AGT.GCC.ATC.TGG.AAC.AT-3′

Nested PCR generates a DNA fragment ranging between 13 bp and 170 bp in length.

- Primers for the cDNA amplification of the *E2A-HLF* fusion transcript generated by the *t(17;19)* translocation (nested PCR):
  –Outer primers:
  FP: 5′-TCG.CCC.AGC.TAC.GAC.GGG.GGT.CTC-3′
  RP: 5′-GAG.GCC.CGG.ATG.GCG.ATC.TG-3′
  –Nested primers:
  FP: 5′-GGC.CTG.CAG.AGT.AAG.ATA.GAA.GAC.C-3′
  RP: 5′-CGT.CGC.GGG.AGC.GCT.TGG.CT-3′

PCR generates a 441-bp DNA fragment in type I translocation and a 294-bp DNA fragment in type II translocation.

3.3.3.6 **Molecular Detection of Acute Myeloid Leukemia**

Acute myeloid leukemia (AML) is a neoplastic proliferation of myeloid hematopoietic stem cells or myeloid precursors including myeloblasts, monoblasts, erythroblasts and megakaryoblasts. According to the FAB classification, AML is classified into eight subtypes (M0–M7) depending on the cell type involved in the neoplastic process.

M0: minimally differentiated acute myeloblastic leukemia.
M1: acute myeloblastic leukemia without maturation.
M2: acute myeloblastic leukemia with maturation.
M3: promyelocytic leukemia.
M4: acute myelomonocytic leukemia.
M5: M5a: acute monoblastic leukemia; M5b: acute monocytic leukemia.
M6: acute erythroblastic leukemia.
M7: acute megakaryoblastic leukemia.

Different cytogenetic abnormalities are found in many types of AML, which also include non-random chromosomal translocations specific for many AML subtypes and associated with different biological behavior and clinical outcome. The following translocations are the most common specific genetic abnormalities associated with different AML subtypes:

t(15; 17)(q22; q12–21), t(8; 21)(q22; q22), t(9; 22), t(9; 11)(p21–23; q23), t(11; 19)(p22, p23), t(3; 5)(q25.1; q34), t(11; 17)(q23; q21) and inv.(16)(p13; q23). The chromosomal abnormalities are clinically divided into three groups according to the clinical behavior of leukemia type bearing these abnormalities:

- Low-risk AML subtypes associated with the t(15; 17), t(8; 21) and abnormal 16q22.
- Intermediate subtypes, usually with normal karyotype.
- High-risk subtypes associated with other genetic abnormalities.

- *t(15; 17)(q22; q12–21) translocation* is a reciprocal translocation usually associated with the acute promyelocytic leukemia (M3) causing the fusion of the PML gene to the RAR-α gene, generating the PML-RAR-α fusion transcript. The RAR-α (*r*etinoic *a*cid *r*eceptor α) gene is located on chromosome 17q11–12, composed of nine exons and encodes a member of the steroid/thyroid hormone receptor family of transcription factors namely the retinoic acid receptor α, which binds retinoic acid. The breakpoints inside the RAR-α gene are constantly clustered within a 16-kb region through intron 2. The RAR-α gene is also involved in other two but rare translocations associated with AML, the t(11; 17)(q23; q21) translocation associated with the M3 subtype of AML and the t(5;17)(q32;q12) translocation associated with atypical acute promyelocytic leukemia characterized by the predominance of hypergranular promyelocytes

  The second gene involved in the t(15; 17) translocation is the PML (*pro*m*y*e-locytic *l*eukemia, also known as MYL) gene, located on chromosome 15q22. This gene encodes a phosphorprotein associated with the nuclear matrix, functioning as a transcription factor and as a tumor suppressor, which also regulates the P53 response to oncogenic signals. The breakpoints within the PML gene are mainly clustered within three regions and include intron 6 (bcr1), intron 3 (bcr3) and less common exon 6 (bcr2). Depending on the size of generated transcript, breakpoints within the bcr1 gene forms the long (L)-type, breakpoints within the bcr3 gene forms the short (S)-type and breakpoints within the bcr2 gene forms the V-type of the PML-RAR-α fusion transcript. Some studies reported that the S-type fusion transcript signifies a worse prognosis than the other types. The generated PML-RAR-α-mRNA is transcriptionally active and encodes a 106-kDa protein, consists of the majority of the PML sequences and large parts of the RAR-α sequences including the retinoic acid-binding domain, whereas the activity of the RAR-α gene under the transcriptional control of the PML gene. The PML-RAR-α fusion protein seems to play a key role in the pathogenesis of the acute promyelocytic leukemia by blocking the terminal differentiation and inhibiting the apoptotic cell death. The detection of this translocation is important for the diagnosis, monitoring and treatment of the acute promyelocytic leukemia as the presence of this fusion transcript predicts a good response to the all-trans-retinoic acid (ATRA) therapy, which acts as a ligand for the encoded fusion protein stimulating the expression of RARA-α and maturation of leukemic cells. Studies and clinical practice show that the presence of PML-RAR-α fusion transcript predicts a high probability of disease-free survival after adequate therapy.

- *t(8; 21)(q22; q22) translocation* is one of the most frequent cytogenetic abnormalities found in both pediatric and adult AML and mainly associated with the acute myeloblastic leukemia (M2) subtype. The t(8; 21) translocation involves the AML1 and the ETO genes. The AML1 gene (*a*cute *m*yeloid *l*eukemia *1* gene, also known as the RUNX1 gene or the runt related transcription factor 1) located on chromosome 21q22 encodes a hematopoietic transcription factor. This transcription factor takes part in the regulation of many genes important for the myeloid differentiation such as myeloperoxidase, GCF, IL-3, neutrophil elastase and T-cell receptors. AML1 gene is also involved in other human leukemia types such as the translocations t(16; 21), t(3; 21) and t(12; 21) found in AML, ALL, CML and MDS. The second gene involved in the t(8; 21) translocation is the ETO gene (*e*ight *t*wenty-*o*ne gene, also known as CBFA2T1 or MTG8 gene) located on chromosome 21q22. This reciprocal translocation generates the transcriptionally active AML1-ETO fusion transcript taking part in the regulation and maturation of the myelopoiesis. The presence of this translocation in AML predicts a good prognosis, good response to chemotherapy and long-term disease-free survival after adequate therapy.

- *inv.(16)(p13; q22)* and related *t(16; 16)(p13; q22) translocation* are recurrent chromosomal rearrangements associated with AML, primarily with the acute myelomonocytic leukemia with increased marrow eosinophils (AML-M4Eo). Both chromosomal abnormalities involve the MYH11 and the CBFβ genes. The MYH11 gene located on chromosome 16 p13 encodes a smooth muscle myosin heavy polypeptide functioning as a contractile protein. The second CBFβ gene located on chromosome 16q22 encodes the beta subunit of the core-binding transcription factor taking part in the regulation of hematopoiesis. The pericentric inversion inv.(16) (p13; q22) and the t (16; 16)(p13; q22) translocation disrupt both genes and generate the transcriptionally active CBFβ-MYH11 fusion transcript, which found in four subtypes (A, B, C and D) depending on the location of the breakpoint within the MYH11 gen. Both cytogenetic abnormalities are associated with a relatively good prognosis and long-term disease-free survival after adequate therapy.

The molecular detection of tumor-specific translocations in hematological malignancies is now an important additional assay for the diagnosis of these malignancies and the estimation of therapy response; and it is an essential method for the detection of residual disease after treatment predicting possible relapse. However, some studies report the persistence of these translocations during complete remission.

Method The method used is mRNA extraction from blood, bone marrow, followed by RT-PCR

- Primers for the cDNA amplification of the *PML-RAR-α* fusion transcript generated by the *t(15; 17)* translocation (nested PCR):

–Outer primers:
FP (PML): 5′-AGC.TGC.TGG.AGG.CTG.TGG.ACG.CGC.GGT.ACC-3′
RP (RAR): 5′-AGG.GCT.GGG.CAC.TAT.CTC.TTC-3′
–Nested primers:
FP (PML): 5′-TGT.GCT.GCA.GCG.CAT.CCG.CA-3′
RP (RAR): 5′-CAG.AAC.TGC.TGC.TCT.GGG.TCT.CAA.T-3′

PCR generates a 248-bp DNA fragment in the S subtype, a DNA fragment of 300–600 bp in the V subtype and DNA fragment of 300–750 bp in the L subtype.

- Primers for the cDNA amplification of the V and L types of the *PML-RAR-α* fusion transcripts generated by *t(15;17)* translocation (nested PCR):
  –Outer primers:
  FP (PML): 5′-AGT.CAG.TGC.CCG.GGG.CAC.AC-3′
  RP (RAR): 5′-AGG.GCT.GGG.CAC.TAT.CTC.TTC-3′
  –Nested primers:
  FP (PML): 5′-AGT.GTA.CGC.CTT.CTC.CAT.CAA.AG-3′
  RP (RAR): 5′-CAG.AAC.TGC.TGC.TCT.GGG.TCT.CAA.T-3′

PCR generates a 324-bp DNA fragment in the L type and about 180-bp DNA fragment in the V type.

- Alternative primers for the cDNA amplification of the L type of the *PML-RAR-α* fusion transcript (nested PCR):
  –Outer primers:
  FP (PML): 5′-GTC.ATA.GGA.AGT.GAG.GTC.TTC-3′
  RP (RAR): 5′-CTC.ACA.GGC.GCT.GAC.CCC.AT-3′

PCR generates a 204-bp DNA fragment.
  –Nested primers:
  FP (PML): 5′-AAC.AGC.AAC.CAC.GTG.GCC.AG-3′
  RP (RAR): 5′-AGC.CTG.AGG.ACT.TGT.CCT.GA-3′

Nested PCR generates a 151-bp DNA fragment.

- Alternative primers for the cDNA amplification of the S type of the *PML-RAR-α* fusion transcript (nested PCR):
  –Outer primers:
  FP (PML): 5′-CCG.ATG.GCT.TCG.ACG.AGT.TC-3′
  RP (RAR): 5′-CTC.ACA.GGC.GCT.GAC.CCC.AT-3′
  PCR generates a 206-bp DNA fragment.
  –Nested primers:
  FP (PML): 5′-TTC.AAG.GTG.CGC.CTG.CAG.GA-3′
  RP (RAR): 5′-AGC.CTG.AGG.ACT.TGT.CCT.GA-3′

Nested PCR generates a 163-bp DNA fragment.

- Primers employed as an internal control for the amplification of the *RAR-α* template:
  FP: 5′-GAC.CAC.TCT.CCA.GCA.CCA-3′
  RP: 5′-GCT.GGG.CAC.TAT.CTC.TTC.AGA.ACT-3′

PCR generates a 100-bp DNA fragment.

- Primers for the cDNA amplification of the *AML1-ETO* fusion transcript generated by the *t(8; 21)* translocation (nested PCR):

–Outer primers:
FP: 5′-AGC.TTC.ACT.CTG.ACC.ATC.AC-3′
RP: 5′-TCA.GCC.TAG.ATT.GCG.TCT.TC-3′
First PCR generates a 166-bp DNA fragment.
–Nested primers:
FP: 5′-TGT.CTT.CAC.AAA.CCC.ACC.GC-3′
RP: 5′-ACA.TCC.ACA.GGT.GAG.TCT.GG-3′
Nested PCR generates a 126-bp DNA fragment.

- Alternative primers for the cDNA amplification of the *AML1-ETO* fusion transcript (nested PCR):
–Outer primers:
FP: 5′-AGC.CAT.GAA.GAA.CCA.GG-3′
RP: 5′-AGG.CTG.TAG.GAG.AAT.GG-3′
First PCR generates a 338-bp DNA fragment.
–Nested primers:
FP: 5′-TAC.CAC.AGA.GCC.ATC.AAA-3′
RP: 5′-GTT.GTC.GGT.GTA.AAT.GAA-3′
Nested PCR generates a 185-bp DNA fragment.

- Alternative primers for the cDNA amplification of the *AML1-ETO* fusion transcript (nested PCR):
–Outer primers:
FP: 5′-GCA.GCC.ATG.AAG.AAC.CA-3′
RP: 5′-GCT.GTA.GGA.GAA.TGG.CT-3′
–Nested primers:
FP: 5′-AGC.TTC.ACT.CTG.ACC.ATC-3′
RP: 5′-TGA.ACT.GGT.TCT.TGG.AGC-3′
Nested PCR generates a 222-bp DNA fragment.

- Primers for the cDNA amplification of the *CBFB/MYH11* fusion transcript (nested PCR):
–Outer primers:
FP: 5′-GCA.GGC.AAG.GTA.TAT.TTG.AAG.G-3′
RP: 5′-CTC.TTC.TCC.TCA.TTC.TGC.TC-3′
First PCR generates a 420-bp DNA fragment.
–Nested primers:
FP: 5′-ACA.CGC.GAA.TTT.GAA.GAT.AGA.G-3′
RP: 5′-TTC.TCC.AGC.TCA.TGG.ACC.TCC-3′
Nested PCR generates a 65-bp DNA fragment.

- Alternative primers for the cDNA amplification of the *CBFB/MYH11* fusion transcript (nested PCR):
–Outer primers:
FP: 5′-CAG.GCA.AGG.TAT.ATT.TGA.AGG-3′
RP: 5′-CTC.CTC.TTC.TCC.TCA.TTC.TGC.TC-3′
First PCR generates a 414-bp DNA fragment in the A subtype of the fusion transcript and various large fragments in the B, C and D subtypes of the same fusion transcript.

–Nested primers:

FP: 5′-GTC.TGT.GTT.ATC.TGG.AAA.GGC.TG-3′

RP: 5′-CCC.GCT.TGG.ACT.TCT.CCA.GC-3′

Nested PCR generates a ~209-bp DNA fragment.

- Alternative nested primers for the amplification of type B, C and D fusion transcripts:

  FP: 5′-GTC.TGT.GTT.ATC.TGG.AAA.GGC.TG-3′

  RP: 5′-CGT.ACT.GCT.GGG.TGA.GGT.TCT-3′

Nested PCR generates different sized DNA fragments, generally long than 210 bp.

3.4 Diagnosis of Lymphoma and Residual Tumor Cells by Molecular Detection of Gene Rearrangements

3.4.1 Introduction

Lymphomas are monoclonal neoplastic proliferation arising from T- or B- or natural killer (NK)-lymphocytes at various stages of differentiation and usually associated with different chromosomal abnormalities and genetic alterations or rearrangements. As lymphoma/leukemia cells are derived from a single transformed precursor cell, accordingly all neoplastic cells have the same genetic changes and gene rearrangements as the precursor neoplastic cell. Consequently, the presence of a monoclonal proliferation pattern is a major diagnostic criterion for the diagnosis of lymphoid and hematopoietic neoplasia and can be detected by different immunological or molecular methods.

All immunoglobulin classes and T-cell receptors are the main molecules responsible for the immune response of both B- and T-lymphocytes. Genes encoding immunoglobulins (Ig) and T-cell receptors (TCR) are located on different chromosomes, have an analogous structure and are composed of the following subunits:

- Sequences encoding the variable segments (V);
- Sequences encoding the joining segments (J);
- Sequences encoding the constant segments (C);
- Sequences encoding the diversity segments (D), diversity segments are only a part of the IgH, TCR β and TCR δ chains.

The rearrangement of different gene segments for the synthesis of the immunoglobulin molecules or T-cell receptors takes place during lymphocyte differentiation generating polyclonal populations of lymphocytes. The gene rearrangement is defined as the process of juxtaposition of different randomly selected gene segments (V, D and J segments) coupled with insertion of random base sequences (N sequences) with looping out and excision of intervening DNA forming a circular excision product called B-cell receptor excision circle or TCR excision circle. The rearrangement process results different sized cell-specific V–D–J sequences. It is important to note that the rearrangement process is a physiological process takes place in normal lymphocytes and not related to the malignant transformation. Noteworthy is the phenomenon of dual rearrangement (inappropriate rearrangement or cross-lineage rearrangement), in which B- or T-cell lymphomas demonstrate the rearrangement of both immunoglobulin heavy chain and T-cell receptor genes and known as lineage infidelity. This phenomenon is mainly noted in lymphomas/leukemias arising from precursor B- and T-lymphocytes and rarely mature B-cell neoplasia and acute myeloid leukemias (AML) that can exhibit TCR rearrangement, consequently the IgH and TCR gene rearrangements are not an absolute lymphocyte-lineage criteria. The inappropriate rearrangement or cross-lineage rearrangement is noted more frequently when multiple primer panels matching different IgH/IgL or TCR regions are used like the BIOMED-2 primer panel, therefore if the rearrangement is used for the primary diagnosis of lymphoma, additional immunophenotyping is necessary to determine the lineage of the lymphoid neoplasia.

Immunoglobulin heavy-chain (IgH) gene and κ light-chain (Igκ) gene rearrangements in B-cell lymphoma and T-cell receptor (TCR) gene rearrangement in T-cell lymphoma are important targets for clonality analysis and widely used to confirm the diagnosis of lymphoma using DNA extracted from broad range of specimens of different origin. Southern blot hybridization and PCR are the most common used methods for clonality analysis. Southern blot hybridization is considered being the gold standard for this assay, but it is labor intensive and requires a large amount of intact high molecular DNA.

The clonality assessment by PCR-based methods is a rapid and inexpensive assay and now becoming the standard method for the diagnosis of lymphoproliferative disorders and requires only small quantities of DNA extracted from frozen or formalin-fixed paraffin-embedded tissue including biopsies, FNA and different body fluids. IgH and TCR γ genes are the most suitable targets for clonality analysis by PCR. In doubtful cases the analysis of TCR δ, TCR β, Igκ and Igλ can give further useful information. After the amplification of rearranged genes by PCR, monoclonal cells generate one or two well defined sharp bands, different from the non-rearranged germ line cells containing a polyclonal cell population, which generate an undefined smear or ladder-like pattern on electrophoresis gel. The presence of more than one sharp band on electrophoresis gel may indicate the presence of more than one neoplastic clone (subclones); these may have different response to the specific therapy.

A substantial percentage of lymphomas and leukemias relapse after chemotherapy, indicating that patients still harbor residual neoplastic cells after therapy. The sensitivity of clonality analysis makes it a useful tool to screen residual tumor cells in the graft to prevent the reinfusion of tumor cells or to detect minimal residual neoplastic cells or the recurrence of lymphoma/leukemia after chemotherapy and bone marrow transplantation. Monitoring of minimal residual disease is found to be a major predictor for post-transplant outcome in children receiving stem cell transplantation or T-cell depleted bone marrow as studies show that minimal residual disease at levels of one malignant cell in 103 background cells or greater is associated with poor outcome and known as a molecular relapse. Furthermore, the formation of neoplastic subclones (known as clonal diversity) during progression of the disease is found to be coupled with a high relapse risk.

An additional useful aspect for the clonality analysis is the screening of cerebrospinal fluid to exclude the central nervous system involvement in patients with acute lymphoblastic leukemia of both B- and T-cell lineage. This method is more accurate than conventional cytology.

Although monoclonality is a fundamental feature of lymphomas and leukemias, monoclonal population of lymphocytes is not equivalent to malignancy but can also occur in physiologic or benign reactive disorders and the diagnosis of malignancy most be done depending on the results of the histopathology, immunohistochemistry, flow cytometry, genetic analyses and clinical data. However, the absence of monoclonality using molecular methods does not necessarily exclude lymphoma.

False-positive results are an important aspect, which we have to consider in clonality analysis as false-positive results are reported in some autoimmune lymphoproliferative disease, immunodeficiency syndrome and lymphoid hyperplasia associated with viral or bacterial infections such as HBV, HCV and helicobacter pylori infection. Moreover the use of very small amount of DNA (<50 ng) can also mimic monoclonal results. Conversely, false-negative results are reported in lymphomas associated with extensive reactive nonneoplastic lymphoid infiltrate or extensive somatic mutations. Somatic hypermutations take place in a 1.5–2.0 kb segment through the entire length of the V–D–J region of the IgH genes and through the entire length of the V–J region of both Igκ and Igλ genes. Somatic hypermutations are an important and frequent cause of false-negative results in the clonality analysis targeting IgH or TCR, especially if simple primer panels are used. Somatic hypermutations of the Ig gene frequently take place in lymphomas of germinal center and post-germinal center origin. Generally, lymphomas associated with low level of somatic mutations such as B-cell chronic lymphocytic lymphoma, prolymphocytic B-cell leukemia, mantle cell lymphoma, hairy cell leukemia, intravascular large B-cell lymphoma and enteropathy type T-cell lymphoma, unspecified peripheral T-cell lymphoma, anaplastic ALK-positive and ALK-negative large cell lymphoma, show optimal results in the clonality analysis in more than 90% of cases, while lymphomas with frequent and extensive somatic mutations such as follicular lymphoma, lymphoplasmacytic lymphoma, marginal zone B-cell lymphoma and marginal zone B-cell lymphoma of MALT type, diffuse large B-cell lymphoma including plasmablastic, immunoblastic and centroblastic lymphoma, primary mediastinal (thymic) large B-cell lymphoma, plasmacytoma and angioimmnoblastic T-cell lymphoma show optimal results in only 50–75% of cases. In such lymphoma types, modified primers and different primer panels to different gene regions (such as the BIOMED-2 primer panel) are recommended to obtain acceptable results. NK cell lymphomas are usually not accompanied with IgH or TCR rearrangements.

A further interesting aspect is the gene rearrangement in Hodgkin's lymphoma. Using single cell PCR, most of Hodgkin and Reed–Sternberg (H&RS) cells demonstrate an immunoglobulin gene rearrangement, which points out the B-cell origin of these cells and only in minor cases is the rearrangement of the TCR is demonstrated. The high rate of somatic mutations found in these neoplastic cells signal to B-cells of germinal center origin. DNA extracted from tissue sections with Hodgkin's lymphoma is not suitable for clonality analysis as neoplastic cells account less than 1% of the whole cell population.

In conclusion, the molecular assessment of monoclonality for the diagnosis of lymphoid neoplasia is a precise, reproducible, rapid and inexpensive method, but must be evaluated with caution taking in consideration possible false-positive and false-negative results in addition to other clinical and hematological data.

3.4.2
Detection of Immunoglobulin Heavy Chain (IgH) Gene Rearrangement in B-Cell Lymphoma

B-cell lymphoma and leukemia are clonal proliferation of neoplastic cells derived from B-lymphocytes at different stages of differentiation. B-lymphocytes are immunocompetent cells, whose main function to produce different types of immunoglobulins (Ig).

Immunoglobulins are a family of proteins with the same basic architecture, divided into five classes: IgM, IgG, IgA, IgD and IgE. Immunoglobulins are composed of different segments encoded by different genes located on different chromosomes. The unit of each immunoglobulin is composed of two identical light chains (IgL) and two identical heavy chains (IgH) (Figure 3.3). The light chains divided into two subclasses kappa (κ) and lambda (λ) light chains. The heavy chains are individual for each immunoglobulin class and divided into five subclasses (α, δ, ε, γ, μ).

The immunoglobulin heavy chain gene extends on a ~1250-kb segment on chromosome 14q32.3. The rearrangement of the immunoglobulin heavy chain gene (IgH) takes place in all normal developing B-lymphocytes before immunoglobulin synthesis. The chromosomal region encoding IgH contains the variable, joining and diversity segments. There are up to 250 variable (V) functional and nonfunctional segments, grouped into six or seven families (VH1–VH7), whereas the functional segments make about 20% of the total amount of the variable segments. The diversity (D) segments are about 30, grouped into 7 families (DH1-DH7) and the joining (J) region, which includes 6 segments. The rearrangement begins with the D and J segments and followed by the V segment, via moving one D, one J, and one V region into close proximity to each other. Essential role in the rearrangement plays the

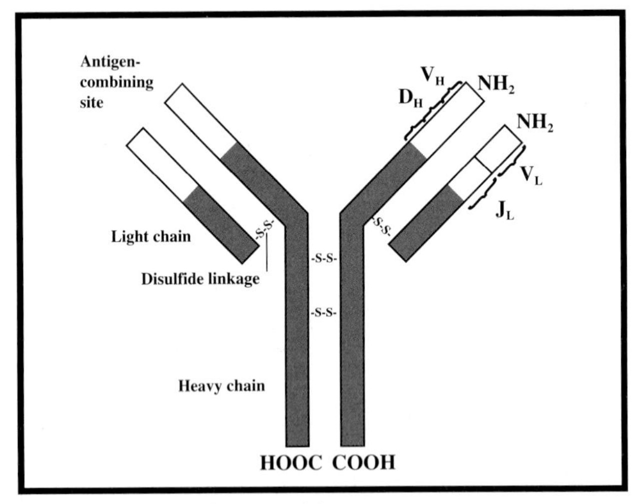

Figure 3.3 Structure of immunoglubin protein.

recombinase enzyme complex that includes the protein products of the *RAG-1* and *RAG-2* genes, which are unique to vertebrates and are active at very early stages of lymphoid development. The recombinase enzyme complex recognize and cut the DNA inside the IgH gene (also inside the TCR gene) at the recombination signal sequences (RSS) located at both sides of the D regions of the IgH gene forming hairpin structures at the RSS coding sequence. At this location, the terminal deoxynucleotidyl transferase (TdT) inserts different nucleotides between the D and J segments to build the hypervariable N region, composed of a random sequence of basses. In the second stage, the terminal deoxynucleotidyl transferase (TdT) builds a second N region between the V and D regions (Figure 3.4). The removed sequences between the rearranged genes are deleted in form of circular excision product (B-cell receptor excision circle). If the rearrangement occurs between the diversity and joining segments only, incomplete D–J joints are generated (incomplete rearrangement), which occurs in precursor B-cells with an immature CD10+/CD19− phenotype.

The terminal deoxynucleotidyl transferase (TdT) is a 58-kDa DNA nuclear polymerase, which catalyzes the template independent polymerization of deoxynucleotidyl triphosphates to double-stranded gene segment DNA. TdT is mainly expressed in primitive B- and T-lymphocytes. Antibodies to TdTare used as a specific marker for the diagnosis of precursor cell lymphomas, namely acute lymphoblastic leukemia and lymphoma.

After the successful rearrangement of the immunoglobulin heavy chain, the rearrangement of the light chain at the kappa locus on chromosome 2p11.2 or at the lambda locus on chromosome 22q11.2 takes place (see below). As a result of the rearrangement, cell-specific DNA sequences unique to that cell occur in the location of gene rearrangement

IgH gene structure on chromosome 14.q32

5'-V_1=V_2=~=$V_{\sim 250}$=//=D_1=D_2=D_3=~=$D_{\sim 30}$=//=J_1=J_2=J_3=J_4=J_5=J_6=//=C==3'

↑ Recombinase enzyme complex and TdT ↑

↓ D–J rearrangement

5'- V_1=V_2=~=$V_{\sim 250}$=//=D=N=J=//=C== -3'

↓ V–D–J rearrangement

5'- V=N=D=N=J=C== -3' (IgH gene components after rearrangement)

V: Variable segments, D: Diversity segments, J: Joining segments, C: Constant segments, N: Hypervariable regin

Figure 3.4 Immunoglobulin heavy chain gene rearrangement in the pre-B stage.

FR1	CDR1	FR2	CDR2	FR3	CDR3			FR4
V_H					N	D	N	J_H

Figure 3.5 Structure of IgH gene.

The mature rearranged IgH gene includes the following three highly variable complementarily determining regions: CDR1 encoding 5–7 amino acids, CDR2 encoding 16–18 amino acids and CDR3 encoding 6–19 amino acids. The three CDR regions are separated by four relatively conserved framework (FR) regions: FR1 encoding 30 amino acids, FR2 encoding 14 amino acids and FR3 encoding 29 amino acids and FR4, where FR1, CDR1, FR2, CDR2 and FR3 regions corresponding to the IgH V region, CDR3 region corresponding to both N regions and IgH D region. The CDR3 is characterized by the highest degree of sequence diversity. The last region is the FR4 region and includes the IgH J region (Figure 3.5). The fact that the CDR3 region represents the most hypervariable segment in the IgH gene makes it the most suitable and informative target for clonality analysis of B-cell lymphomas. In order to analyze the CDR3 gene, consensus primer sets to both adjacent FR3 and FR4 (JH) regions are used. The generated PCR product includes the CDR3 region, which can be analyzed by electrophoresis or by sequencing.

After the rearrangement of the immunoglobulin heavy chain gene (IgH), the immunoglobulin kappa light chain (Ig κ) region locating on chromosome 2p11.2 undergoes rearrangement in a similar fashion with the exception that it does not contain the diversity region (D). The Ig κ gene includes 5 Jκ segments and many Vκ segments, about 33 of them are functional segments, grouped in seven families. The variable region (V) is constantly coupled to a leader region (L) encoding the signal polypeptide, regulating the secretion of the gene product (Figure 3.6). If the rearrangement of the kappa light chain gene is not productive in either allele, which happens in approximately one-third of cases, the kappa light chain constant and/or variable regions are deleted and the immunoglobulin lambda light chain (Ig λ)

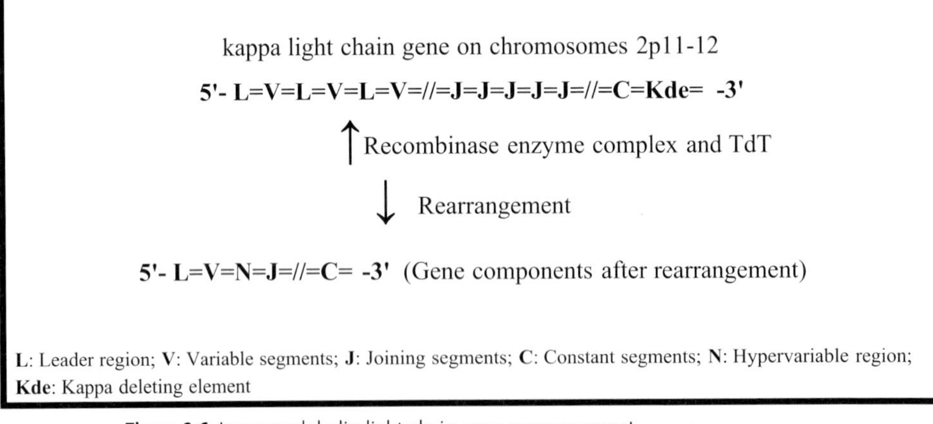

Figure 3.6 Immunoglobulin light chain gene rearrangement.

region undergoes rearrangement. The Ig λ region is located on chromosome 22q11.2. Igλ locus consists of 31 active segments grouped in ten Vλ families and four Jλ segments. As a result of this mechanism, B-lymphocytes express only one type of light chains, phenomenon known as the principle of allelic exclusion. Both Ig κ and Ig λ genes have a similar structure and the rearrangement takes place in a similar pattern. The immunoglobulin light chain rearrangements are not reported in any type of T-lineage lymphomas, whereas the IgH rearrangements are found in some types of T-cell lymphoma (so-called dual or inappropriate rearrangements or lineage infidelity).

Because each normal B-cell undergoes a unique individual rearrangement, there are constantly genetic differences between B-lymphocytes, which results a polyclonal B-cell population and each B-cell is able to make only one type of antibody out of a possible repertoire of over ten0 million possibilities, each antibody recognizes only one antigenic shape.

Since lymphomas are clonal neoplasia, immunoglobulin heavy chain as well as kappa or lambda light chain rearrangements are detectable in essentially most of the neoplastic cases. However, malignancies derived from precursor-B-cells (lymphoblastic lymphomas and leukemia) reveal only immunoglobulin heavy chain rearrangements as the neoplastic transformation occurs before the rearrangement of the immunoglobulin kappa light chain region takes place. The fact that lambda light chain rearrangements do not always occur or occur in late stages of B-cell development makes this region an unsuitable initial target for clonality testing. Nevertheless it can be used as a complementary assay in some cases.

A significant aspect in clonality analysis is the extensive somatic mutations appearing within the immunoglobulin heavy chain gene of some neoplastic B-cell disorders, especially these lymphomas arising from post-germinal center cells and memory B-cells such as follicular lymphoma, Marginal zone B-cell lymphoma, hairy cell leukemia, Burkitt's lymphoma and plasma cell malignancies. These somatic mutations extensively alter the sequence of the immunoglobulin loci amplified by designed primers so that the primer hybridization is incomplete or does not take place, causing false-negative clonality results in about 30% of these lymphoma types. Consequently, a negative PCR result does not exclude the presence of a monoclonal B-cell proliferation and the failure to demonstrate a monoclonal population in a suspicious specimen may needs further genetic analysis such as the detection of specific translocations associated with different types of lymphoma as t(14; 18) in follicular lymphoma, t(2; 5) in anaplastic large cell lymphoma or trisomy 3 in MALT lymphoma. Optimal results in clonality analysis are noted in lymphomas arising from pre-germinal B-cells as lymphoblastic lymphoma, mantle cell lymphoma and chronic lymphocytic leukemia.

Recently few interesting observations show that clonality analyses may be also useful to analyze some types of classical Hodgkin's lymphoma, namely nodular sclerosis and lymphocyte predominance type, as the vast majority of the neoplastic L&H cells (about 95%) proved to be of B-cell origin, i.e. from germinal center B-lymphocytes. The rest (~5%) of L&H cells seems to be of T- or null-cell origin. The IgH rearrangement is found in up to 20% of Hodgkin's lymphoma types.

Additional and significant factor, which we have also to take in consideration in interpretation of clonality results, is the possibility of false-positive results due to different immunopathologic disorders such as the monoclonal gammopathy of undetermined significance. Furthermore, few viral and bacterial infections such as HIV, HBV, HCV, EBV and Helicobacter pylori infections may be associated with benign but clonal inflammatory infiltrate and may mimic a monoclonal proliferation pattern on the gel electrophoresis. Representative example for this phenomenon is the observations demonstrating the loss of monoclonality in lymphoma suspected gastric biopsies after eradication of Helicobacter pylori. Consequently, the presence of IgH clonal rearrangements in gastric biopsies should not be considered as an absolute criterion for MALT lymphoma, while the persistence of monoclonality in chronic gastritis should be an indication for further follow-up as MALT lymphoma commonly derived from a monoclonal lymphoid background. The same circumstances must be considered in lymphoma appearing in the thyroid gland or salivary glands after Hashimoto thyroiditis and Sjögren syndrome.

In conclusion, clonality analysis is fundamental criteria for the diagnosis of many types of lymphoma and an essential tool for the differentiation between lymphoma and other benign or unclassified hyperplasic lymphoid lesions or pseudolymphoma. Additionally clonality analysis is also useful for the screening of bone marrow before autologous bone marrow or stem cell transplantation or for the detection of minimal residual disease in bone marrow and peripheral blood.

The use of PCR-based methods for the detection of immunoglobulin heavy chain gene rearrangements allows the use of small amounts of DNA and high-molecular-weight DNA is not essential. Moreover DNA extracted from formalin-fixed paraffin-embedded tissue is suitable for this assay, which makes this method a valuable method in histopathology. However, observations show that the use of very small amount of DNA can mimic monoclonal results by generating pseudo-monoclonal bands on electrophoresis. The amount of required DNA must be more than 50 ng, 5 ng of which are of lymphocytic origin. This amount of DNA can be extracted from 5 μm paraffin sections (either three sections 2×1 cm or seven sections 3×3 mm). It is also important to mention that the highest sensitivity rate is obtained when the monoclonal cells make not less than 15% of the total polyclonal background cells.

Method The method used is DNA extraction from blood, bone marrow, fresh or formalin-fixed paraffin-embedded tissues, followed by PCR or multiplex PCR. To provide a better discrimination of the monoclonal bands it is advisable to run out the PCR product on a polyacrylamide gel electrophoresis. Heteroduplex assay for the analysis of the PCR product improves the results and minimizes the possibility of false-positive or false-negative results due to improper primer annealing.

- *Optimal primer selection*: for optimal clonality screening it is recommended to begin with one or two simple primer sets, each is composed of two primers. First and most effective primer set found to be the one composed of forward primer matching the FR3 region, as this region represents the most constant VH framework, combined with a reverse primer matching the FR4 (JH) region. This initial primer set shows to have the highest efficiency in clonality analysis and able to detect up to 95% of low-grade B-cell lymphoma and up to 75% of intermediate and high-grade lymphoma in addition to plasma cell malignancies. If the initial primer set fails to detect a suspicious lymphoproliferative disorder, further primer sets can be used which include primers matching other conserved framework regions (FR1, FR2) of the IgH gene or κ and λ light chain genes. The BIOMED-2 primer panels listed below are efficient standardized primer panels and can be successfully used for clonality study of B- and T-cell malignancies. Doubtful cases with suspicious histology can also be the subject of southern blot hybridization. In unclear cases, the use of other primer sets to detect frequent lymphoma-associated translocations can be helpful. This approach is cost and labor effective and help to minimize false-positive and false-negative results caused by frequent somatic mutations occurring in some lymphomas such follicular lymphoma. It is also important to evaluate the integrity of the extracted DNA by the simultaneous amplification of one of the housekeeping genes.

- Primer sets for simple screening of IgH clonality consensus to the *FR3* and *Ig J* regions: best results are obtained using the first two primer sets (1 and 2).

Primer set 1:
FP (FR3-V670): 5′-ACG.GCC.GTG.TAT.TAC.TG-3′
RP (Ig J-OL-4): 5′-ACC.TGA.GGA.GAC.GGT.GAC.C-3′
Primer set 2:
FP (FR3 A): 5′-ACA.CGG.C(C/T)(G/C).TGT.ATT.ACT.GT-3′
RP (VLJH): 5′-GTG.ACC.AGG.GT(A/G/C/T).CCT.TGG.CCC.CAG-3′
Both primers mentioned above are degenerate primers, composed of a mixture of two or more primers that differs in base composition at the positions placed in brackets.
Primer set 3:
FP (Ig V): 5′-CTG.TCG.ACA.CGG.C(C/T)(G/C).TGT.ATT.ACT-3′
RP (JH): 5′-AAC.TGC.AGA.GGA.GAC.GGT.GAC.C-3′
The FR primer is a degenerate primer, composed of a mixture of two or more primers that differs in base composition at the positions placed in brackets.
Primer set 4:
FP (IgH2-VH26): 5′-AAC.AGC.CTG.AGA.GCC.GAG.GA-3′
RP (OL-4): 5′-ACC.TGA.GGA.GAC.GGT.GAC.C-3′

Specimens containing a monoclonal B-cell population generate equal DNA fragments exhibiting the same size, which are demonstrated as additional sharp one or two bands on electrophoresis gel (ranging between 70 bp and 150 bp in size). The presence of more than one band seems to be as a result of formation of tumor subclones (clonal diversity) during tumor progression. The generated monoclonal bands give distinctive DNA fragments after digestion by restriction enzymes, usually different from the non-rearranged germ line cells. Specimens containing a polyclonal B-cell population generate an undefined smear or ladder-like pattern on electrophoresis gel.

- Primer set consensus to the *FR2* gene family and *JH* region:

  FP FR2: 5′-TGG.(A/G)TC.CG(A/C).CAG.(G/C)C(T/C).(T/C)C(A/C/GGT).GG-3′

  RP JH: 5′-AAC.TGC.AGA.GGA.GAC.GGT.GAC.C-3′

The forward primer is a degenerate primer, composed of a mixture of two or more primers that differs in base composition at the positions placed in brackets; monoclonal cell population generates one or two bands of 200–250 bp in size.

- Primer set consensus to the *FR1* gene family and *JH* region:
  1. FP (VH1): 5′-CCT.CAG.TGA.AGG.TCT.CCT.GCA.AGG-3′
  2. FP (VH2): 5′-TCC.TGC.GCT.GGT.GAA.AGC.CAC.ACA-3′
  3. FP (VH3): 5′-GGT.CCC.TGA.GAC.TCT.CCT.GTG.CA-3′
  4. FP (VH4a): 5′-TCG.GAG.ACC.CTG.TCC.CTC.ACC.TGC.A-3′
  5. FP (VH4L): 5′-ATG.AAA.CAC.CTG.TGG.TTC.TT-3′
  6. FP (VH5): 5′-GAA.AAA.GCC.CGG.GGA.GTC.TCT.GAA-3′
  7. FP (VH6): 5′-CCT.GTG.CCA.TCT.CCG.GGG.ACA.GTG-3′
  8. RP (JH): 5′-ACC.TGA.GGA.GAC.GGT.GAC.CAG.GGT-3′

PCR is carried out in seven tubes, each contain one VH primer and the JH primer. Monoclonal lymphoid cell population generates one or two bands ranging between 250 bp and 500 bp in size.

- *BIOMED-2* primer sets consensus to the *FR1*, *FR2* and *FR3* gene families and to the *JH* region gives generally optimal results in the vast majority of cases. Multiplex PCR is carried out in three tubes (A, B, C), each contains the reverse primer for the JH region and the forward primers for one FR gene family. The following reaction conditions are recommended.

Reaction volume:	50 μl
Template DNA quantity:	~100 ng
Primer quantity:	20 pmol of each primer (concentration in reaction volume 100 nM)
Initial denaturation:	95 °C; 2 min
Denaturation:	94 °C; 40 s
Primer annealing:	60 °C; 40 s
Primer extension:	72 °C; 40 s
Final extension:	72 °C; 10 min
Number of cycles:	35 cycles

For the analysis of the PCR product, polyacrylamide gel and heteroduplex analysis is recommended.

–*Tube A*:

FP (VH1-FR1): 5′-GGC.CTC.AGT.GAA.GGT.CTC.CTG.CAA.G-3′
FP (VH2-FR1): 5′-GTC.TGG.TCC.TAC.GCT.GGT.GAA.ACC.C-3′
FP (VH3-FR1): 5′-CTG.GGG.GGT.CCC.TGA.GAC.TCT.CCT.G-3′
FP (VH4-FR1): 5′-CTT.CGG.AGA.CCC.TGT.CCC.TCA.CCT.G-3′
FP (VH5-FR1): 5′-CGG.GGA.GTC.TCT.GAA.GAT.CTC.CTG.T-3′
FP (VH6-FR1): 5′-TCG.CAG.ACC.CTC.TCA.CTC.ACC.TGT.G-3′

–*Tube B*:
FP (VH1-FR2): 5′-CTG.GGT.GCG.ACA.GGC.CCC.TGG.ACA.A-3′
FP (VH2-FR2): 5′-TGG.ATC.CGT.CAG.CCC.CCA.GGG.AAG.G-3′
FP (VH3-FR2): 5′-GGT.CCG.CCA.GGC.TCC.AGG.GAA-3′
FP (VH4-FR2): 5′-TGG.ATC.CGC.CAG.CCC.CCA.GGG.AAG.G-3′
FP (VH5-FR2): 5′-GGG.TGC.GCC.AGA.TGC.CCG.GGA.AAG.G-3′
FP (VH6-FR2): 5′-TGG.ATC.AGG.CAG.TCC.CCA.TCG.AGA.G-3′
FP (VH7-FR2): 5′-TTG.GGT.GCG.ACA.GGC.CCC.TGG.ACA.A-3′
–*Tube C*:
FP (VH1-FR3): 5′-TGG.AGC.TGA.GCA.GCC.TGA.GAT.CTG.A-3′
FP (VH2-FR3): 5′-CAA.TGA.CCA.ACA.TGG.ACC.CTG.TGG.A-3′
FP (VH3-FR3): 5′-TCT.GCA.AAT.GAA.CAG.CCT.GAG.AGC.C-3′
FP (VH4-FR3): ′-GAG.CTC.TGT.GAC.CGC.CGC.GGA.CAC.G-3′
FP (VH5-FR3): 5′-CAG.CAC.CGC.CTA.CCT.GCA.GTG.GAG.C-3′
FP (VH6-FR3): 5′-GTT.CTC.CCT.GCA.GCT.GAA.CTC.TGT.G-3′
FP (VH7-FR3): 5′-CAG.CAC.GGC.ATA.TCT.GCA.GAT.CAG-3′

–*General reverse primer*:
–RP (JH): 5′-CCA.GTG.GCA.GAG.GAG.TCC.ATT.C-3′

Monoclonal lymphoid cell population generates one or two clonal bands ranging between 150 bp and 400 bp in size.

- *BIOMED-2* primers for the detection of *incomplete IgH (D–J) rearrangement* can be used when the above primer sets fail to give successful results. Multiplex PCR is carried out in two tubes (A and B). The first tube (A) contains the DH1–6 primers in addition to the JH reverse primer. The second tube (B) contains both DH7 and JH primers. The multiplex PCR has the same conditions as the above reaction.

–*Tube A*:
FP (DH1): 5′-GGC.GGA.ATG.TGT.GCA.GGC-3′
FP (DH2): 5′-GCA.CTG.GGC.TCA.GAG.TCC.TCT-3′
FP (DH3): 5′-GTG.GCC.CTG.GGA.ATA.TAA.AA-3′
FP (DH4): 5′-AGA.TCC.CCA.GGA.CGC.AGC.A-3′
FP (DH5): 5′-CAG.GGG.GAC.ACT.GTG.CAT.GT-3′
FP (DH6): 5′-TGA.CCC.CAG.CAA.GGG.AAG.G-3′
–*Tube B*:
FP (DH7): 5′-CAC.AGG.CCC.CCT.ACC.AGC.-3′
–*General reverse primer*:
RP (JH): 5′-CCA.GTG.GCA.GAG.GAG.TCC.ATT.C-3′

- Primers for the detection of the *kappa light chain* (*Ig* κ) gene rearrangement:

FP (FR3 κ): 5′-TTC.AG(C/T).GGC.AGC.GG(A/G).TCT.GGG-3′
RP (Jκ): 5′-CA(G/C).CTT.(G/T)GT.CCC.(C/T)TG.GCC.GAA-3′

Both primers are degenerate primers, composed of a mixture of two or more primers that differs in base composition at the positions placed in brackets. Monoclonal cell population generates a clonal band 120–150 bp in size.

- Alternative primers for the detection of the *kappa light chain* (*Ig* κ) gene rearrangement. The reaction is carried out in three tubes; Tube 1 includes the primers 1, 2 and 4 (Vκ I/VI, Vκ II and Jκ); Tube 2 includes the primers 3 and 4 (Vκ III/IV and Jκ); Tube 3 includes the primers 1, 3 and 5 (Vκ I/VI, Vκ III/IV and KDE):
  1. FP (Vκ I/VI): 5′-TTC.AG(T/C).GGC.AGT.GGA.TCT.GG-3′
  2. FP (Vκ II): 5′-TTC.AGT.GGC.AGT.GGG.TCA.GG-3′
  3. FP (Vκ III/IV): 5′-TTC.AGT.GGC.AG(T/C).GGG.TCT.GG-3′
  4. RP (Jκ): 5′-GTT.TGA.TCT.CCA.CCT.TGG.TCC.C-3′
  5. RP (KDE): 5′-CCC.TTC.ATA.GAC.CCT.TCA.GGC.AC-3′

  In case of monoclonal cell population, the first two reactions generate a band 130–150 bp in size; the third reaction generates a monoclonal band 250–300 bp in size.

- *BIOMED-2* primers for the detection of the *kappa light chain* (*Ig* κ) gene rearrangement. The multiplex PCR is carried out in two tubes (A andB). The first (A tube) with the first six V κ forward primers (FP) and both A (J κ) reverse primers (RP), the second (B tube) with all seven forward primers and the B (Kde region) reverse primer.

  –Forward primers:
  FP (V κ 1f/6): 5′-TCA.AGG.TTC.AGC.GGC.AGT.GGA.TCT.G-3′
  FP (V κ 2f): 5′-GGC.CTC.CAT.CTC.CTG.CAG.GTC.TAG.TC-3′
  FP (V κ 3f): 5′-CCC.AGG.CDC.CTG.ATC.TAT.GAT.GCA.TCC-3′
  FP (V κ 4): 5′-CAA.CTG.CAA.GTC.CAG.CCA.GAG.TGT.TTT-3′
  FP (V κ5): 5′-CCT.GCA.AAG.CCA.GCC.AAG.ACA.TTG.AT-3′
  FP (V κ 7): 5′-GAC.CGA.TTT.CAC.CCT.CAC.AAT.TAA.TCC-3′
  FP (J κ – C κ intron): 5′-CGT.GGC.ACC.GCG.AGC.TGT.AGA.C-3′
  –Reverse primers (B tube):
  RP (J κ1–4): 5′-CCC.TGG.TTC.CAC.CTC.TAG.TTT.GCA.TTC-3′
  RP (J κ5): 5′-CCC.TGT.GCT.GAC.CTC.TAA.TTT.GCA.TTC-3′
  –B reverse primers:
  RP (Kde): 5′-ATC.CTG.TTG.GAC.GAG.ACT.GGA.GAC.TCC-3′

The monoclonal cell population generates a clonal band. The length of the PCR product depends on the type of the rearrangement.

- *BIOMED-2* primers for the detection of the lambda light chain (Ig λ) gene rearrangement. The multiplex reaction is carried out with both V λ foreword primers (FR) and the J λ reverse primer (RP):

  FP (V λ1/2): 5′-ATT.CTC.TGG.CTC.CAA.GTC.TGG.C-3′
  FP (V λ3): 5′-GGA.TCC.CTG.AGC.GAT.TCT.CTG.G-3′
  RP (J λ): 5′-CCC.TGG.TTC.GAG.TGG.CAG.GAT.C-3′

Monoclonal cell population generates a clonal band 100–200 bp in size.

3.4.3 Detection of T-Cell Receptor Gene Rearrangement in T-Cell Neoplasia

T-cell lymphoma is a clonal proliferation of neoplastic cells originated from different stages of T-lymphocytes. Thymic and prethymic T-cell lymphomas derive from the early T-cell precursors whereas peripheral T-cell lymphomas from more mature T-lymphocytes. T-lymphocytes are characterized by the expression of T-cell receptors (TCR) on the membrane surface. These receptors are essential for the recognition of antigens and different Immunologic reactions. The T-cell receptor (TCR) is a heteromer with two subtypes:

- TCR1 composed of α (alpha) and β (beta) chains (TCR alpha/beta type), presents on about 95% of circulating T-lymphocytes and on the majority of TCR-positive thymocytes.
- TCR2 composed of γ (gamma) and δ (delta) chains (TCR2 gamma/delta type) presents on T-cells concentrated in various mucosa types such as intestinal, uterine and tongue mucosa.

All α β γ δ chains are composed of the V, J and C segments, but both β and δ chains also include the D segment similar to the IgH chain. The T-cell receptor (TCR) genes undergo V–D–J or V–J rearrangements at an early stage of T-lymphocyte development in a similar way as the immunoglobulin heavy chain and kappa light chain genes in B-lymphocytes and plasma cells in the following sequential order: TCR δ on chromosome 14q11, TCR γ on chromosome 7q15, TCR β on chromosome 7q34 and TCR α on chromosome 14q11. About 95% of circulating T-lymphocytes undergoes all four rearrangements and only a small percentage of the T-lymphocytes undergo only the first two rearrangements, namely TCR δ and TCR γ rearrangements. These cells (γ/δ lymphocytes) are primarily located in the red pulp of the spleen and are the origin of the hepatosplenic γ/δ T-cell lymphoma.

Both TCR γ and TCR β chains exhibit an unsophisticated structure and are usually used for the rearrangement study of T-cell lymphoma using PCR-based methods, whereas the TCR α gene locus exhibit more complicated structure and rarely used for clonality analysis. Some studies reported benefits using consensus primer sets to the TCR δ gene, specifically in the cases of precursor lymphoblastic leukemia of B- or T-cell origin. As the TCR δ gene components are situated within the TCR α gene on chromosome 14q11 between the V and J regions, the TCR α gene rearrangement is associated with the deletion of the TCR δ gene (Figure 3.7). In such cases, the rearrangement study of the TCR δ gene is unproductive.

The TCR γ chain is less complex than the other chains, spans a segment of about 160 kb on chromosome 7p14, and is constantly rearranged earlier than both TCR β and TCR α chains. The gene encoding the TCR γ chain is located on chromosome 7p15 and is composed just of the variable (V), joining (J) and constant (C) segments but not the diversity (D) segments. The human TCR γ gene contains the following segments:

- 14 V γ gene segments, most of them are included in 4 V families:
  –V γ I family: V γ1, V γ2, V γ3, V γ4, V γ5, V γψ5, V γψ6, V γψ7 and V γ8

$5'\text{-}V_{\alpha 1}=V_{\alpha 2}=\sim\sim=V_{\alpha\sim 70}=//=V_{\delta}=V_{\delta}=V_{\delta}=D_{\delta}=D_{\delta}=D_{\delta}=//=J_{\delta}=J_{\delta}=J_{\delta}=C_{\delta}=//=J_{\alpha}=J_{\alpha}=J_{\alpha}=//=C_{\alpha}=3'$

V: Variable segments; D: Diversity segments; J: Joining segments; C: Constant segments

Figure 3.7 Structure of TCR α and TCR δ genes on chromosome 14q11.

–V γ II family: V γ9
–V γ III family: V $\gamma\psi$10
–V γ IV family: V $\gamma\psi$11

- 5 J γ gene segments: J γP1, J γP, J γ1, J γP2 and J γ2
- 2 C γ gene segments: C γ1 and C γ2

Studies show that only ten of the 14 V γ segments (V γ 2, V γ3, V γ4, V γ5, V $\gamma\psi$7, V γ8, V γ9, V $\gamma\psi$10, and V $\gamma\psi$11) are functional segments capable to undergo the rearrangement. The TCR γ rearrangement takes place in a single allele by recognition and cutting the DNA at the recombination signal sequences, followed by adding of the hypervariable (N) sequence to form heterogeneous V–N–J sequences (Figure 3.8), which varies by about 20 bp in a polyclonal cell population. The rearrangement process is mediated by the recombinase enzyme complex and the terminal deoxynucleotidyl transferase (TdT) as described in the IgH rearrangement. The second allele remains as a backup for unsuccessful initial rearrangement.

For optimal TCR γ rearrangement analysis by PCR, consensus primers to the V γ2–11 regions (all four V γ families) together with consensus primers to the J region are used. As no somatic hypermutations occur in the TCR γ locus the use of such primer panels is able to detect up to 90% of clonal T-cell neoplasia and needs only small amounts of DNA and even low molecular DNA extracted from formalin-fixed paraffin-embedded tissue can be successfully used. This makes this method ideal in

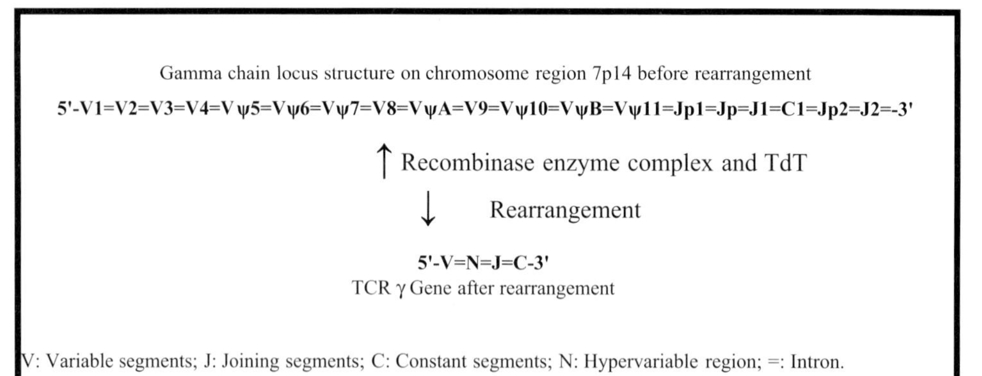

Figure 3.8 Structure and rearrangement of the TCR γ gene.

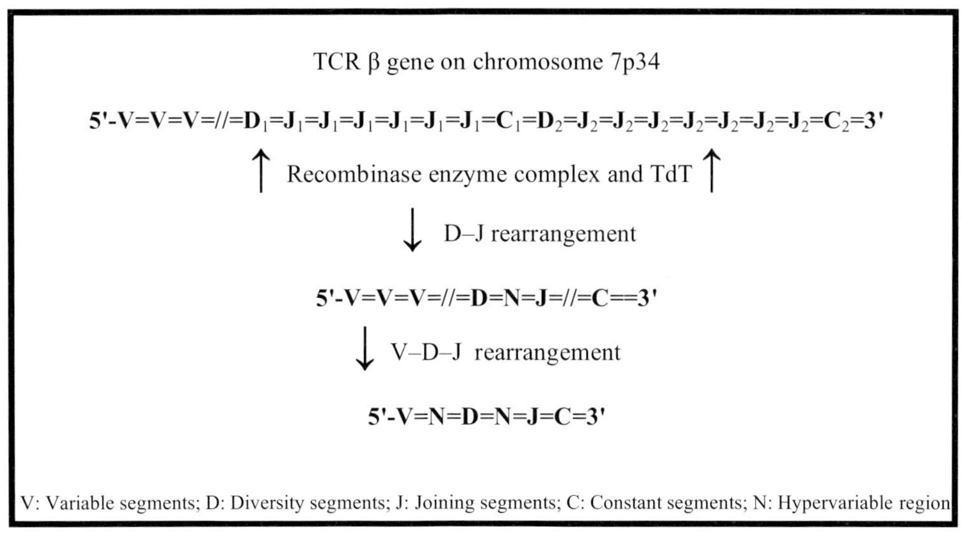

Figure 3.9 Structure and rearrangement of the TCR β gene.

routine histopathology for the diagnosis of T-cell lymphoma and to differentiate between reactive T-cell infiltrate and T-cell neoplasia. Furthermore this method can be used for the detection of minimal residual disease after specific therapy.

The TCR β gene is an additional target used for clonality analysis. The structure of the TCR β gene is more sophisticated than the TCR γ gene, it spans a sector of about 600 kb on chromosome 7q34 and is composed of the following segments (see Figure 3.9):

- 75–100 V β gene segments;
- two D genes: D β1 and D β2 gene segments;
- two J genes: J β1, composed of six segments and J β2, composed of seven segments;
- two C β gene segments: C β1 and C β2.

Rearrangement of the TCR β gene is similar to rearrangement of the IgH gene and includes the D–J and V–D steps. Because of the distribution of the gene segments within the TCR β locus, the D1 segment may join either J1 or J2 segments, whereas the D2 segments join only the J1 segment (Figure 3.9).

Special caution must be taken when evaluating cutaneous lymphoid lesions, as TCR clonal rearrangements may appear in different benign inflammatory cutaneous lesions, such as chronic dermatitis and many other reactive skin lymphoid infiltrate; consequently T-cell clonality cannot be taken as an absolute criterion for cutaneous lymphoma. Nevertheless benign monoclonal cutaneous lymphocytic infiltrates are more likely to develop cutaneous lymphoma than polyclonal infiltrates, which usually continue to have a benign clinical course.

Another factor to consider in clonality analysis of T-cell lymphoma is that the monoclonal proliferation of T-lymphocytes can be observed as a physiologic condition in healthy persons, especially individuals older than 70 years (mainly CD8+ phenotype) or as an immunological reaction due to viral infection (so-called T-cell expansion of undetermined significance).

Method The method used is DNA extraction from blood, bone marrow or tissue (fresh or formalin-fixed paraffin-embedded tissues), followed by PCR (multiplex PCR) using specific primers. To provide a better discrimination of the monoclonal bands it is recommended to run out the PCR product by polyacrylamide gel electrophoresis. High-resolution electrophoresis is essential to analyze small fragments of the rearranged of TCR γ chain with slight variations. A heteroduplex assay for the analysis of the PCR product improves the results and minimizes the possibility of false-positive and false-negative results due to improper primer annealing.

Specimens containing a monoclonal T-cell population generate equal DNA fragments exhibiting the same size, which demonstrates one or more additional sharp bands on the electrophoresis gel. Generated monoclonal bands present distinctive DNA fragments after digestion by restriction enzymes, usually different from the non-rearranged germ line cells. Specimens containing a polyclonal T-cell population generate an undefined smear or ladder-like pattern on electrophoresis gel.

Optimal primer selection: For the TCR rearrangement, analysis different forward primers designed to match some or all-11 variable regions and reverse primers designed to match the joining regions are used. It is recommended to begin with a simple screening primer set, which includes one or two forward and one or two reverse primers. In case of negative results, a complete primer set containing variable primers matching different V and J regions can be helpful to indicate possible monoclonality. Remarkable is the increased number of false-positive results using primers matching the V γ9 region only, probably due to the presence of cell clones with minimal junctional diversity. The BIOMED-2 primer panels listed below is an efficient standardized method, which can be successfully used for the clonality study of related lymphoid malignancies. Another factor to consider targeting the TCR γ is the generation of monoclonal bands in rare cases of B-cell lymphoma or ALL (so-called dual or inappropriate rearrangements or lineage infidelity). This phenomenon is described in both B- and T-cell lymphomas and makes it necessary to carry out the monoclonality analysis of both IgH and TCR simultaneously. Finally, it is also important to use the adequate quantity and quality of DNA as described in the previous section.

- Primer set for the simple amplification of *TCR* γ gene:
  FP (V γ): 5′-AGG.GTT.GTG.TTG.GAA.TCA.GG-3′
  RP (J γ): 5′-CGT.CGA.CAA.CAA.GTG.TTG.TTC-3′

PCR generates a clonal band 160–190 bp in size on electrophoreses gel.

- Primer set for the amplification of multiple regions within the *TCR* γ gene:

  1. FP (V γ11): 5′-TCT.GG(G/A).GTC. TAT.TAC.TGT.GC-3′

2. FP (V γ101): 5′-CTC.ACA.CTC.(C/T)CA.CTT.C̀-3
3. RP (J γJ12): 5′-CAA.GTG.TTG.TTC.CAC.TGC.C-3′
4. RP (J γJp2): 5′-GTT.ACT.ATG.AGC.(T/C)TA.GTC.C-3′

V γ 11 consensus primer to V γ1–8 regions; V γ101 consensus primer to V γ10–11 regions;

J γ J12 consensus primer to J1 and J2 regions; J γ Jp2 consensus primer to Jp1 and Jp 2.

PCR carried out in two reactions, first reaction using both V forward primers (primer 1 and 2) with one J reverse primer (primer 3). This reaction generates a band on electrophoreses 75–95 bp in size, second reaction using the same V forward primers with the other J reverse primer (primer 4). This reaction generates a monoclonal band (or two bands) on electrophoreses 80–110 bp in size.

- Complete primer set matching different V and J regions of the *TCR* γ gene. This primer set is usually used as multiplex PCR and recommended if simple screening primer sets reveal insufficient results.

1. FP (V γ2): 5′-AGC.GTC.TTC.AGT.ACT.ATG.ACT.CCT.ACA.ACT.CCA.A-3′
2. FP (V γ3): 5′-CCA.CAG.CGT.CTT.CTG.TAC.TAT.GAC.GTC.TCC.ACC.G-3′
3. FP (V γ4): 5′-GCG.TCT.TCT.GTA.CTA.TGA.CTC.CTA.CAC.CTC.CAG.C-3′
4. FP (V γ5): 5′-ATC.TTC.TGC.ACT.ATG.AAG.TCT.CCA.ACT.CAA.GGG.A-3′
5. FPV γ8): 5′-CTT.CTG.TAC.TAT.GAC.TCC.TAC.AAC.TCC.AGG.GTT.G-3′
6. FP (V γ9): 5′-GAC.GGC.ACT.GTC.AGA.AAG.GAA.TCT.GGC.ATT.CCG-3′
7. FP (V γ10): 5′-GTC.TCA.ACA.AAA.TCC.GCA.GCT.CGA.CGC.AGC.ATG.G-3′
8. FP (V γ11): 5′-TCT.TGA.CAA.TCT.CTG.CTC.AAG.ATT.GCT.CAG.GTG.G-3′

1. RP (J γ12): 5′-AAG.TGT.TGT.TCC.ACT.GCC.AAA-3′
2. RP (J γ3): 5′-AGT.TAC.TAT.GAG.C(T/C)T.AGT.CCC-3′
3. RP (J γ4): 5′-TGT.AAT.GAT.AAG.CTT.TGT.TCC.

J γ 12 primer consensus to J 1 and J 2 regions; J γ 3 primer consensus to Jp 1andJp 2 regions; J γ 4 primer consensus to Jp 2 region. Multiplex PCR is carried out with the 8 forward primers and the two J γ 12 and J γ 3 reverse primers. In case of negative or doubtful rearrangement results the additional J γ 4 reverse primer consensus to the rarely rearranged Jp 2 region can be used in a second multiplex PCR. In the case of a monoclonal cell population, PCR generates one or two monoclonal bands 160–240bp in size.

- *BIOMED-2* primers consensus to the *V* and *J* regions of the *TCR* γ gene. Multiplex PCR is carried out in two tubes (A and B) each contains two forward primers for the V regions and both J reverse primers.

  –*Tube A:*
  FP (V γfl): 5′-GGA.AGG.CCC.CAC.AGC.(A/G)TC.TT-3′
  FP (V γ10): 5′-AGC.ATG.GGT.AAG.ACA.AGC.AA-3′
  –*Tube B:*
  FP (V γ9): 5′-CGG.CAC.TGT.CAG.AAA.GGA.ATC-3′
  FP (V γ11): 5′-CTT.CCA.CTT.CCA.CTT.TGA.AA-3′

–*Reverse primers:*

RP J γ1.1/2.1 (J p1/2): 5′-CGA.GTA.TCA.TTG.AAG.CGG.ACC.ATT-3′
RP J γ1.3/2.3 (J γ1/2): 5′-GAG.AAA.CCG.TCA.CCT.TGT.TGT.G-3′

The V γ fl FR is a degenerate primer. A monoclonal cell population generates a clonal band 100–300 bp in size.

- Primers for the detection of *TCR* β gene rearrangement:
  FP (D β1): 5′-CAA.AGC.TGT.AAC.ATG.TGG.GGA.C-3′
  RP (J β2): 5′-AGC.ACC.GTG.AGC.CTG.GTG.GC-3′

If a monoclonal cell population present, PCR generates a clonal band 80–120 bp in size.

- Primers for the detection of *TCR* δ gene rearrangement. Multiplex PCR is carried out in two different tubes. The first reaction using the first primer set (primers 1 and 2). The second reaction with the second primer set (primers 3 and 4).

1. FP (V δ2): 5′-TGG.CCC.TGG.TTT.CAA.AGA.CAA.TTT.CCA-3′
2. RP (D δ3.2): 5′-TGA.GGA.TAT.CCC.AGG.GAA.ATG.GCA.CTT-3′
3. FP (V δ1): 5′-CGC.CTT.AAC.CAT.TTC.AGC.CTT.ACA.GCT.AGA.A-3′
4. RP (J δ1): 5′-CCT.TAA.CCT.TAA.ACT.TCA.GAT.AAA.TAA.ATG.AGT.TAC-3′

If a monoclonal cell population is present, PCR generates a 280-bp clonal band as a monoclonal electrophoresis pattern.

- *BIOMED-2* primers consensus to the *V* and *J* regions for the detection of the *TCR* δ gene rearrangement. PCR is carried out in one tube with all the V and J primers and able to screen the complete V–D–J rearrangements but not the partial V–D, D–D and D–J rearrangements.
  FP (V δ1): 5′-ATG.CAA.AAA.GTG.GTC.GCT.ATT-3′
  FP (V δ2): 5′-ATA.CCG.AGA.AAA.GGA.CAT.CTA.TG-3′
  FP (V δ3): 5′-GTA.CCG.GAT.AAG.GCC.AGA.TTA-3′
  FP (V δ4): 5′-ATG.ACC.AGC.AAA.ATG.CAA.CAG-3′
  FP (V δ5): 5′-ACC.CTG.CTG.AAG.GTC.CTA.CAT-3′
  FP (V δ6): 5′-CCC.TGC.ATT.ATT.GAT.AGC.CAT-3′
  RP (J δ1): 5′-CTT.GGG.CAC.ACT.GAC.ACC.TTG-3′
  RP (J δ2): 5′-CTT.GTG.TTG.AGT.AGC.ACC.TTG-3′
  RP (J δ3): 5′-GAG.AAG.CAC.CTC.GGG.GCA.CTC-3′
  RP (J δ4): 5′-CCT.TGG.ATA.GAC.CTC.CAT.GTT-3′

A monoclonal electrophoresis with monoclonal bands is generated if a monoclonal cell population is present.

3.5 Diagnosis of Tumors, Micrometastases, Isolated and Circulating Tumor Cells by Molecular Detection of Tissue-Specific Gene Expression

3.5.1 Introduction

Another approach for the diagnosis and classification of solid tumors and their metastases is the detection of the tissue-specific gene expression encoding tissue-specific antigens and oncofoetal antigens by molecular detection of selected complementary mRNA. The concentration of many tissue-specific gene products can be measured in serum and in many cases they are used as tumor markers, well known examples are PSA, AMV, NSE and CEA, which are now widely used for monitoring of tumors. Table 3.13 includes the most common tissue-specific molecular markers used for the diagnosis of related tumors.

These tumor markers are transcriptional active in the original tissue and usually in the malignant cells originated from this tissue but not in the surrounding tissue harboring metastatic tumor tissue. The molecular detection of complementary mRNA is more sensitive than traditional antibody-based methods such as ELISA and immunohistochemistry and can be used as a sensitive molecular tool, ideal for the detection of minimal residual tumor cells, micrometastases and circulating tumor cells originated from primary tumors with tissue-specific antigens. The detection of complementary mRNA can be also used as an additional marker in the primary histopathologic diagnosis of tumors and for the identification of metastases of unknown origin.

The molecular detection of micrometastases is more sensitive and selective than the classic histopathologic examination of lymph nodes. Many studies showed that about one-third of breast cancer patients with histological negative lymph nodes develop recurrent disease due to the presence of lymph node micrometastases or isolates tumor cells.

Lymph node staging and mapping of lymphatic routes by examination of the sentinel lymph nodes using a RT-PCR-based method found to be a better predictor of possible recurrence of various tumors and fundamental for the success of the therapy. Molecular screening of sentinel lymph nodes is now frequently used for the staging of some tumors like melanoma and breast cancer.

Finally, PCR- and RT-PCR-based methods are very sensitive methods and a minimal contamination of samples can cause false-positive results. For example, the contamination of blood or bone marrow samples by epidermal epithelial cells during sample collection can cause false-positive results when targeting squamous cell carcinoma. The contamination of the samples by non-epithelial cells of the skin such as melanocytes and Merkel cells can also cause false-positive results in melanoma, different neuroendocrine tumors and neuroblastoma. Other reasons for false-positive cells can be caused by ectopic tissue, which may be presented in lymph nodes or other locations. A further possible cause for false-positive results is the illegitimate transcription of genes (illegitimate transcription is caused by the possibility of the

Table 3.13 Tissue-specific molecular markers used for the diagnosis and monitoring of tumors.

Thyroid carcinoma (papillary and follicular carcinoma)	Thyroglobulin (Tg), thyroid peroxidase, thyroid transcription factor-1 (TTF-1)
Thyroid carcinoma (medullary carcinoma)	Calcitonin, thyroid transcription factor-1 (TTF-1), chromogranin (CgA), neuron-specific enolase (NSE), carcinoembryonic antigen (CEA)
Squamous cell carcinoma (head and neck, lung)	Squamous cell carcinoma antigen (SCCA), CK19
Non-small cell lung carcinoma	CK8, CK19, CEA, TTF-1, MUC-1
Breast carcinoma	Carcinoembryonic antigen (CEA), MUC-1, mammaglobin, CK8, CK19
Adenocarcinoma of gastrointestinal tract	Guanylyl cyclase C (GCC), carcinoembryonic antigen (CEA), MUC-1, CK19, CK20, CDX-2
Adenocarcinoma of pancreas and bile ducts	Carcinoembryonic antigen (CEA), CK19, CK20
Hepatocellular carcinoma	Alpha fetoprotein (AFP), albumin, MAGE-1
Gynecological adenocarcinoma (uterus and ovaries)	MUC-1, CK8, CK19, Carcinoembryonic antigen (CEA)
Squamous cell carcinoma of uterine cervix	Squamous cell carcinoma antigen (SCC A), HPV
Carcinoma of prostatic gland	Prostate-specific antigen (PSA), prostate-specific membrane antigen (PSMA), PTI-1, CK19
Transitional cell carcinoma of urinary tract	CK19, CK20, uroplakin
Malignant melanoma	Tyrosinase, P97, MUC 18, MART-1 (melan A), MAGE-3
Neuroblastoma	Tyrosine hydroxylase, PGP 9.5, MAGE, GAGE
Neuroendocrine tumors and small cell carcinoma of the lung	Chromogranin (CgA), neuron-specific enolase (NSE), choline acetyltransferase (CAT), secretogranin (SgII), protein gene product 9.5 (PGP9.5)
Rhabdomyosarcoma	Specific translocations, MyoD1, myogenin, myoglobin, fetal type acetylcholine receptor
Germ cell tumors (embryonal carcinoma, yolk sac tumor)	Alpha fetoprotein (AFP)

transcription of any gene in any cell type) or by the presence of pseudogenes in the background cells. To minimize the error incidence, it is recommended to use more than one molecular marker (primer panel with multiplex PCP), followed by second nested PCR. The immediate denaturation of the samples and inactivation of cell enzymes is an important step to prevent the illegitimate transcription. The use of real-time PCR is also another way to minimize false-positive results.

3.5.2 Detection of Micrometastases and Circulating Tumor Cells of Prostatic Carcinoma by Amplification of Prostate-Specific Antigen, Prostate-Specific Membrane Antigen and PTI-1 mRNA

Adenocarcinoma of the prostatic gland is one of the most common carcinomas of adult males diagnosed with increasing frequency because of public screening by measuring of the prostate-specific antigen (PSA) levels in serum. Prostate-specific antigen (PSA) and prostate-specific membrane antigen (PSMA) are two important antigens specific for the prostatic gland, which are widely used for the diagnosis and screening of prostatic malignancies.

- *Prostate-Specific Antigen*: Prostate-specific antigen (PSA) is a single chain protein of the kallikrien family of serine proteases with a molecular mass of 34 kDa, mainly expressed by prostatic epithelial cells. The conventional measuring of PSA serum levels is an approved assay for the diagnosis and monitoring of prostate carcinoma but PSA levels may also show abnormal elevation by inflammatory processes of the prostatic gland or after digital examination.

  Adenocarcinoma of the prostate metastasizes to the draining pelvic lymph nodes after invasion of lymph vessels and to bone or bone marrow after invasion of blood vessels of the vertebral plexus situated near the prostatic gland, initiating systemic hematogenous dissemination. Neither of the clinically common used methods for early staging of prostatic carcinoma like CT, MRI and transrectal sonography are accurate enough to detect microscopic capsular penetration or micrometastases after confirmation of the diagnosis by needle biopsy, which may be the reason for under-staging of the prostatic cancer leading to an inadequate therapy or unnece-ssary radical prostatectomy, as a result 20–50% of patients found to develop an extraprostatic recurrence despite radical prostatectomy. Furthermore, the Gleason score for grading of prostatic adenocarcinoma, which is also an important prognostic factor, evaluated on needle biopsy is a subject to upgrading in up to 50% of cases after radical operation (upgrading from <7 to ≥ 7). All these factors signify that more accurate pretherapeutic staging tests are needed.

  Many studies show that the use of PCR-based molecular methods to amplify PSA-mRNA as molecular marker offer a sensitive assay for the detection of extraprostatic PSA synthesizing cells suitable for monitoring and detection of micrometastases and circulating tumor cells originated from prostatic carcinoma. The use of PSA as a molecular target proved to be more sensitive than the amplification of other prostate-specific tissue markers like prostate-specific membrane antigen (PSMA) or human glandular kallikrien (a member of the kallikrien family of serine proteases with trypsin like activity).

- *Prostate-Specific Membrane Antigen*: Prostate-specific membrane antigen (PSMA) is a type II integral membrane glycoprotein expressed by prostatic epithelial cells.

Ectopic (illegitimate) expression of PSMA-mRNA in background blood cells makes PSMA an inaccurate molecular marker for prostate carcinoma, but it can be used in combination with PSA.

- *Prostate Tumor-Inducing Gene-1*: Prostate tumor-inducing gene-1 (PTI-1) is an additional marker for prostate carcinoma. The PTI-1 gene is proved to be a fraction of the human genome, a part of which found to have a high similarity to the mycoplasma hypopneumoniae 23S ribosomal RNA and seems to be an oncogene involved in the malignant transformation and growth control of prostatic carcinoma. The PTI-1 gene encodes a 398-amino-acid protein expressed in the majority of prostatic carcinoma (about 90%) but not in normal prostatic tissue or in benign prostatic hyperplasia. The presence of PT-1 mRNA is also discovered in some cell lines of breast, lung and colon carcinoma. Since only few studies reported the used of PTI-1 as a suitable marker for prostatic carcinoma, further investigations are needed to decide the specificity of this gene for prostate cancer or other malignancies.

The presence of circulating tumor cells in peripheral blood or bone marrow arising from a carcinoma of the prostate correlates with capsular penetration, metastatic spread or positive surgical margins after surgical treatment. The molecular amplification of PSA mRNA or other prostate-specific mRNA can be used as an accurate method for the preoperative staging. For this purpose EDTA blood samples must be collected from patient suspicious to have prostatic cancer few days before digital rectal examination or needle biopsy. To avoid false-positive results, blood samples can be suspended in guanidine thiocyanate solution and conserved for later RNA extraction after pathological confirmation of neoplasia. This method is also useful for the screening of pelvic surgical margins after radical prostatectomy taking in consideration that about 20% of patients with histologically negative surgical margins have local relapse. For molecular screening of surgical margins, tissue specimens are obtained from four different areas of prostatic fossa including urethra, bladder neck, left and right lateral lobes. The presence of tumor cells within the surgical margins can be an indication for early radiotherapy after radical prostatectomy. The molecular method is also valuable as a postoperative method to monitor tumor progression or to detect micrometastases in pelvic lymph nodes. This assay is especially informative when the serum PSA is not significantly elevated due to chemotherapy or antiandrogen hormone treatment, which may reflect that a PSA synthesis is an androgen depending process. Studies reported that up to 10% of patients with progressive prostate cancer show no elevation of PSA serum levels, possibly because of the syntheses of defective PSA protein not recognized by immunological methods (e.g. ELISA) using specific antibodies to PSA molecules.

Finally, it must be mentioned that different published studies reported different and some times controversial results about the prognostic value of the PSA mRNA detection in blood, bone marrow and lymph nodes, which indicates that this method needs further standardization. However, despite conflicting results found by different authors, a positive molecular test is generally strongly linked to the tumor stage and a predictor of upstaging of a clinically prostate confined early stage cancer.

Method The method used is mRNA extraction from various tissue types (lymph node, surgical margins), peripheral blood, bone marrow or body fluids, followed by RT and nested PCR.

- Primers for the cDNA amplification of the *PSA* mRNA template:
  FP (exon 2): 5′-TCT.GCG.GCG.GTG.TTC.TG-3′
  RP (exon 3): 5′-GCC.GAC.CCA.GCA.AGA.TCA-3′

PCR generates a 68-bp DNA fragment.

- Alternative primers for the cDNA amplification of the *PSA* mRNA template:
  FP: 5′-AGG.CTG.GGG.CAG.CAT.TGA.ACC.AGA.GGA-3′
  RP: 5′-GTC.CAG.CGT.CCA.GCA.CAC.AGC.ATG.AAC.T-3′

PCR generates a 144-bp DNA fragment.

- Alternative primers for the cDNA amplification of the *PSA* mRNA template:
  FP: 5′-CCT.CCT.GAA.GAA.TCG.ATT.CCT-3′
  RP: 5′-CGT.CCA.GCA.CAC.AGC.ATG.AA-3′

PCR generates a 291-bp DNA fragment.

- Alternative primers for the cDNA amplification of the *PSA* mRNA template (nested PCR):
  –Outer primers:
  FP: 5′-GCC.TCT.CGT.GGC.AGG.GCA.GTC-3′
  RP: 5′-CAT.CAC.CTG.GCC.TGA.GGA.ATC-3′

PCR generates a 216-bp DNA fragment.

  –Nested primers:
  FP: 5′-CAC.TGC.ATC.AGG.AAC.AAA.AGC.GT-3′
  RP: 5′-CAT.CAC.CTG.GCC.TGA.GGA.ATC-3′

PCR generates a final 156-bp DNA fragment.

- Alternative primers for the cDNA amplification of the *PSA* mRNA template (nested PCR):
  –Outer primers:
  FR: 5′-GAT.GAC.TCC.AGC.CAC.GAC.CT-3′
  RP: 5′-CAC.AGA.CAC.CCC.ATC.CTA.TC-3′

PCR generates a 710-bp DNA fragment.

  –Nested primers:
  FP: 5′-GAT.ATG.TCT.CCA.GGC.ATG.GC-3′
  RP: 5′-GCA.AGT.TCA.CCC.TCA.GAA.GG-3′

Nested PCR generates a 455-bp DNA fragment.

- Primers for the cDNA amplification of the *PSMA* mRNA template (nested PCR):
  FP: 5′-AAA.AGT.CCT.TCC.CCA.GAG.TTC.AGT-3′
  RP: 5′-ACT.GTG.ATA.CAG.TGG.ATA.GCC.GCT-3′

PCR generates a 165-bp DNA fragment.

- Alternative primers for the cDNA amplification of the *PSMA* mRNA template (nested PCR):
  –Outer primers:
  FP: 5′-CAG.ATA.TGT.CAT.TCT.GGG.AGG.TC-3′
  RP: 5′-AAC.ACC.ATC.CCT.CCT.CGA.ACC-3′

PCR generates a 647-bp DNA fragment.

–Nested primers:
FP: 5′-CCT.AAC.AAA.AGA.GCT.GAA.AAG.CCC-3′
RP: 5′-ACT.GTG.ATA.CAG.TGG.ATA.GCC.GCT-3′
PCR generates a 234-bp DNA fragment.

- Alternative primers for the cDNA amplification of the *PSMA* mRNA template (nested PCR):
  –Outer primers:
  FR: 5′-ATG. GGT GTT.TGG.TGG.TAT.TGA.CC-3′
  RP: 5′-TGC.TTG.GAG.CAT.AGA.TGA.CAT.GC-3′
  PCR generates a 947-bp DNA fragment.
  –Nested primers:
  FP: 5′-ACT.CCT.TCA.AGA.GCG.TGG.CG-3′
  RP: 5′-AAC.ACC.ATC.CCT.CCT.CGA.ACC-3′

Nested PCR generates a 434-bp DNA fragment.

- Primers for the cDNA amplification of the *PTI-1* mRNA template:
  FP (UU): 5′-ACC.CGA.GAG.GGG.AGT.GAA.ATA-3′
  RP (UL): 5′-TGC.CGC.CAT.TCC.ACA.TTC.AGT-3′

PCR generates a 424-bp DNA fragment.

- Primers for the cDNA amplification of the *PTI-1* mRNA template (Bridge region):
  FP (BU): 5′-ATG.GGG.GTA.GAG.CAC.TGA.ATG-3′
  RP (BL): 5′-AAC.ACC.AGC.AGC.AAC.AAT.CAG-3′PCR generates a 252-bp DNA fragment.
- Alternative primers for the cDNA amplification of the *PTI-1* mRNA template:
  FR: 5′-GAG.CAC.GTT.AAA.GTG.TGA.TGG-3′
  RP: 5′-TAA.AGT.CTC.TGT.GTC.CTG.GGG-3′

PCR generates a 507-bp DNA fragment.

3.5.3 Detection of Micrometastases and Circulating Tumor Cells of Follicular and Papillary Thyroid Carcinoma by Amplification of Thyroglobulin, Thyroid Peroxidase and TTF-1 mRNA

Follicular and papillary thyroid carcinomas are the most common endocrine malignancies. Follicular, papillary and anaplastic carcinoma are derived from follicular epithelium of thyroid gland, but characterized by different histological appearance, different biological and clinical behavior and associated with different genetic abnormalities.

Papillary thyroid carcinoma is a tumor of thyroid follicular epithelium with papillary and follicular growth pattern and the characteristic nuclear features. Papillary thyroid carcinoma makes about 70% of all thyroid carcinomas, often associated with radiation exposure or high iodine intake, and mainly metastasizes by invasion of lymph vessels. Papillary thyroid carcinoma frequently associated with different rearrangements mainly involving the RET/PTC gene discussed in a previous section.

Follicular thyroid carcinoma is also a tumor of thyroid follicular epithelium exhibiting pure follicular differentiation and lacking the nuclear features characteristic for papillary carcinoma. Follicular thyroid carcinoma makes about 20% of all thyroid carcinomas, usually associated with endemic goiter or low iodine intake and mainly metastasizes by hematogenous spread. Follicular thyroid carcinoma frequently associated with genetic abnormalities involving the chromosomes 2, 3, 7 and 8 in addition to the t(2; 3)(q13; p25) translocation, discussed in a previous section.

The most important prognostic factor for all types of thyroid carcinoma is the TNM status including the invasion of blood and lymph vessels and metastatic spread. For the monitoring of thyroid carcinoma two methods are usually used: the nuclear body-scan with radioactive iodine and measuring of serum thyroglobulin, but we have to consider that both methods can reliably detect systemic metastatic spread or relapse of thyroid carcinoma only, if thyroid tissue has been completely removed.

Other important monitoring parameter for thyroid carcinoma is the presence of tumor cells in bone marrow or the circulation of tumor cells in peripheral blood. For the detection of these tumor cells, molecular methods based on the detection of tissue-specific gene expression are developed. The most efficient molecular marker is mRNA encoding to the thyroid-specific hormone thyroglobulin (Tg), a further target is mRNA encoding specific enzymes involved in the synthesis of thyroid hormones such as thyroid peroxidase (TPO). Thyroid transcription factor-1 (TTF-1) is another thyroid-specific marker, which can be used for the diagnosis of thyroid tumors.

- *The thyroglobulin gene*: the thyroglobulin gene is a large gene composed of 40 exons encoding a large glycosylated protein consisting of two identical subunits ~339 kDa each, and is a prohormone synthesized by thyroid follicular epithelial cells and their neoplasia but not in medullary carcinoma originated from C cells of thyroid gland.

- *Thyroid peroxidase*: thyroid peroxidase (TPO) is a membrane bounding protein catalyzing iodide oxidation and iodination of tyrosine residues to generate T3 and T4. This enzyme is encoded by a gene consists of 17 exons.

- *Thyroid transcription factor-1*: thyroid transcription factor-1 (TTF-1) is a tissue-specific transcription factor and a member of the homeodomain transcription family, expressed selectively in thyroid follicular epithelial cells in addition to thyroid C cells, diencephalon cells and epithelial cells of the lung, namely in type II alveolar cells and in non-ciliated bronchiolar epithelial cells (Clara cells). TTF-1 is a transcription factor regulating the transcription of thyroglobulin and thyroid peroxidase in follicular epithelium and surfactant in alveolar cells as well as Clara cell protein.

The amplification of thyroglobulin, thyroid peroxidase and TTF-1 mRNA can be a valuable sensitive method for the detection of micrometastases, minimal residual cancer cells and circulating tumor cells of papillary and follicular thyroid carcinoma. TTF-1 can be also used for the detection of thyroid medullary carcinoma and different lung neoplasia derived from alveolar cells type II and Clara cells (mainly adenocarcinoma). The presence of circulating tumor cells is associated with poor prognosis and can be used for the post-operative follow-up of patients with thyroid cancer. Finally, it is important to distinguish follicular, papillary or anaplastic thyroid carcinoma from medullary thyroid carcinoma. For this purpose the simultaneous amplification of calcitonin mRNA can be helpful. Calcitonin is a 32-amino-acid polypeptide functioning as a hormone, exclusively expressed by the C thyroid cells and related tumors.

Method The method used is mRNA extraction from different tissue types, blood and bone marrow, followed by RT-PCR.

- Primers for the cDNA amplification of the *Tg* mRNA template:
  FP: 5′-GTG.CCA.ACG.GCA.GTG.AAG.T-3′
  RP: 5′-TCT.GCT.GTT.TCT.GTA.GCT.GAC.AAA-3′

PCR generates an 87-bp DNA fragment.

- Alternative primers for the cDNA amplification of the *Tg* mRNA template:
  FP: 5′-CTG.CTG.GCT.CCA.CCT.TGT.TT-3′
  RP: 5′-CAG.GGC.GTG.GGG.CAT.TTC.TT-3′

PCR generates a 160-bp DNA fragment.

- Primers for the cDNA amplification of the *TPO* mRNA template:
  FP: 5′-CAC.CAA.AGG.CGG.CTT.CCA-3′
  RP: 5′-CAG.AGC.AGC.CAG.CGA.CAT-3′

PCR generates a 124-bp DNA fragment.

- Primers for the cDNA amplification of the *TTF-1* mRNA template (exon 1 and 2):
  FP: 5′-CCA.GGA.CAC.CAT.GAG.GAA.CA-3′
  RP: 5′-TGC.ACT.CGT.TCT.TGT.ACC.GG-3′

PCR generates a 149-bp DNA fragment.

- Alternative primers for the cDNA amplification of the *TTF-1* mRNA template:
  FP: 5′-CCA.GGA.CAC.CAT.GAG.GAA.CA-3′
  RP: 5′-GTC.GCT.CCA.GCT.CGT.ACA.C-3′

PCR generates a 236-bp DNA fragment.

- Primers for the cDNA amplification of the *calcitonin* mRNA template:
  FP: 5′-GGC.AGC.CTC.CAT.GCA.GCA.CC-3′
  RP: 5′-CCA.GGT.GCT.CCA.ACC.CC-3′

PCR generates a 291-bp DNA fragment.

- Alternative primers for the cDNA amplification of the *calcitonin* mRNA template:
  FP: 5′-GGA.CTA.TGT.GCA.GAT.GAA.GG-3′
  RP: 5′-CCT.CCC.ATC.TGA.AGT.TTG.AGA.C-3′

PCR generates a 401-bp DNA fragment.

3.5.4
Detection of Micrometastases and Circulating Tumor Cells of Hepatocellular Carcinoma by Amplification of Alpha-Fetoprotein and Albumin mRNA

Hepatocellular carcinoma (HCC) is a malignant tumor derived from hepatocytes or their precursors and one of the common malignant tumors, usually associated with HBV or HCV infection, liver cirrhoses, nutritional or storage disorders. Hepatocellular carcinoma is characterized by intrahepatic and hematogenic metastatic spread via venous invasion and systemic dissemination. Hepatocellular carcinoma shows a high incidence of relapse after curative resection or liver transplantation because of the presence of undetectable metastases before surgery, especially after immunosuppression to avoid rejection of transplanted liver. Alpha-fetoprotein (AFP) and albumin mRNA are two markers specific for liver tissue and hepatocellular neoplasia and can be used for the diagnosis and monitoring of hepatocellular carcinoma. MAGE-1 is a further marker useful for the diagnosis of hepatocellular carcinoma.

- *Alpha fetoprotein*: Alpha fetoprotein (AFP) is an oncofoetal 70-kDa glycoprotein encoded by a 17-kb gene located on chromosome 4. AFP is mainly expressed in fetal yolk sac, fetal liver and intestines but rapidly reduced after birth. Some tumors such as hepatocellular carcinoma and few other germ cell tumors such as yolk sac tumor and embryonal carcinoma exhibit a high AFP expression. Measuring of AFP levels in serum is an important marker for the monitoring of these tumors and has a half-life of 5–7 days. The detection of AFP mRNA expressing cells in different human samples can be employed as a molecular marker for these tumors, helpful for tumor diagnosis and tumor follow-up.
- *Albumin*: albumin is an important protein, mainly expressed by hepatocytes, encoded by a 19-kb gene located on chromosome 4q11–22. The amplification of albumin mRNA can also be used as a marker for hepatocytes and for tumors of hepatocytic origin.
- *MAGE-1 (melanoma antigen)*: MAGE-1 is a member of the MAGE family known as a tumor rejection antigen, expressed in many tumors and in more than 50% of hepatocellular carcinoma and malignant melanoma. Antibodies to MAGE-1 are used as an immunohistochemical marker for these tumors. The amplification of MAGE-1 mRNA can be used as a molecular marker for tumors expressing this antigen (further members of this tumor family and the sequences of corresponding primers are listed and discussed in the melanoma section). Note that antibodies targeting the MAGE-1 molecule were recently used as a specific immunotherapy against these tumors.

The preoperative molecular detection of neoplastic cells from hepatocellular carcinoma in peripheral blood by the amplification of the AFP and albumin mRNA can be a useful method for the selection of transplant recipients so as to minimize the frequency of relapse after liver transplantation. This method can also be used for the primary diagnosis and monitoring of this malignancy. The amplification of MAGE-1 antigen can also be useful for the selection of patients for immunotherapy.

Method The method used is mRNA extraction from tissue, blood, bone marrow and body fluids, followed by RT-PCR.

- Primers for the cDNA amplification of the *AFP* mRNA template (nested PCR):
  –Outer primers:
  FP: 5′-ACT.GAA.TCC.AGA.ACA.CTG.CAT.AG-3′
  RP: 5′-TGC.AGT.CAA.TGC.ATC.TTT.CAC.CA-3′

First PCR generates a 176-bp DNA fragment.

–Nested primers:

FP: 5′-TGG.AAT.AGC.TTC.CAT.ATT.GGA.TTC-3′

RP: 5′-AAG.TGG.CTT.CTT.GAA.CAA.ACT.GG-3′

Nested PCR generates a 101-bp DNA fragment.

- Alternative primers for the cDNA amplification of the *AFP* mRNA template (nested PCR):

–Outer primers:

FP: 5′-AAC.TAG.CAA.CCA.TGA.AGT.GG-3′

RP: 5′-CAT.GGC.AAA.GTT.CTT.CCA.GA-3′

–Nested primers:

FP: 5′-CCA.ATG.TAC.TGC.AGA.GAT.AAG-3′

RP: 5′-TCT.AAA.CAC.CCT.GAA.GAC.TG -3′

Nested PCR generates a 159-bp DNA fragment.

- Primers for the cDNA amplification of the *albumin* mRNA template (nested PCR):

–Outer primers:

FP: 5′-AGA.AAG.TAC.CCC.AAG.TGT.CAA-3′

RP: 5′-AGC.TGC.GAA.ATC.ATC.CAT.AAC-3′

–Nested primers:

FP: 5′-ACT.ATC.TAT.CCG.TGG.TCC.TGA-3′

RP: 5′-TCT.TGA.TTT.GTC.TCT.CCT.TCT-3′

Nested PCR generates a 222-bp DNA fragment.

- Alternative primers for the cDNA amplification of the *albumin* mRNA template:

FP: 5′-CTT.GAA.TGT.GCT.GAT.GAC.AGG-3′

RP: 5′-GCA.AGT.CAG.CAG.GCA.TCT.CAT.C-3′

PCR generates a 157-bp DNA fragment.

3.5.5
Detection of Micrometastases and Circulating Tumor Cells of Malignant Melanoma by Amplification of mRNA of Melanoma-Associated Antigens: Tyrosinase, P97, MUC-18, HMB45, Melan-A, MAGE and GAGE Family Members

Malignant melanoma is a heterogeneous highly malignant tumor arising from melanocytes derived from the neural crest and migrating to the periphery during fetal life. Cutaneous melanoma arises from neoplastic proliferation of melanocytes in the dermo-epidermal junction, able to develop systemic metastases during its vertical growth. Other types of malignant melanoma, like intraocular and mucosal melanoma, and including oral and genital melanoma, have the same histogenetic origin and identical clinical behavior.

The most important prognostic factors for malignant melanoma considered being the tumor thickness i.e. the vertical growth classified by CLARK level, histological type, number of positive lymph nodes and the presence of distant metastases.

Another prognostic factor corresponding to the other factors is the systemic dissemination of tumor cells including the presence of tumor cells in peripheral blood or bone marrow, which is an important feature for the majority of melanoma types correlated with bad prognosis and poor clinical outcome.

Malignant melanoma is associated with a broad spectrum of cytogenetic abnormalities, which frequently involve chromosomes 1, 5 and 6. The t(12; 22) translocation is not a feature for malignant melanoma but is a characteristic genetic abnormality for clear cell sarcoma described in a prior section.

The detection of melanoma cells in peripheral blood or bone marrow can be managed by the conventional histopathology supported by immunohistochemistry or by the molecular amplification of mRNA of melanoma-associated markers. Most of the melanoma-associated markers are enzymes involved in the melanin synthesis pathway or melanoma-associated antigens such as tyrosinase, P97, HMB45, MUC-18, Melan A (MART-1), MITF (microphthalmia transcription factor), MAGE and GAGE family. The majority of these antigens are also expressed in clear cell sarcoma and the antibodies to these antigens are important markers used for the immunohistochemical diagnosis. The antigens listed below are the most common targets for molecular diagnosis of melanoma.

- *Tyrosinase*: tyrosinase is an important protein, specific for all melanocytes and melanocytic tumors and is one of the enzymes involved in the syntheses of melanin catalyzing the conversion of tyrosine to dopa and of dopa to dopaquinone.

- *HMB45*: HMB45 is a melanosomal matrix glycoprotein, which is an intracytoplasmic melanoma-associated antigen encoded by the gp100 gene (also known as Pmel 17) located on chromosome 12q13–14. The HMB45 antigen is highly specific for melanocytes and melanocytic tumors and the expression of this antigen indicates a melanocytic differentiation. Specific monoclonal antibodies to HMB45 are widely used in the immunohistochemistry for the diagnosis melanoma.

- *P97*: P97 is a glycoprotein encoded by the P97 gene. The P97 glycoprotein is a melanoma-associated antigen known as melanotransferrin.

- *MUC-18*: MUC-18 is a glycoprotein encoded by the MUC-18 gene. The expression of MUC-18 shows a direct correlated with the metastatic activity of melanoma.

- *Melan A*: Malan A is a protein encoded by the MART-1 gene. Melan A is expressed in melanin producing cells and considered as a specific melanocyte differentiation and melanoma marker.

- *MAGE and GAGE*: MAGE is a gene family consists of more than 12 closely related genes located on chromosome Xq and transcribed in many malignant tumors including hepatocellular carcinoma, germ cell tumors and melanoma. The GAGE gene family encodes another tumor-associated antigens expressed in many tumors such as malignant melanoma, neuroblastoma, and other carcinomas and sarcomas but not in normal tissue with the exception of testicular tissue. Both of MAGE and GAGE gene families encode different tumor-associated antigenic peptides recognized by the autologous cytotoxic T-lymphocytes.

MAGE-3 and GAGE are transcribed in 50–70% of melanoma cells but usually not in normal background cells or melanocytes and reported as specific melanoma markers. The MAGE proteins expressed by tumor cells are potential targets for specific immunotherapy, which may be effective in some malignancies such as melanoma and hepatocellular carcinoma.

The molecular detection of above-mentioned melanoma-associated antigens can be used as a complementary method in the primary histopathologic diagnosis of the malignant melanoma or for the classification of metastases of unknown origin. It is also useful for detection of submicroscopic micrometastases in bone marrow, regional or sentinel lymph nodes and even in surgical margins using mRNA extracted from frozen or formalin-fixed specimens, followed by RT-PCR. The amplification of melanoma-associated antigens mRNA is described by many researchers to be a specific and sensitive molecular test used for the primary diagnosis, early detection and monitoring of malignant melanoma in addition to the follow-up of hematogenous dissemination of melanoma cells, even in early stages of the disease. It can be also used for monitoring of tumor progression after surgical treatment and during therapy for the evaluation of treatment efficiency.

The early dissemination of melanoma cells, even in superficial melanomas makes the detection of melanoma cells by the molecular amplification of melanoma markers an unsuitable method for an accurate melanoma staging, however the presence of circulating melanoma cells correlates with the clinical stage and considered as an independent prognostic factor for different tumors including the majority types of malignant melanoma. Clinical observations and molecular analysis reported that the presence of melanoma cells in bone marrow and peripheral blood is a risk factor for short-term recurrence, rapid progression and associated with poor prognosis.

Despite the high sensitivity and specificity of PCR-based molecular methods, false-positive and false-negative results are reported. Source for false-positive results can be the contamination by skin melanocytes during blood or bone marrow collection or the temporary circulation of skin melanocytes after surgical intervention or due to the presence of nodal melanocytic nevi, when regional or sentinel lymph nodes are examined.

Malignant melanoma is a highly malignant heterogeneous tumor that can shows different genetic features and different antigen expression pattern, which may lead to false-negative results. To avoid possible false-negative or false-positive results it is recommended to collect two venous blood samples and to use the second sample for the mRNA extraction as the first potion may contain skin melanocytes (see also the other notes mentioned in the introduction of the next chapter). The use of different primer sets matching more than one melanoma-associated marker is recommended to minimize false results; this assay can also be carried out as multiplex PCR. Best results are achieved by the amplification of tyrosinase mRNA simultaneously with melanoma-associated markers as MAGE-3 and GAGE.

Method The method used is mRNA extraction from different tissue types, peripheral blood or bone marrow, followed by RT-PCR.

- Primers for the cDNA amplification of the *tyrosinase* mRNA template (nested PCR):
  –Outer primers:
  FP: 5′-TTG.GCA.GAT.TGT.CTG.TAG.CC-3′
  RP: 5′-AGG.CAT.TGT.GCA.TGC.TGC.T-3′

First PCR generates a 284-bp DNA fragment.
  –Nested primers:
  FP: 5′-GTC.TTT.ATG.CAA.TGG.AAC.GC-3′
  RP: 5′-GCT.ATC.CCA.GTA.AGT.GGA.CT-3′

Nested PCR generates a 207-bp DNA fragment.

- Primers for the cDNA amplification of the *HMB45* mRNA template:
  FP: 5′-CTG.TGC.CAG.CCT.GTG.CTA.C-3′
  RP: 5′-CAC.CAA.TGG.GAC.AAG.AGC.AG-3′

PCR generates a 334-bp DNA fragment.

- Primers for the cDNA amplification of the *P 97* mRNA template (nested PCR):
  –Outer primers:
  FP: 5′-TAC.CTG.GTG.GAG.AGC.GGC.CGC.CTC-3′
  RP: 5′-AGC.GTC.TTC.CCA.TCC.GTG.T-3′

First PCR generates a 286-bp DNA fragment.
  –Nested primers:
  FP: 5′-CGG.TGA.TGG.GCT.GCG.ATG.TA-3′
  RP: 5′-CCA.GTA.CCG.TGC.TGT.GCT.TC-3′

Nested PCR generates a 239-bp DNA fragment.

- Primers for the cDNA amplification of the *MUC-18* mRNA template (nested PCR):
  –Outer primers:
  FP: 5′-CCA.AGG.CAA.CCT.CAG.CCA.TGT.C-3′
  RP: 5′-CTC.GAC.TCC.ACA.GTC.TGG.GAC.GAC.T-3′

First PCR generates a 437-bp DNA fragment.
  –Nested primers:
  FP: 5′-GTC.ATC.TTC.CGT.GTG.CGC.CA-3′.
  RP: 5′-GTA.GCG.ACC.TCC.TCA.GGC.TCC.TTA.C-3′

Nested PCR generates a 264-bp DNA fragment.

- Primers for the cDNA amplification of the *MART-1* mRNA template (nested PCR):
  –Outer primers:
  FP: 5′-ACT.GCT.CAT.CGG.CTG.TTG-3′
  RP: 5′-TCA.GCA.TGT.CTC.AGG.TG-3′

First PCR generates a 268-bp DNA fragment.
  –Nested primers:
  FP: 5′-GGA.TAC.AGA.GCC.TTG.ATG.G-3′
  RP: 5′-TCT.CGC.TGG.CTC.TTA.AGG-3

Nested PCR generates a 213-bp DNA fragment.

- Primers for the cDNA amplification of the *MAGE-1* mRNA template (nested PCR):

–Outer primers:
FP: 5′-GTA.GAG.TTC.GGC.CGA.AGG.AAC-3′
RP: 5′-CAG.GAG.CTG.GGC.AAT.GAA.GAC-3′

First PCR generates a 149-bp DNA fragment.

–Nested primers:
FP: 5′-TAG.AGT.TCG.GCC.GAA.GGA.AC-3′
RP: 5′-CTG.GGC.AAT.GAA.GAC.CCA.CA-3

Nested PCR generates a 143-bp DNA fragment.

- Alternative primers for the cDNA amplification of the *MAGE-1* mRNA template:
FP: 5′-CGG.CCG.AAG.GAA.CCT.GAC.CCA.G-3′
RP: 5′-GCT.GGA.ACC.CTC.ACT.GGG.TTG.CC-3′

PCR generates a 421-bp DNA fragment.

- Primers for the cDNA amplification of the *MAGE-2* mRNA template:
FP: 5′-AAG.TAG.GAC.CCG.AGG.CAC.TG-3′
RP: 5′-GAA.GAG.GAA.GAA.GCG.GTC.TG-3′

PCR generates a 230-bp DNA fragment.

- Primers for the cDNA amplification of the *MAGE-3* mRNA template:
FP: 5′-GAA.GCC.GGC.CCA.GGC.TCG-3′
RP: 5′-GGA.GTC.CTC.ATA.GGA.TTG.GCT.CC-3′

PCR generates a 423-bp DNA fragment.

- Primers for the cDNA amplification of the *MAGE-4* mRNA template:
FP: 5′-GAG.CAG.ACA.GGC.CAA.CCG-3′
RP: 5′-AAG.GAC.TCT.GCG.TCA.GGC-3′

PCR generates a 446-bp DNA fragment.

- Primers for the cDNA amplification of the *GAGE* mRNA template:
FP: 5′-AGA.CGC.TAC.GTA.GAG.CCT-3′
RP: 5′-CCA.TCA.GGA.CCA.TCT.TCA-3′

PCR generates a 239-bp DNA fragment.

3.5.6
Detection of Micrometastases and Circulating Tumor Cells of Adenocarcinomas by Amplification of Carcinoembryonic Antigen mRNA

Carcinoembryonic antigen (CEA, also clustered as CD66e) is an oncofoetal heavily glycosylated protein with a molecular weight of 180 kDa, encoded on chromosome 19q13.2. CEA is expressed in both normal mucosa and epithelial tumors arising from various mucosa types, including adenocarcinomas of the gastrointestinal tract, especially adenocarcinoma of colon in addition to many other different adenocarcinomas such as adenocarcinoma of the breast, endometrium, ovaries, lung, bile ducts and pancreas.

CEA level in serum is a widely accepted tumor marker used for the diagnosis of various types of adenocarcinoma and for the follow-up of tumor progression and metastatic activity as well as the monitoring of the efficiency of chemotherapy. CEA serum level is elevated in the majority of colorectal adenocarcinoma and in 40–75% of gastric, pancreatic, lung and breast adenocarcinoma. CEA is not expressed by prostatic glandular epithelia, thyroid follicular cells, mesothelial cells or tumors derived from these tissue types.

The molecular detection of CEA mRNA expressing cells is more sensitive than the conventional measuring of CEA levels in serum for the monitoring of tumor progression and metastatic activity. It is also successfully used for the detection of tumor cells in body fluids, micrometastases and minimal residual cells of related tumors. It is also more efficient than the immunohistochemical detection of CEA- or cytokeratin-positive cells in bone marrow or peripheral blood. Many studies showed that the molecular amplification of CEA mRNA can detect one CEA-expressing cell in up to 10^5 background cells, which leads to an earlier diagnosis, improved tumor staging and early specific treatment of tumors with subclinical and submicroscopic tumor manifestation.

Few cases of false-positive results were reported targeting CEA mRNA, may be due to the presence of migrating macrophages containing phagocytosed cancer cells or due to the presence of mRNA of highly homologous genes expressed in myeloid cells. Second factor can be avoided by the construction of specific primers, which are not complementary with any other homologous genes. The use of nested PCR is also another way to avoid non-specific target amplification.

Method The method used is mRNA extraction from different tissue types, peripheral blood, body fluids or bone marrow, followed by PCR and confirmed by nested PCR.

- Primers for the cDNA amplification of the *CEA* mRNA template:
  FP: 5′-CTG.GCC.GCA.ATA.ATT.CCA.TAG -3′
  RP: 5′-CCA.GCT.GAG.AGA.CCA.GGA.GAA -3′

PCR generates a 73-bp DNA fragment.

- Alternative primers for the cDNA amplification of the *CEA* mRNA template:
  FP: 5′-AAC.TGG.TGT.CCC.GGA.TAT.CA-3′
  RP: 5′-ATA.TTC.TTT.GCT.CCT.TGC.CA-3′

PCR generates a 138-bp DNA fragment.

- Alternative primers for the cDNA amplification of the *CEA* mRNA template (nested PCR):
  –Outer primers:
  FP: 5′-TCT.GGA.ACT.TCT.CCT.GGT.CTC-3′
  RP: 5′-TGT.AGC.TGT.TGC.AAA.TGC.TTT.AAG-3′

PCR generates a 160-bp DNA fragment.

–Nested primers:
FP: 5′-GGG.CCA.CTG.TCG.GCA.TCA.TGA.TTG.G-3′
RP: 5′-TGT.AGC.TGT.TGC.AAA.TGC.TTT.AAG.GAA.GAA.GC-3′

Nested PCR generates a 131-bp DNA fragment.

- Alternative primers for the cDNA amplification of the *CEA* mRNA template:
  FP: 5′-ACC.TTC.TCC.TGG.TCT.CTC-3′
  RP: 5′-GCA.AAT.GCT.TTA.AGG.AAG.AAG-3′

PCR generates a 146-bp DNA fragment.

3.5.7
Detection of Micrometastases and Circulating Tumor Cells of Neuroblastoma by Amplification of Tyrosine Hydroxylase and PGP 9.5 mRNA

Neuroblastoma is the most common extracranial solid tumor in children and is a tumor of neural crest origin, mainly arising in the adrenal medulla or sympathic ganglia. Histologically, the tumor consists of neuroblasts derived from primitive sympathogonia. Neuroblastoma is characterized by the secretion of catecholamines, which can be detected by measuring of urinary vanillyl mandelic and homovanillic acid. Further molecular methods for the diagnosis and monitoring of neuroblastoma are recently developed. These methods target the mRNA corresponding to the enzymes and antigens specific for neuroblastoma such as tyrosine hydroxylase, PGP 9.5 and other neuroendocrine markers.

- *Tyrosine hydroxylase*: tyrosine hydroxylase (TH) is the first enzyme involved in the process of catecholamine syntheses. Some clinical and molecular studies found that the molecular detection of TH mRNA could be more sensitive than the monitoring of urinary vanillyl mandelic and homovanillic acid taking in consideration that more than 5% of neuroblastoma shows no elevation of urinary vanillyl mandelic and homovanillic acid because of incomplete catecholamine syntheses.
- *Protein gene product 9.5*: The protein gene product 9.5 (PGP 9.5) neuron gene product is another marker found to be specific for neuroblastoma as well as for other tumors originated from the neural crest. PGP 9.5 is a cytosolic protein belongs to the ubiquitin-carboxy-terminal hydrolase subclass with a molecular weight of 27 kDa, highly expressed in the neurons and neuronal tumors including K cells of the diffuse neuroendocrine cell system of the lung and in both small and non-small cell lung cancer. PGP 9.5 is also expressed in different neuroendocrine tumors and in malignant nerve sheet tumor. Specific antibodies to PGP 9.5 are also used as an immunohistochemical marker for the discrimination of neural and neuroendocrine differentiation.

The molecular detection of PGP 9.5 and TH mRNA and other neuroblastoma-associated antigens such as the members of the MAGE and GAGE family (discussed in the melanoma section) reported to be a sensitive and specific method for the early detection and monitoring of neuroblastoma. This method is also efficient for the detection of micrometastases or circulating tumor cells in peripheral blood in addition to disseminated tumor cells in bone marrow. The molecular methods based on RT-PCR are able to detect one neuroblast in up to one million-background bone marrow or nucleated cells of peripheral blood and found to be more sensitive than the immunohistochemical staining of tumor cells in bone marrow samples.

The involvement of bone marrow is an important feature of neuroblastoma. The presence of neuroblasts in bone marrow considered to be a major component of clinical staging of the disease and consequently have an important therapeutic and prognostic relevance as the staging system recommends the screening of bone marrow biopsies and aspirates to rule out marrow involvement. Studies showed that the presence of neuroblastoma cells in bone marrow is correlated with poor prognosis irrespective of tumor stage, however the molecular detection of neuroblasts does not automatically means the upstaging of patient from stages I–III to stage IV as the dissemination of tumor cell seems to occur in the early stages of the disease. Further useful application of the molecular detection method is the screening of bone marrow before autologous bone marrow or stem cell transplantation.

False-negative results can appear during tumor progression due the presence of primitive undifferentiated tumor cells with abnormal genetic transcription and incomplete catecholamine synthesis pathway. Using a primer panel for both tyrosine hydroxylase and PGP 9.5 in addition to nested PCR can noticeably diminish the frequency of false-negative results.

Method The method used is mRNA extraction from tissue, blood and bone marrow, followed by RT-PCR.

- Primers for the cDNA amplification of the *TH* mRNA template (nested PCR):
  –Outer primers:
  FP: 5′-TGT.CAG.AGC.TGG.ACA.AGT-3′
  RP: 5′-TAT.TGT.CTT.CCC.GGT.AGC-3′
  –Nested primers:
  FP: 5′-CAC.CTG.GTC.ACC.AAG.TTC-3′
  RP: 5′-GGT.GTA.GAC.CTC.CTT.CCA-3′

Nested PCR generates a 180-bp DNA fragment.

- Alternative primers for the cDNA amplification of the *TH* mRNA template:
  FP: 5′-GTG.TGC.GCC.AGG.TGT.CAG.AG-3′
  RP: 5′-GGC.CCT.TCA.GCG.TGG.TGT.AG-3′

PCR generates a 256-bp DNA fragment.

- Alternative primers for the cDNA amplification of the *TH* mRNA template:
  FP: 5′-TGT.CAG.AGC.TGG.ACA.AGT.GT-3′
  RP: 5′-GAT.ATT.GTC.TTC.CCG.GTA.GC-3′

PCR generates a 299-bp DNA fragment.

- Primers for the cDNA amplification of the *PGP* 9.5 mRNA template:
  FP: 5′-CTG.TGG.CAC.AAT.CGG.ACT.TA-3′
  RP: 5′-TGT.CAT.CTA.CCC.GAC.ATT.GG-3′

PCR generates a 203-bp DNA fragment.

- Alternative primers for the c DNA amplification of the *PGP* 9.5 mRNA template (nested PCR):
  –Outer primers:
  FP: 5′-AGA.TGT.CTG.AAC.AAA.GTC.TG-3′
  RP: 5′-CAG.AGA.GCC.ACG.GCA.GAG.AA-3′
  –Nested primers:
  FP: 5′-TGG.GGC.TGG.AAG.AGG.AGT.C-3′
  RP: 5′-GAA.CTG.GCG.CCA.TGG.TTC.A-3′

Nested PCR generates a 472-bp DNA fragment.

- Alternative primers for the cDNA amplification of the *PGP* 9.5 mRNA template:

PGP 9.5 RT-specific primer: 5′-TGT.TTC.ACA.AGT.ACT-3′
FP: 5′-AGA.TCA.ACC.CCG.AGA.TGC.TGA.ACA.AAG.TGC.TG-3′
RP: 5′-ATT.AGG.CTG.CCT.TGC.AGA.GAG.CCA.CGG.CAG.AGA.A-3′

PCR generates a 653-bp DNA fragment.

3.5.8 Detection of Micrometastases and Circulating Tumor Cells of Tumors with Neuroendocrine Differentiation, PNET and Small Cell Carcinoma of the Lung by Amplification of Neuroendocrine-Specific Markers mRNA: Chromogranin A, Neuron-Specific Enolase, Choline Acetyltransferase, Secretogranin II and Somatostatin Receptor Type 2

Chromogranin A (CgA), secretogranin II (SgII), neuron-specific enolase (NSE), choline acetyltransferase (CAT) and somatostatin receptor type 2 (SST2) are widely expressed proteins in different human neuroendocrine tissue and in various tumors with neuroendocrine differentiation such as primitive neuroectodermal tumors (PNET), Ewing's sarcoma, neuroblastoma, small cell lung carcinoma, carcinoids and many other neuroendocrine tumors including tumors of the endocrine glands like parathyroid, pituitary and adrenal gland tumors. The above-listed neuroendocrine markers are now used as biochemical, immunological and molecular markers for the diagnosis of these tumors.

- *Chromogranin A*: chromogranin A (CgA) is a member of the chromogranin/secretogranin family, which includes chromogranin A, chromogranin B (also known as secretogranin I) and chromogranin C (also known as secretogranin II). Chromogranins are expressed in almost all neuroendocrine cells and neuroendocrine tumors. Chromogranin A is a non-glycosylated protein encoded on chromosome 14. Chromogranin A is found in secretory granules of many types of neuroendocrine cells and in the majority of endocrine and neuroendocrine tumors. Specific antibodies to chromogranin A are significant immunohistochemical markers for the diagnosis of neuroendocrine differentiation and neuroendocrine tumors.

- *Secretogranin II*: secretogranin II (SgII; chromogranin C) is another member of the chromogranin/secretogranin family, an acid protein found in most of endocrine cells. Both chromogranin A and secretogranin II are considered reliable markers for neuroendocrine differentiation and tumors derived from neuroendocrine cells.

- *Neuron-specific enolase*: neuron-specific enolase (NSE) is a glycolytic enzyme catalyzing the reaction pathway between 2-phospho-glycerate and phosphophenol pyruvate. Enolases are homo or heterodimers composed of the three subunits: 46-kDa alpha (α)-subunit, 44-kDa beta (β)-subunit and 46-kDa gamma (γ)-subunit. The α-subunit is expressed in most tissue types while the expression of the β-subunit is restricted mainly to the striated muscle including skeletal muscle and myocardium. The γ-subunit is expressed primarily in neurons, in normal and in neoplastic neuroendocrine cells, but it can be also expressed in megakaryocytes, T-lymphocytes in addition to striated and smooth muscle cells. The γ-dimeric (γ-γ) enolase iso-enzyme is the neuron-specific enolase (NSE), a cytoplasmic enzyme of the glycolysis pathway predominantly found in neuronal and neuroendocrine tissues. The γ-γ enolase dimeric form is found at elevated concentrations in small cell lung carcinoma and neuroblastoma.

- *Choline acetyltransferase*: choline acetyltransferase (CAT) is an enzyme catalyzing the syntheses of acetylcholine from choline and acetyl coenzyme A. Choline acetyltransferase is expressed in neural and neuroendocrine cells and their tumors.
- *Somatostatin receptor type 2*: somatostatin receptor type 2 (SST2) is a member of the somatostatin family, commonly detected in neuroendocrine tumors including small cell carcinoma of the lung. Some poorly differentiated non-neuroendocrine lung cancers show also weak expression of SST2.

The expression of the above-mentioned markers is diagnostic for neuroendocrine differentiation and the molecular detection of CgA, NSE, CAT, SgII and SST2 mRNA can be used as a sensitive molecular assay for the primary diagnosis and monitoring of different neuroendocrine tumors, even by negative immunohistochemical results. They are also an important indicator for neuroendocrine differentiation in non-neuroendocrine tumors of different origin.

Small cell carcinoma of the lung contains abundant CgA mRNA but very little stored chromogranin A protein whereas Ewing's sarcoma expresses SgII but no CgA and NSE can be expressed other non-neuroendocrine cells. This difference in the expression pattern of these neuroendocrine markers makes it important to use a primer panel (multiplex PCR) complementary to different neuroendocrine markers, this approach helps to minimize false-negative results and can be successfully used for the detection and monitoring of micrometastases and circulating cells of different neuroendocrine tumors.

Method The method used is mRNA extraction from tissue, peripheral blood and bone marrow, followed by RT-PCR.

- Primers for the cDNA amplification of the *CgA* mRNA template:
  FP: 5′-GCT.CCA.AGA.CCT.CGC.TCT.CC-3′
  RP: 5′-GAC.CGA.CTC.TCG.CCT.TTC.CG-3′

PCR generates a 583-bp DNA fragment.

- Primers for the cDNA amplification of the *SgII* mRNA template:
  FP: 5′-GAT.TCA.CTG.AGT.GAA.GAA.GAC-3′
  RP: 5′-GAC.CAC.ATC.TTC.ATA.GGC.AAT-3′

PCR generates a 429-bp DNA fragment.

- Primers for the cDNA amplification of the *SST2* mRNA template:
  FP: 5′-CAG.TCA.TGA.GCA.TCG.ACC.GA-3′
  RP: 5′-GCA.AAG.ACA.GAT.GAT.GGT.GA-3′

PCR generates a 284-bp DNA fragment.

- Primers for the cDNA amplification of the *NSE* mRNA template:
  FP: 5′-TCT.GAA.CGT.CTG.GCT.AAA.TAC.AAC-3′
  RP: 5′-GAA.AAG.GAA.GAA.GTG.GAA.AGT.GCG.G-3′

PCR generates a 333-bp DNA fragment.

- Primers for the cDNA amplification of the *CAT* mRNA template:
  FP: 5′-GAC.CAT.CCT.TTT.CTG.CAT.CT-3′
  RP: 5′-GGC.TTG.CTC.TCA.GTA.GGC-3′

PCR generates a 135-bp DNA fragment.

3.5.9
Molecular Diagnosis of Rhabdomyosarcoma and Detection of Micrometastases and Circulating Tumor Cells by Amplification of Muscle-Specific Markers mRNA: Myo D1, Myogenin, Myoglobin and Fetal Acetylcholine Receptor

Rhabdomyosarcoma is a highly malignant sarcoma derived from primitive mesenchymal cells exhibiting striated muscle differentiation. Rhabdomyosarcoma considered being the most common soft tissue sarcoma of childhood and young adults, accounting for about 10% of all pediatric tumors and characterized by a poor prognosis. Based on the morphological features, cytogenetic properties and clinical behavior, the international classification of rhabdomyosarcoma includes the following three subtypes:

1. Embryonal rhabdomyosarcoma: spindle cell and botryoid types;
2. Alveolar rhabdomyosarcoma;
3. Pleomorphic (anaplastic) rhabdomyosarcoma.

All rhabdomyosarcoma types show morphologic, immunophenotypic and ultrastructural aspects of skeletal (striated) muscle differentiation and resemble the morphology of skeletal muscle in different embryonal stages as the tumor cells fail to reach the terminal differentiation and are insufficient to fuse into myotubes. In contrast, different rhabdomyosarcoma types have different cytogenetic features and different clinical behavior.

- *Embryonal rhabdomyosarcoma* is the most common type of rhabdomyosarcoma exhibiting the morphological features of a primitive small round cell tumor with botryoid and spindle cell types, mainly presents in head and neck, genitourinary and retroperitoneal regions. The morphology of embryonal rhabdomyosarcoma resembles the morphology of skeletal muscle in 7–10 gestation weeks. Cytogenetically, embryonal rhabdomyosarcoma has no constant or specific genetic abnormality, but usually associated with loss of heterozygosity and allelic deletions of chromosome 11p15.5. Gain of chromosomes is also a common cytogenetic abnormality observed in association with this tumor type, mainly chromosomes 2, 7, 8, 11, 12 and 13.

- *Alveolar rhabdomyosarcoma* is the second common type rhabdomyosarcoma, exhibiting a different morphological appearance, different biological behavior and different prognosis than the other types, and usually more aggressive than embryonal rhabdomyosarcoma. The morphology of alveolar rhabdomyosarcoma resembles the morphology of skeletal muscle in 10–12 gestation weeks. Alveolar rhabdomyosarcoma presents typically in the trunk and extremities and commonly associated with one of the specific translocations t(2; 13)(q35; q14) or t(1; 13)(p36; q14) discussed in a previous section.

- *Pleomorphic rhabdomyosarcoma* makes less than 5% of all rhabdomyosarcoma and mainly affect patients over 40 years, primarily in the region of large muscles of extremities. All types of rhabdomyosarcoma show the similar immunohistochemical phenotype demonstrating positive immunohistochemical reactions with antibodies specific to different myogenic antigens such as desmin, sr-Actin, myoglobulin, myosin and antibodies to myogenic regulatory proteins including Myo D1, Myf-3, myogenin, Myf-5 and Myf-6. Some of the available immunohistochemical markers such as myoglobin, fetal acetylcholine receptor (AchR) and myogenic regulatory proteins are specific markers for skeletal muscle and/or rhabdomyoma and rhabdomyosarcoma and not detected in any other tissue type like smooth muscle and smooth muscle tumors like leiomyosarcoma. The specific mRNAs encoding the myogenic regulatory factors such as Myo D1, Myf-3, myogenin (Myf-4) and Myf-6 in addition to the muscle proteins such as myoglobin and the acetylcholine receptor are used as specific molecular markers for the diagnosis of rhabdomyosarcomas.
- *Myo D1 gene*: Myo D1 is a myogenic regulatory factor encoded by a gene located on 11p15 chromosome, encoding a member of the basic helix-loop-helix protein family. Myo D1 phosphoprotein (with great homology to Myf-3), is a transcriptional factor involved in the regulation of the skeletal muscle differentiation and maintenance of myogenic program. Myo D1 is found in skeletal muscle and almost in all rhabdomyosarcoma types. Monoclonal antibody to Myo D1 is used as a specific immunohistochemical marker for the diagnosis of rhabdomyosarcoma and the amplification of the Myo D1 mRNA can be also used as a sensitive molecular marker for this tumor group.
- *Myogenin*: myogenin is a myogenic transcription factor involved in the regulation of muscle development and encoded by a member of the myogenic regulatory gene family, which includes Myo D1, Myf-3, Myf-4 (myogenin), Myf-5 and Myf-6. Myogenin is a specific transcriptional factor for skeletal muscle differentiation and the detection of the complementary mRNA can be used as a specific marker for tumors with myogenic differentiation, especially rhabdomyosarcoma. Antibodies specific for this transcription factor are also used as sensitive and specific markers for the diagnosis of rhabdomyosarcoma.
- *Myoglobin*: myoglobin is a cytoplasmic polypeptide composed of 153 amino acids expressed in skeletal and cardiac muscle, acting as oxygen transporting pigment. Antibodies to myoglobin are also specific immunohistochemical markers for myogenic tumors and the amplification of the myoglobin mRNA is an additional sensitive molecular marker.
- *Acetylcholine receptor*: acetylcholine receptor (AChR) is a nicotinic muscle receptor, a pentamer ion channel with a prominent extracellular domain, mediating the

synaptic transmission at the skeletal neuromuscular junction. There are two types of the acetylcholine receptor, each composed of four subunits. The fetal type composed of the $\alpha 2\beta\gamma\delta$ subunits and the adult type composed of $\alpha 2\beta\varepsilon\delta$ subunits. The adult type occurs by replacement of the γ subunit by the ε subunit during the development of the neuromuscular junction, which makes the γ subunit specific for the fetal type of the acetylcholine receptor and usually not found in the adult receptor type. The γ to ε switch takes place rapidly in the first postnatal weeks, whereas α, β and δ subunits persist throughout the differentiation of the receptor. In adults, the expression of the fetal type is limited to the extraocular muscles, denervated striated muscles and thymic myoid cells. The thymic myoid cells are thought to be epithelial cells with features of striated muscle cells. The AchR γ subunit and mRNA encoding this molecule are detected in almost all types of rhabdomyosarcoma and can be used as an immunohistochemical and molecular markers for the diagnosis of all types of rhabdomyosarcoma. Furthermore, the fact that the acetylcholine receptor has a prominent extracellular domain makes the γ subunit molecule a potential target for specific antibodies for immunotherapy of rhabdomyosarcoma.

The molecular amplification of mRNA complementary to the above-mentioned myogenic markers can be used not only as sensitive and specific molecular markers for rhabdomyosarcoma but also for the detection of micrometastases or disseminated tumor cells in peripheral blood, bone marrow as well as a tool for precise screening of surgical margins if these margins free of skeletal muscle.

Method The method used is mRNA extraction from tissue, peripheral blood and bone marrow, followed by RT-PCR.

- Primers for the cDNA amplification of the *Myo D1* mRNA template:
  FP: 5′-AGC.ACT.ACA.GCG.GCG.ACT-3′
  RP: 5′-GCG.ACT.CAG.AAG.GCA.CGT.C-3′

PCR generates a 266-bp DNA fragment.

- Primers for the cDNA amplification of the *myogenin* mRNA template:
  FP: 5′-TAA.GGT.GTG.TAA.GAG.GAA.GTC-3′
  RP: 5′-CCA.CAG.ACA.CAT.CTT.CCA.CTG.T -3′

PCR generates a 438-bp DNA fragment.

- Primers for the cDNA amplification of the *AChR* γ subunit mRNA template:
  FP: 5′-ATC.TCA.GTC.ACC.TAC.TTC.CCC-3′
  RP: 5′-TAC.TTG.CTG.ATT.AAG.GTG.TGT.AAG.AGG.AAG.TC-3′

PCR generates a 496-bp DNA fragment.

- Alternative primers for the cDNA amplification of the *AChR* γ subunit mRNA template:
  FP: 5′-GAA.GCC.CTC.ACC.ACC.AAT.GT-3′
  RP: 5′-GTA.GGT.GAA.GGA.AGG.ACG.GT-3′

PCR generates a 513-bp DNA fragment.

- Primers for the cDNA amplification of the *myoglobin* mRNA template (nested PCR):

–Outer primers:
FP: 5′-TCG.GAA.TGC.ATC.ATC.CAG.GT-3′
RP: 5′-GTT.GGA.GGC.CAT.GTC.CTT.CC-3′
–Nested primers:
FP: 5′-TCG.GAA.TGC.ATC.ATC.CAG.GT-3′
RP: 5′-AAC.AGC.TCC.AGG.GCC.TTG.TT-3′
PCR generates a 114-bp final DNA fragment.

3.5.10
Detection of Micrometastases and Circulating Tumor Cells of Squamous Cell Carcinoma by Amplification of Squamous Cell Carcinoma Antigen mRNA

Squamous cell carcinoma is a common malignancy arising in different organs and systems including skin, respiratory, gastro-intestinal- and genito-urinary systems, derived from stratified squamous epithelia or from metaplastic epithelia.

Squamous cell carcinoma antigen (SCCA) is a group of cytoplasmic proteins with a molecular weight of approximately 45 kDa, which is a serine and cysteine protease inhibitor and a subfraction of the tumor-associated antigen TA4. Squamous cell carcinoma antigen (SCCA) includes two highly homologous fractions, a neutral fraction (SCCA 1) and an acidic fraction (SCCA 2), both fractions are encoded by two genes located on chromosome 18q21.3. Both genes are members of the high-molecular-weight *ser*ine *p*roteinase *in*hibitor (serpin) superfamily encoding the SCCA1 and SCCA2 antigens that are 98% and 92% homologous at the nucleotide and amino acid levels, respectively.

Squamous cell carcinoma antigen is released from various epidermal and squamous cells and their neoplasia with a marked elevation in squamous cell carcinoma, especially non-keratinizing squamous cell carcinoma. The neutral fraction expressed in both benign and malignant squamous cells, whereas the acidic fraction is mainly expressed in malignant cells and significantly increased in squamous cell carcinoma and easily released extracellularly to appear in the circulation. Serum levels of SCCA (mainly SCCA2) can be employed as a tumor marker for the monitoring of squamous cell carcinoma. Serum SCCA levels correlate with the stage and clinical course of the carcinoma and can predict the relapse after treatment.

The amplification of SCCA mRNA may be a useful method for the detection of micrometastases and the monitoring of squamous cell carcinoma such as squamous cell carcinoma of uterine cervix, head and neck and esophagus. The simultaneous amplification of other epithelial antigens such as CK19 as multiplex PCR is recommended for precise results.

Method The method used is mRNA extraction from tissue, blood and bone marrow, followed by RT-PCR.

- Primers for the cDNA amplification of the *SCCA* mRNA template. The reverse primer is complementary to both SCCA 1 and 2 fractions:

FP (SCCA1): 5′-CGCGGTCTCGTGCTATCTGG-3′
FP (SCCA2): 5′-CACGGTCTCTCAGTATCTAA-3′
RP (common): 5′-AGAAGAGGATGCTGTTGGTC-3′

PCR generates a 191-bp DNA fragment.

- Alternative primers for the cDNA amplification of the *SCCA* mRNA template (nested PCR). The primers are complementary to both SCCA 1 and 2 genes:
  –Outer primers:
  FP: 5′-GAG.TTC.CAG.ATC.ACA.TCG-3′
  RP: 5′-CTT.CTG.GAG.CAT.TTG.CA-3′

PCR generates a 450-bp DNA fragment.

  –Nested primers:
  FP: 5′-TCA.GTG.AAG.CCA.ACA.CCA.AG-3′
  RP: 5′-TTG.TTG.GCG.ATC.TTC.AGC.TC-3′

PCR generates a 307-bp DNA fragment.

- Alternative primers for the cDNA amplification of the *SCCA* mRNA template (nested PCR):
  –Outer primers:
  FP (exon 1): 5′-GCC.CAC.CTC.TGC.TTC.CTC.TA-3′
  RP (exon 8): 5′-GCT.TCT.GCT.CCC.TCC.TCT.GT-3′

PCR generates a 1083-bp DNA fragment.

  –Nested primers:
  FP (exon 5): 5′-GCA.AAT.GCT.CCA.GAA.GAA.AG-3′
  RP (exon 7): 5′-CGA.GGC.AAA.ATG.AAA.AGA.TG-3′

Nested PCR generates a 261-bp DNA fragment.

3.5.11
Detection of Micrometastases and Circulating Tumor Cells of Adenocarcinoma by Amplification of MUC-1 mRNA

Mucins are a group of mucopolysaccharides composed of high-molecular-weight glycosylated proteins consisting of core protein linked to carbohydrate side-chains. Mucins are synthesized and expressed by glandular epithelial cells of different locations including glandular epithelium of breast, gastrointestinal and gynecological organs. Different mucins are also expressed in various types of adenocarcinoma and borderline lesions of various origins. There are two main gene families encoding the epithelial mucins:

1. Gel-forming secreted mucins, encoded by genes located at chromosome 11p15 and include MUC-2, MUC-5AC, MUC-5B and MUC-6.
2. Membrane-bound mucins, encoded by genes located on the chromosomes 7q22, 3q and 1q2. This mucin family includes MUC-1, MUC3A, MUC-3B, MUC-4, MUC-12, MUC-13 and MUC-17.

MUC-1 is a glycosylated mucin expressed by epithelial cells and different carcinoma types, positioned as a transmembrane molecule with a large extracellular domain.

MUC-1 has protective and regulatory functions and found to be overexpressed in some carcinoma types.

The amplification of MUC-1-mRNA may be of value for the detection of micrometastases and the monitoring of glandular malignancies such as gynecological and gastrointestinal adenocarcinomas in addition to breast carcinoma. Some papers reported a low specificity but high sensitivity of MUC-1 molecular detection method because of possible illegitimate transcription in nucleated blood cells and CD34+ cells; consequently, it is recommended to use the MUC-1 marker in combination with other markers more specific for epithelial tumors like CK19 and CEA

Method The method used is mRNA extraction from tissue, bone marrow, peripheral blood or body fluids, followed by RT-PCR.

- Primers for the cDNA amplification of the *MUC-1* mRNA template:
  FP: 5′-GTG.CCC.CCT.AGC.AGT.ACC.G-3′
  RP: 5′-GAC.GTG.CCC.CTA.CAA.GTT.GG-3′

PCR generates a 100-bp DNA fragment.

- Alternative primers for the cDNA amplification of the *MUC-1* mRNA template:
  FP: 5′-ACC.TAC.CAT.CCT.ATG.AGC.GAG-3′
  RP: 5′-GGT.TTG.TGT.AAG.AGA.GGC.TGC-3′

PCR generates a 132-bp DNA fragment.

- Alternative primers for the cDNA amplification of the *MUC-1* mRNA template:
  FP: 5′-GGT.ACC.TCC.TCT.CAC.CTC.CTC.CAA-3′
  RP: 5′-CGT.CGT.GGA.CAT.TGA.TGG.TAC.C-3′

PCR generates a 288-bp DNA fragment.

- Alternative primers for the cDNA amplification of the *MUC-1 mRNA* template (nested PCR):
  –Outer primers:
  FP: 5′-GCC.AGT.AGC.ACT.CAC.CAT.AGC.TCG-3′
  RP: 5′-CTG.ACA.GAC.AGC.CAA.GGC.GAG-3′

First PCR generates a 516-bp DNA fragment.

–Nested primers:
FP: 5′-TTC.TCG.GAA.GGC.CAG.AGT.CAA.TTG.T-3′
RP: 5′-TCA.CAG.CAC.TTC.TCC.CCA.GTT.GTC.T-3′

Nested PCR generates a 248-bp DNA fragment.

3.5.12 Detection of Micrometastases and Circulating Tumor Cells of Colorectal Adenocarcinoma by Amplification of Guanylyl Cyclase C and CDX-2 mRNA

Colorectal adenocarcinoma is one of the most common malignancies in humans and a major cause of cancer related mortality. Colorectal adenocarcinoma develops from intestinal mucosal cells after a cascade of genetic alterations. The most important prognostic parameter for all types of carcinoma is the TNM status including tumor spread, the presence of lymph node and organ metastases and the invasion of blood and lymph vessels in addition to the status of the surgical margins. Because of the high recurrence incidence between the pN0-operated patients, we have to think about the additional TNM parameters including the pN0 (i+), pN0 (mol+), pM0 (i+) and pM0 (mol+) tumor cases. These parameters reflect the presence of micrometastases or circulating tumor cells in peripheral blood or in bone marrow as additional important monitoring parameter and prognostic factor for the treatment and follow-up of these patients. For the detection of isolated tumor cells or disseminated and circulated tumor cells, molecular methods based on the detection of tissue-specific gene expression are developed. For colorectal adenocarcinoma, efficient molecular markers are mRNAs encoding tissue-specific antigens or transcription factors or cytokeratins such as CK20, CEA (both discussed in other sections), guanylyl cyclase C (GCC) and CDX-2

- *Guanylyl cyclase C*: Guanylyl cyclase C (GCC) is a member of the guanylyl cyclase family of the transmembrane signaling proteins and the receptor for the endogenous peptides guanylin and uroguanylin and it is the receptor for bacterial diarrheagenic heat-stable enterotoxins. Guanylyl cyclase C is selectively expressed in the apical membranes of the intestinal mucosal cells from duodenum to the rectum as well as adenoma and adenocarcinoma neoplastic cells arising from the intestinal mucosa. In some studies, low expression level of GCC mRNA in blood was reported, most likely due to the ectopic expression of guanylyl cyclase C in CD34-positive progenitor cells. Guanylyl cyclase C is not expressed in the normal esophageal or gastric mucosa but expressed in areas of intestinal metaplasia in esophagus and stomach.
- *CDX-2*: CDX-2 is a caudal type homeobox gene, which encodes a 33-kDa 311-amino-acid nuclear protein. CDX-2 is an intestine-specific transcription factor expressed in the early intestinal development stage supposed to be involved in the differentiation, adhesion, proliferation and apoptosis of intestinal epithelial cells.

Clinical observations showed that up to 30% of patients with T1 and up to 50% of T2 colorectal adenocarcinoma develops lymph node and organ metastasis, probably as a result of the presence of micrometastases in regional lymph nodes and/or the inadequate lymph node sampling and histopathologic examination. The under staging of there tumors can be the cause of under treatment of these patients. The amplification of mRNA encoding related tissue markers such as GCC, CDX-2 or CK20 can be a valuable sensitive method for the detection of micrometastases, minimal residual cancer cells and circulating tumor cells of colorectal adenocarcinoma as these factors are associated with poor prognosis. This approach also helps precision in the postoperative tumor classification, which improves the postoperative follow-up and treatment of patients with colorectal carcinoma.

Method The method used is mRNA extraction from different tissue types, blood and bone marrow, followed by RT-PCR.

- Primers for the cDNA amplification of the *GCC* mRNA template:
  FP (exon 6): 5′-ATT.CTA.GTG.GAT.CTT.TTC.AAT.GAC.CA-3′
  RP (exon 7): 5′-CGT.CAG.AAC.AAG.GAC.ATT.TTT.CAT-3′

PCR generates an 84-bp DNA fragment.

- Alternative primers for the cDNA amplification of the *GCC* mRNA template:
  FP: 5′-ACG.TCT.GCA.AAA.TGC.TGG.C -3′
  RP: 5′-GGC.AGT.CGC.CTG.AGT.TAT.GAA -3′

PCR generates a 86-bp DNA fragment.

- Alternative primers for the cDNA amplification of the *GCC* mRNA template (nested PCR):
  –Outer primers:
  FP: 5′-TGG.ATT.TCA.GTC.GCA.GA-3′
  RP: 5′-ATG.TAG.GGT.TAG.GTC.ATC.AAA.G-3′

First PCR generates a 341-bp DNA fragment.

  –Nested primers:
  FP: 5′-AAT.CAG.CGT.CCT.GAT.GAT.GGG.CAA.C-3′
  RP: 5′-ATG.AGG.ACA.CAG.CCC.ATC.CGT.TGT.G-3′

Nested PCR generates a 252-bp DNA fragment.

- Primers for the cDNA amplification of the *CDX-2* mRNA template:
  FP: 5′-AAAGTGAGCTGGCTGCCACACTTG-3′
  RP: 5′-TCCATCAGTAGATGCTGTTCGTGG-3′

PCR generates a 426-bp DNA fragment.

- Alternative primers for the cDNA amplification of the *CDX-2* mRNA template:
  FP: 5′-GCTTTGGGAACATTTCCCAAACTCAGTG-3′
  RP: 5′-TTCTCGCAGCGTCCATACTCCTCAT-3′

PCR generates a 361-bp DNA fragment.

3.5.13 Detection of Micrometastases and Circulating Tumor Cells of Breast Carcinoma by Amplification of Mammaglobin mRNA

Breast cancer is the most frequent malignancy in women with increased frequency. The breast is composed of heterogeneous tissue, with epithelial, mesenchymal, endothelial and lymphatic components. Different genes are predominantly expressed by the breast epithelial cells (breast parenchyma) and by tumors derived from the breast epithelial cells. These genes and the mRNA encoding them can be used as a molecular marker for the diagnosis and follow-up of different types of breast carcinoma. Good examples are the genes encoding the members of the secretoglobin superfamily. This superfamily includes the uteroglobin family, mammaglobin B, lipophilin A and lipophilin B. These genes are located as a gene cluster on chromosome 11q12–13. Additional molecular markers for breast carcinoma such as MUC1, maspin, cytokeratins and CEA were discussed in previous sections.

Mammaglobin A (*SCGB2A2*) is a low molecular protein, a member of the uteroglobin family, which belongs to the secretoglobin gene superfamily. Mammaglobin A consists of a 93 amino acid polypeptide chain homologous to the human Clara cell protein. Due posttranslational modifications mammaglobin A exists in the breast in two main forms with molecular masses of ~18 kDa and 25 kDa. Mammaglobin A is a specific breast tissue protein, usually found in breast tissue as a complex with lipophilin B (*SCGB1D2*) and encoded by a gene located on chromosome 11q12.1–13.1 composed of three exons and two introns. This gene is frequently activated and modified during the genesis of breast carcinoma. Mammaglobin A is expressed only in adult breast tissue and is found in 80–90% of primary breast carcinoma and lymph node metastases. This protein is overexpressed in about 50% of breast carcinoma with a potential correlation between mammaglobin A expression, the expression of steroid hormone receptors and the overexpression of HER-2. Poorly differentiated, estrogen receptor-negative invasive breast carcinomas are frequently mammaglobin-negative. Furthermore, there is a positive correlation between the expression of mammaglobin A and the expression of other members of the uteroglobin family, namely lipophilin A, lipophilin B and the small mucin like protein, all of them encoded by genes located near to the gene encoding mammaglobin A. Interesting that few types of lymphoid malignancies such as mantle cell lymphoma, associated with the t(11; 14)(q13; q32) translocation also involves the mammaglobin A gene, as this gene is located close to the breakpoints caused by this translocation. This mechanism may cause the activation of this gene and the transcription of mammaglobin mRNA.

Mammaglobin B is another protein highly homologous to mammaglobin A, but less specific to breast tissue and breast tumors than mammaglobin A.

A further interesting protein characteristic for breast tissue is NY-BR-1. NY-BR-1 is a differentiation antigen expressed in normal breast tissue and epithelial breast tumors in addition to testicular parenchyma and brain tissue. NY-BR-1 is a nuclear polypeptide functioning as transcription factor encoded by a gene located on chromosome 10p11–12.

The specificity of both mammaglobins and NY-BR-1 for adult breast tissue and breast cancers makes these proteins and their complementary mRNA-specific markers for breast carcinomas. Mammaglobin and NY-BR-1 mRNA can be the subject of molecular amplification as a precise method for the diagnosis of breast cancer and for the detection of submicroscopic metastases or circulating tumor cells as well as tumor staging. This method is useful to detect micrometastases and isolated tumor cells in lymph nodes including sentinel lymph nodes. Using molecular methods, different studies reported the detection of tumor cells in 5–15% of histologically negative sentinel lymph nodes, which can change the staging of the tumor and perhaps the therapeutic strategy. This method can be also used for the detection of tumor cells in bone marrow and leukaphereses for autologous stem cell transplantation after high-dose chemotherapy, peripheral blood and body fluids such as pleural effusion. It can also be used for the discrimination of metastatic tissue of an unknown tumor.

Studies show that the presence of mammaglobin mRNA in peripheral blood of breast cancer patients correlates with tumor stage, disease progression and metastatic spread. Recent studies showed that the detection of mammaglobin mRNA in the bone marrow of breast cancer patients is highly prognostic of relapse and overall survival. The molecular staging of breast cancer can be an important indicator for monitoring of therapy success. Finally, the unique expression of mammaglobins in adult breast tissue and breast tumors makes these polypeptides a suitable molecular target for specific therapeutic agents.

For more precise diagnostic approach the mammaglobins A and B and NY-BR-1 may be used in combination with other breast tissue-specific molecular markers such as maspin, lipophilin B, B305D, B726P, γ-aminobutyrate type A receptor π subunit (GABAπ), CK19, CEA and MUC-1 in a multiplex PCR.

Method The method used is mRNA extraction from tissue, peripheral blood or body fluids and bone marrow, followed by RT-PCR.

- Primers for the cDNA amplification of the *mammaglobin A* mRNA template:
  FP: 5′-TCT.CCC.AGC.ACT.GCT.ACG.C-3′
  RP: 5′-TTA.GAC.ACT.TGT.GGA.TTG.ATT.GTC.TTG-3′

PCR generates a 82-bp DNA fragment.

- Alternative primers for the cDNA amplification of the *mammaglobin A* mRNA template:
  FP: 5′-AGA.ACT.GCA.GGG.ATG.GTG.AGA.A -3′
  RP: 5′-ACA.TGT.ATA.GCA.GGT.TTC.AAC.AAT.TGT-3′

PCR generates a 114-bp DNA fragment.

- Alternative primers for the cDNA amplification of the *mammaglobin A* mRNA template (nested PCR):
  –Outer primers:
  FP: 5′-GAA.GTT.GCT.GAT.GGT.CCT.CAT.GCT.GGC-3′
  RP: 5′-CTC.ACC.ATA.CCC.TGC.AGT.TCT.GTG.AGC-3′

First PCR generates a 324-bp DNA fragment.

  –Nested primers:
  FP: 5′-CTC.CCA.GCA.CTG.CTA.CGC.AGG.CTC-3′
  RP: 5′-CAC.CTC.AAC.ATT.GCT.CAG.AGT.TTC.ATC.CG-3′

Nested PCR generates a 201-bp DNA fragment.

- Alternative primers for the cDNA amplification of the *mammaglobin A* mRNA template (nested PCR):
  –Outer primers:
  FP: 5′-CAG.CGG.CTT.CCT.TGA.TCC.TTG-3′
  RP: 5′-ATA.AGA.AAG.AGA.AGG.TGT.GG-3′

First PCR generates a 402-bp DNA fragment.

  –Inner primers:
  FP: 5′-TGA.ACA.CCG.ACA.GCA.GCA.G-3′
  RP: 5′-TCC.GTA.GTT.GGT.TTC.TCA.CC-3′

Nested PCR generates a 367-bp DNA fragment.

- Alternative primers for the cDNA amplification of the *mammaglobin A* mRNA template:
  FP: 5′-CTC.TGG.CTG.CCC.CTT.ATT.GGA-3′
  RP: 5′-ATA.AGA.AAG.AGA.AGG.TGT.GG-3′

PCR generates a 329-bp DNA fragment.

- Alternative primers for the cDNA amplification of the *mammaglobin A* mRNA template:
  FP: 5′-CCG.ACA.GCA.GCA.GCC.TCA.C -3′
  RP: 5′-TCC.GTA.GTT.GGT.TTC.TCA.C -3′

PCR generates a 361-bp DNA fragment.

- Primers for the cDNA amplification of the *mammaglobin B* mRNA template:
  FP: 5′-ACT.CCT.GGA.GGA.CAT.GGT.TGA-3′
  RP: 5′-TCT.GAG.CCA.AAC.GCC.TTG.GGT-3′

PCR generates a 245-bp DNA fragment.

- Primers for the cDNA amplification of the *NY-BR-1* mRNA template:
  FP: 5′-CAA.AGC.AGA.GCC.TCC.CGA.GAA.G -3′
  RP: 5′-CCT.ATG.CTG.CTC.TTC.GAT.TCT.TCC -3′

PCR generates a 573-bp DNA fragment.

- Primers for the cDNA amplification of the *lipophilin B* mRNA template:
  FP: 5′-CTG.AGC.TCA.CAG.CAA.AAC-3′
  RP: 5′-GAG.CTG.GGC.AGA.AC-3′

PCR generates a 105-bp DNA fragment.

- Alternative primers for the cDNA amplification of the *lipophilin B* mRNA template:
  FP: 5′-TTG.CAG.CCA.AGT.TAG.GAG.TG-3′
  RP: 5′-AAG.ACA.GTG.GAA.ACC.AGG.ATG-3′

PCR generates a 160-bp DNA fragment.

3.5.14
Detection of Micrometastases and Circulating Tumor Cells of Transitional Cell Carcinoma by Amplification of Uroplakins mRNA

Transitional cell carcinoma is one of the most common malignancies in humans and a major case of mortality throughout the world. Transitional cell carcinoma develops from the urothelium (transitional cell mucosa) of the urinary tract mainly the urinary bladder. Generally, the most important prognostic parameter for all malignancies is the differentiation grade of the tumor (G) and the TNM status. An additional important monitoring and prognostic factor for malignancies is the presence of isolated tumor cells and micrometastases in addition to the presence of tumor cells in bone marrow or circulating tumor cells in peripheral blood. For the detection of the disseminated or circulated tumor cells, molecular methods targeting related tissue-specific antigens and mRNA encoding these antigens are now used. Efficient molecular markers for transitional cell carcinoma are mRNA encoding urothelium-specific cytokeratins (such as CK7 and CK20) and specific antigens such as the different members of the uroplakin family.

The uroplakins are urothelium-specific transmembrane proteins expressed as rigid 0.2–0.5 μm plaques on the apical surface of mammalian urothelium. They are important for the differentiation of urothelium and take part in strengthening of the urothelial apical surface during distention of urinary bladder and urinary tract. Uroplakins are divided in four subtypes Ia, Ib, II, and III all of then are expressed by the urothelium of the urinary tract and tumors originated from the urothelium. Uroplakin subtypes Ia and II are specific for urothelium and were not detected in any other tissue or carcinoma type other than transitional cell carcinoma. Both uroplakins are also negative in primary squamous cell carcinoma and adenocarcinoma of the urinary bladder. The uroplakin subtype Ib was detected in other epithelial cells such as tracheal and bronchial epithelia and in mucosa exhibiting squamous metaplasia, whereas uroplakin subtype III was also detected in prostatic glandular epithelium. Antibodies specific for different uroplakins are used as immunohistochemical markers for the diagnosis of transitional cell carcinoma.

Since the uroplakins Ia and II are highly specific for urothelium and urothelial tumors, the mRNA encoding these uroplakin subtypes can be employed as specific molecular marker for the diagnosis and monitoring of transitional cell carcinoma including the examination of regional lymph nodes, surgical margins and perivesical tissue after radical cystectomy. The molecular detection of mRNA encoding uroplakin II in perivesical tissue is a risk factor for local recurrence, which can improve the postoperative follow-up of these patients. The detection of circulating tumor cells or tumor cells in bone marrow is also an important risk factor for the development of distant metastases. For accurate results, it is recommended to use uroplakins with other markers such as CK 20.

Method The method used is mRNA extraction from different tissue types, blood and bone marrow, followed by RT-PCR.

- Primers for the cDNA amplification of the *uroplakin Ia* mRNA template:
  FP: 5′-GCTCATCGTCTACATCTTCGA-3′
  RP: 5′-CTGACGTGAAGTTCACCCAG -3′

PCR generates a 221-bp DNA fragment.

- Primers for the cDNA amplification of the *uroplakin Ib* mRNA template:
  FP: 5′-TTCCAGGGCCTGCTGATTTT-3′
  RP: 5′-CATGGTACCCAGGAGAACCC-3′

PCR generates a 726-bp DNA fragment.

- Primers for the cDNA amplification of the *uroplakin II* mRNA template:
  FP: 5′-GATCCTGATTCTGCTGGCTG-3′
  RP: 5′-ATGTAGTATTTGGTTCCTGG-3′

PCR generates a 264-bp DNA fragment.

- Alternative primers for the cDNA amplification of the *uroplakin II* mRNA template (nested PCR):
  –Outer primers:
  FP: 5′-TCC.CCA.GGG.GCT.GCA.GAC.TT-3′
  RP: 5′-GGT.TTG.TCA.CCT.GGT.ATG.CAC.T-3′

First PCR generates a 268-bp DNA fragment.

  –Inner primers:
  FP: 5′-CAT.CTC.AAG.CCT.CTC.TCT.GGT.C-3′
  RP: 5′-AGC.CAG.CAC.CAC.TGT.CCA.CC-3′

Nested PCR generates a 203-bp DNA fragment.

- Alternative primers for the cDNA amplification of the *uroplakin II* mRNA template:
  FP: 5′-AACATCTCAAGCCTCTCTGGTCTG-3′
  RP: 5′-TGTGGACATTGGGATCTCTCTGCTG -3′

PCR generates a 322-bp DNA fragment.

- Primers for the cDNA amplification of the *uroplakin III* mRNA template:
  FP: 5′-GGAGGCATGATCGTCATCAC-3′
  RP: 5′-TCACGGACGTGTAGGAAGAC-3′

PCR generates a 193-bp DNA fragment.

3.6
Diagnosis of Tumors, Micrometastases, Isolated and Circulating Tumor Cells by Molecular Detection of Tissue-Specific Cytokeratins

3.6.1
Introduction

Cytoskeleton of mammalian cells is composed of microtubules, microfilaments and filaments of intermediate size. Filaments of intermediate size consist of five classes including cytokeratin type I and II, vimentin, desmin, glial fibrous acidic protein (GFAP) and neurofilament.

Cytokeratins are a complex family with more than 20 isotypes consists of keratin-like proteins with a molecular weight of 40–68 kDa and are the most important markers for epithelial structures and epithelial differentiation. Cytokeratins 9–20 are type I intermediate filaments whereas cytokeratins 1–8 are type II. Specific antibodies to different cytokeratins are widely used in immunohistochemistry to detect the cytokeratin expression and tumor differentiation as the cytokeratin profile remains usually constant after malignant transformation. Another approach for the detection of cytokeratin expression is the detection of their complementary mRNA using RT-PCR-based molecular methods.

Molecular detection of cytokeratins is an efficient tool to distinguish tumors with epithelial differentiation. It is also a precise method for the detection of epithelial tumor cells in peripheral blood, bone marrow or surgical margins, which is an important prognostic factor for therapy management. Many studies demonstrate that the detection rate of cytokeratins by PCR-based methods in blood and bone marrow correlates with the tumor stage. Furthermore the presence of cytokeratin-positive cells originated from epithelial tumors in bone marrow correlate with a decrease in tumor-free survival period of cancer patients.

Cytokeratin 20 (CK 20), cytokeratin 19 (CK 19) and cytokeratin 8 (CK 8) are the most common cytokeratins used as a target for the molecular detection of epithelial tumors and their metastases. CK 8 and CK 19 are widely expressed in mucosal epithelial cells and highly expressed in breast and lung carcinoma. CK 20 is highly expressed in adenocarcinoma of gastrointestinal tract in addition to pancreas, bile ducts and some gynecological adenocarcinomas. Other cytokeratins like cytokeratin 18 (CK 18) are not optimal for molecular detection due to the presence of pseudogenes or ectopic gene expression causing false-positive RT-PCR results.

An important aspect targeting cytokeratins and many other tissue-specific antigens using molecular methods are the contradictory results reported in the literature about recurrent false-positive results. To minimize the possibility of false-positive results we have to take the following factors in consideration:

- The contamination of blood and bone marrow samples during sample collection by epidermal epithelial cells or epithelial cells of skin appendages in addition to other nonepithelial cells such as Merkel cells and melanocytes, which can be the cause of false-positive results. To avoid this contamination it is recommended to aspirate two sample tubes and to use only the second portion for molecular analysis as the first portion may contain the above-mentioned contaminating cells.

- The presence of pseudogenes in the cells or the illegitimate expression of some cytokeratins in non-epithelial cells, especially in the nucleated and CD34+ positive blood and bone marrow cells, which can be the cause of false-positive results. Helpful to avoid this factor is the optimal primer construction and the use of nested primers. The simultaneous use of a primer panel complementary to two or more cytokeratins is also useful. Immediate denaturation of collected blood or bone marrow samples is also an effective method for the preservation of RNA and inactivation of cellular enzymes, which is important to prevent the illegitimate transcription of genes including those encoding cytokeratins. Methods to remove the non-specific cells from samples such as immunobead capture and gradient centrifugation can reduce the amount of the targeted cells and diminish the sensitivity of PCR.
- The presence of benign ectopic epithelial structures in lymph nodes and soft tissue parts can be an error source while screening of lymph nodes and surgical margins for micrometastases. These epithelial structures can be of different origin such as Müllerian epithelial inclusions and endometriosis in pelvic and abdominal lymph nodes or heterotopic ducts and glands in abdominal, thoracic and cervical lymph nodes. Microscopic examination of H&E tissue sections prior to molecular examination can be helpful to exclude the presence of such epithelial structures.
- PCR conditions as excess of PCR cycles and low annealing temperature can lead to the amplification of non-specific sequences.

3.6.2
Detection of Micrometastases and Circulating Tumor Cells of Malignant Epithelial Neoplasia by the Amplification of Cytokeratin 8 and Cytokeratin 18 mRNA

Cytokeratin 8 (CK 8) is an intermediate filament type II and one of the first cytokeratins expressed in the embryonal tissue and persists throughout the adult tissue. CK 8 is encoded on chromosome 12 and expressed in the vast majority of adenocarcinomas including lung, breast, gastrointestinal adenocarcinoma in addition to transitional, neuroendocrine and hepatocellular carcinoma, but is not detectable in squamous epithelium. CK 8 is also known as tissue polypeptide antigen (TPA) detectable in serum of cancer patients.

Cytokeratin 18 is an intermediate filament protein, usually co-expressed with cytokeratin 8 and is found in most simple ductal and glandular epithelia.

The molecular detection of mRNA complementary to both CK8 and CK18 in lymph nodes, blood or bone marrow can be used for detection of submicroscopic metastases or circulating tumor cells of different carcinoma types. Some studies demonstrate recurrent false-positive results targeting CK8/18 mRNA in blood samples or bone marrow, most probably due to the presence of pseudogenes or the expression of these cytokeratins in non-epithelial structures. Consequently, it is recommend to target these markers in combination with other epithelium-specific markers such as CEA, MUC-1 using a primer panel in addition to a negative control.

Method The method used is mRNA extraction from peripheral blood, bone marrow, body fluids, or tissue (lymph node), followed by RT-PCR.

- Primers for the cDNA amplification of the *CK 8* mRNA template:
  FP: 5′-AAC.AAC.CTT.AGG.CGG.CAG.CT-3′
  RP: 5′-GCC.TGA.GGA.AGT.TGA.TCT.CG-3′

PCR generates a 244-bp DNA fragment.

3.6.3 Detection of Micrometastases and Circulating Tumor Cells of Malignant Epithelial Neoplasia by the Amplification of Cytokeratin 19 mRNA

Cytokeratin 19 (CK 19) is an intermediate filament type I, and the smallest human keratin filament protein with a molecular mass of 40 kDa, encoded on chromosome 17 and expressed in many types of neoplasia arising from ductal, glandular and non-keratinizing epithelia especially basal layer components.

CK 19 has a high expression level in many types of breast cancer and in adenocarcinomas of the lung. Many reports demonstrated that the amplification of CK 19 mRNA is a sensitive method for the detection of micrometastases and circulating tumor cells in of many carcinoma types specially breast and lung adenocarcinomas, which usually correlates with tumor stage including tumor size and nodal involvement. It is also a valuable method to monitor the circulating tumor cells after surgical excision. The molecular detection of tumor cells in bone marrow is a bad prognostic factor predicting early recurrence and distant metastases.

Some studies reported false-positive results amplifying CK 19 mRNA extracted from blood and bone marrow samples, which may be due to the presence of CK 19 pseudogenes, possible illegitimate expression or due to the contamination of blood samples by different cutaneous cells during blood aspiration or bone marrow collection. To avoid sample contamination by cutaneous epithelial cells, it is important to take two blood samples and to use the second one for the RNA extraction. It must be also mentioned that some lymph nodes harbor benign ectopic epithelial structures such as benign Müllerian inclusions and endometriosis in pelvic and abdominal lymph nodes or heterotopic ducts and glands in axillary lymph nodes, which can be also the source of contaminating cytokeratin mRNA causing false-positive results. Consequently, it is more informative to use the CK 19 mRNA as a molecular target in combination with other epithelial molecular markers such as CEA, CK 20, and MUC-1 in a primer panel, also as multiplex PCR. The use of such panels reduces the incidence of false-negative results.

Method The method used is mRNA extraction from different tissue types such as lymph nodes, peripheral blood, body fluids and bone marrow, followed by RT-PCR.

- Primers for the cDNA amplification of the *CK 19* mRNA template:
  FP: 5′-TCG.ACA.ACG.CCC.GTC.TG-3′
  RP: 5′-CCA.CGC.TCA.TGC.GCA.G-3′

PCR generates a 75-bp DNA fragment.

- Primers for the cDNA amplification of the *CK 19* mRNA template:

FP: 5′-ATGAAAGCTGCCTTGGAAGA-3′
RP: 5′-TGATTCTGCCGCTCACTATCAG-3′

PCR generates a 138-bp DNA fragment.

- Alternative primers for the cDNA amplification of the *CK 19* mRNA template:
  FP: 5′-CCG.CGA.CTA.CAG.CCA.CTA.CTA.CAC-3′
  RP: 5′-GAG.CCT.GTT.CCG.TCT.CAA.A-3′

PCR generates a 169-bp DNA fragment.

- Alternative primers for the cDNA amplification of the *CK 19* mRNA template:
  FP: 5′-CCA.AGA.TCC.TGA.GTG.ACA.TGC.GAA.G-3′
  RP: 5′-TGC.AGC.TCA.ATC.TCA.AGA.CCC.TGA.A-3′

PCR generates a 198-bp DNA fragment.

- Alternative primers for the cDNA amplification of the *CK 19* mRNA template:
  FP: 5′-GCG.GGA.CAA.GAT.TCT.TGG.TG-3′
  RP: 5′-CTT.CAG.GCC.TTC.GAT.CTG.CAT-3′

PCR generates a 214-bp DNA fragment.

- Alternative primers for the cDNA amplification of the *CK 19* mRNA template (nested PCR):
  –Outer primers:
  FP: 5′-TTT.GAG.ACG.GAA.CAG.GCT.CT-3′
  RP: 5′-CAG.CTC.AAT.CTC.AAG.ACC.CTG-3′

First PCR generates a 426-bp DNA fragment.

  –Nested primers:
  FP: 5′-GCA.GAT.CGA.AGG.CCT.GAA-3′
  RP: 5′-TGA.ACC.AGG.CTT.CAG.CAT.C-3′

Nested PCR generates a 209-bp DNA fragment.

- Alternative primers for the cDNA amplification of the *CK 19* mRNA template:
  FP: 5′-AGG.TGG.ATT.CCG.CTC.CGG.GCA-3′
  RP: 5′-ATC.TTC.CTG.TCC.CTC.GAG.CA-3′

PCR generates a 460-bp DNA fragment.

3.6.4 Detection of Micrometastases and Circulating Tumor Cells of Malignant Epithelial Neoplasia by the Amplification of Cytokeratin 20 mRNA

Cytokeratin 20 (CK 20) is a type I intermediate filament protein composed of 424 amino acids with a molecular mass of 46 kDa (low molecular weight cytokeratin) encoded by a 18-kb gene composed of 8 exons and 7 introns located on chromosome 17 and transcribed to a ~1.75-kb mRNA. CK 20 is the major cellular protein of mature enterocytes, goblet cells and expressed in many epithelial glandular tumors especially colorectal adenocarcinomas and in a less degree in gastric adenocarcinomas. Furthermore, CK 20 is also expressed in various other adenocarcinomas including adenocarcinomas of pancreas, gall bladder and bile ducts, mucinous ovarian tumors, transitional cell carcinomas and Merkel cell carcinoma. CK 20 is not expressed by breast, lung, endometrium, neuroendocrine tumors or non-mucinous ovarian carcinomas.

For CK 20-positive tumors, CK 20-mRNA is a suitable molecular marker to follow-up. Various studies showed that patients with RT-PCR-positive results for CK 20 mRNA in both bone marrow and peripheral blood or in bone marrow alone revealed shorter survival than RT-PCR negative patients. These studies demonstrate also that the molecular detection of CK 20 expressing cells is a more discriminatory method than the immunohistochemical or molecular detection of other cytokeratins or tumor markers. However, the dilemma of occasional false-positive results as described for other cytokeratins still restricts the use of CK 20 as an accurate marker. CK 20 is not found in endothelial and lymphoid tissue, but several studies demonstrate the illegitimate expression of CK 20 in granulocytes. To minimize this negative aspect it is essential to achieve the denaturation of the blood or bone marrow samples immediately after the aspiration using a strong denaturing solution as described for RNA extraction. The use of nested PCR in addition to the optimal selection of primers is another important factor to minimize this problem.

In conclusion, the amplification of CK 20 mRNA in blood, bone marrow, lymph nodes or body fluids is an informative method for the detection of micrometastases or circulating tumor cells originated from colorectal and gastric adenocarcinomas.

Method The method used is mRNA extraction from different tissue types such as lymph nodes or peripheral blood, body fluids and bone marrow, followed by RT-PCR or nested PCR. It is necessary to consider the remarks mentioned in the introduction to this chapter in order to avoid false-positive results due to the contamination of blood or bone marrow samples by cytokeratin 20-positive epidermal cells or Merkel cells integrated in the epidermis.

- Primers for the cDNA amplification of the *CK20* mRNA template:
  FP: 5′-GCG.ACT.ACA.GTG.CAT.ATT.ACA.GAC.AA-3′
  RP: 5′-GCA.GGA.CAC.ACC.GAG.CAT.TT-3′

PCR generates an 87-bp DNA fragment.

- Alternative primers for the cDNA amplification of the *CK 20* mRNA template:
  FP: 5′-GGT.CGC.GAC.TAC.AGT.GCA.TAT.TAC.A-3′
  RP: 5′-CCT.CAG.CAG.CCA.GTT.TAG.CAT.TAT.C-3′

PCR generates a 120-bp DNA fragment.

- Alternative primers for the cDNA amplification of the *CK 20* mRNA template:
  FP: 5′-CTG.AAT.AAG.GTC.TTT.GAT.GAC.C-3′
  RP: 5′-ATG.CTT.GTG.TAG.GCC.ATC.GA-3′

PCR generates a 138-bp DNA fragment.

- Alternative primers for cDNA amplification of the *CK 20* mRNA template (nested PCR):
  –Outer primers:
  FP1: 5′-ATG.GAT.TTC.AGT.CGC.AGA-3′
  RP1: 5′-ATG.TAG.GGT.TAG.GTC.ATC.AAA.G-3′

PCR generates a 557-bp DNA fragment.

–Nested primers:

FP2: 5′-TCC.AAC.TCC.AGA.CAC.ACG.GTG.AAC.TAT.G-3′

RP2: 5′-CAG.GAC.ACA.CCG.AGC.ATT.TTG.CAG-3′

Nested PCR generates a final 290-bp DNA fragment.

- Alternative primers for the amplification of the *CK 20* mRNA template:

  FP: 5′-AGA.CCA.AGG.CCC.GTT.ACA.G-3′

  RP: 5′-ACG.ACC.TTG.CCA.TCC.ACT.ACT.TC-3′

PCR generates a 298-bp DNA fragment.

- Alternative primers for the cDNA amplification of the *CK 20* mRNA template:

  FP: 5′-CAG.ACA.CAC.GGT.GAA.CTA.TGG-3′

  RP: 5′-GAT.CAG.CTT.CCA.CTG.TTA.GAC.G-3′

PCR generates a 370-bp DNA fragment.

3.7 Molecular Detection of Tumor-Associated Viruses

3.7.1 Introduction

Only limited numbers of viruses are able to transform human cells to an oncogenic status to induce malignant tumors, while only a few human tumors are associated with chronic viral infection.

During the latent viral infection, the viral genome persists as viral episomes in the infected cells, whereas the direct integration of the viral genome within the cell genome still in most cases controversial. The viral oncogenesis has general mechanisms, which mainly include the deregulation of the cell cycle through the activation of different cellular proto-oncogenes or the inhibition of tumor suppressor genes. The following DNA and RNA virus families are the main viruses reported to be the etiologic factor or the co-factor in human malignancies:

3.7.1.1 **DNA Viruses**

- *Papovaviridae*: papovaviridae is a DNA virus family containing double-stranded DNA. The human papillomavirus is the most important member of this virus family, usually associated with cervical intraepithelial dysplasia and squamous cell carcinoma, skin tumors and tumors of the oropharyngeal region. The second member of this virus family is the polyomavirus associated with other animal tumors.

- *Herpesviridae*: the herpesvirus family consists of a large number of encapsulated DNA viruses and subdivided into three subfamilies: alpha herpesvirinae, beta herpesvirinae and gamma herpesvirinae. This classification was created by the herpesvirus study group of the International Committee on Taxonomy of Viruses according to biological properties of the viruses and it does not always count on the DNA sequence homology of the viruses.

There are eight members of the herpesvirus family able to infect humans: HHV-1 (herpes simplex virus 1, HSV-1); HHV-2 (herpes simplex virus 2, HSV-2); HHV-3 (varicella zoster virus,VZV); HHV-4 (Epstein–Barr virus, EBV); HHV-5 (cytomegalovirus, CMV); HHV-6 (human B-cell lymphotrophic virus or roseolovirus); HHV-7 (closely related to HHV-6); HHV-8 (a type of the rhadinovirus).

The Epstein–Barr virus (EBV; HHV-4) and the human herpesvirus 8 (HHV-8) are two members of the gamma herpesvirus family, able to induce human tumors. EBV (HHV-4) is found in association with different human neoplasia such as Hodgkin's disease, lymphoepithelial carcinoma (nasopharyngeal carcinoma), peripheral T-cell lymphoma and NK/T-cell lymphoma. It is also found in about 90% of endemic Burkitt's lymphoma and in up to 30% of the sporadic cases. HHV-8 is discovered in association with Kaposi's sarcoma, multiple myeloma, primary effusion lymphoma, inflammatory myofibroblastic tumor (inflammatory pseudotumor) and Castleman's disease.

- *Hepadnaviridae*: the hepadnavirus family is represented by the hepatitis B virus (HBV). HBV is one of the few known non-retroviral DNA viruses that use the reverse transcription as a stage of its replication cycle. After the infection of the cell, the viral DNA genome is transferred into the cell nucleus where the viral mRNAs are transcribed. HBV is an important cause of virus hepatitis and frequently associated with chronic liver cell injury, liver cirrhosis and hepatocellular carcinoma.

3.7.1.2 RNA Viruses

- *Retroviridae*: the retrovirus family includes enveloped RNA viruses, which depend on the reverse transcriptase enzymes to achieve the reverse transcription of its RNA into DNA, able to be integrated into the cellular genome by the activity of an integrase enzyme. This viral family includes the human T-cell leukemia virus types I-IV.

Type I and II (HTLV I and II) are two closely related types with a high genomic similarity, tropic for CD4+ T-lymphocytes. HTLV I is associated with the adult T-cell lymphoma and leukemia in addition to some types of myelopathy. HTLV II might be involved in certain demyelinating diseases. HTLV-III has similarity to the Simian T-lymphotropic virus 3 (STLV-III)

- *Flaviviridae*: the hepacivirus family is represented by its single member the hepatitis C virus. This important human hepatitis virus is reported to be associated with hepatocellular carcinoma, lymphoplasmacytic lymphoma, extranodal marginal zone B-cell lymphoma and monoclonal lymphoid proliferation.

3.7.2 Detection of Human Papillomavirus

Human papillomavirus (HPV) is a member of the papovavirus group containing a double stranded DNA genome composed of approximately 8000 bp and containing nine open reading frames. Viruses that differ by more than 10% of their L1 gene homology are categorized as a different virus type, by 2–10% as different subtype and >2% as variant. More than 100 HPV types are reported with different pathogenicity and able to infect various animals from birds to mammals. The viral genome is composed of the following three main regions separated by two early and late polyadenylation sites:

- Upstream regulatory region (URR);
- Early region (ER) includes several genes occupying over 50% of the virus genome, encoding six common viral oncoproteins (E1, E2, E4, E5, E6 and E7);
- Late regions, L1 encoding the major viral capsid protein and L2 encoding the minor viral capsid protein.

Each virus type has different sets of oncogenes, which determine its pathogenicity and the clinical behavior of the infection. The main oncogenes include:

- E1 viral DNA replication factor;
- E2 responsible for DNA replication and RNA transcription;
- E4 encodes the protein causing the collapse of cytoplasmic cytokeratin;
- E5 responsible for EGF signal transmission and growth regulatory mechanisms;
- E6 a cell cycle regulator affecting P53 and telomerase;
- E7 an oncogene reacting with the retinoblastoma (Rb) tumor suppressor gene.

According to the composition of gene sets integrated within the virus type, human papillomavirus is divided into low-risk, probable high-risk and high-risk types. Low-risk HPV types include the virus types 6, 11, 42, 43 and 44. High-risk HPV types include the types 16, 18, 31, 33, 35, 39, 45, 51, 52, 56, 58, 59, 68, 73 and 82. The HPV types 26, 53 and 66 are classified as probable high-risk types. In the case of mixed infection, a potential synergy between different HPV types is possible to cause malignancies.

HPV is characterized by its high tropism for epithelial tissue and subdivided into mucosal and cutaneous types. Different virus types cause different epithelial lesions: HPV type 1, 2 and 4 cause the common and plantar warts, HPV type 6, 10, 11 and 40–54 are responsible for the anogenital warts (condyloma acuminatum). HPV types 16, 18, 31 are able to cause various grades of squamous cell dysplasia including cervical intraepithelial dysplasia (CIN), which may be transformed in to a frank squamous cell carcinoma. Genetic analysis demonstrate that HPV-16 alone is responsible for the development of up to 60% of cervical cancer, it is also found in association with oropharyngeal squamous cell carcinoma and bladder transitional cell papilloma or transitional cell carcinoma. Clinical follow-ups demonstrate that HPV-negative genital lesions are rarely transformed to invasive cervical carcinoma. Other HPV types (especially HPV 18) are able to infect the glandular epithelium causing adenoma, adenocarcinoma *in situ* or adenocarcinoma.

The full replication cycle of human papillomaviruses takes place in epithelial cells. The infection begins after the viral insertion into the nuclei of basal epithelial cells. After the infection, few viral copies survive in the cells or persist as episomes. HPV oncogenes are capable to activate the cellular DNA methylation mechanisms and as a result epigenetically regulate both viral and cellular genes. E6 binds to p53 and causing its degradation. E7 binds to Rb and prevents it from degradation. Consequently, the HPV oncogenes E6 and E7 are able to convert the infected cells into P53- and Rb-negative status, which seems to be one of the major steps into malignant transformation. The later activation of the E2 gene causes down-regulation of E6 and E7 genes and the cells may be able to continue with normal cell cycle and differentiation. In the final stage, the L1 and L2 capsid genes are expressed and the viral particles are assembled in the nucleus, followed by the release of complete viruses in the upper layer of the epithelium without destruction of the cells. The integration of viral oncogenic sequences into human genome is the first step in malignant transformation and seems to be an important mechanism in the progression of dysplasia in to invasive carcinoma. During the integration process, the viral genome undergoes fragmentation. The E2 gene seems to be inactivated and the down regulation of the E6 and E7 gene is removed, as a result the infected cells continue to be in a P53- and Rb-negative status.

The above-described mechanisms make the detection of HPV an important parameter for the follow-up and further treatment of different dysplastic and neoplastic epithelial lesions. *In situ* hybridization and different PCR-based molecular methods are the most common used molecular tools to detect HPV in various specimens.

Method The method used is extraction of viral DNA from different tissue types, swabs or body fluids, followed by PCR.

- Screening primers for the amplification of the *HPV* genome (nested PCR):
  –Outer primers:
    FP (MY11): 5′-GC(A/C).CAG.GG(A/T).CAT.AA(C/T).AAT.GG-3′
    RP (MY09): 5′-CGT.CC(A/C).A(A/G)(A/G).GGA.(A/T)AC.TGA.TC-3′

Both primers are degenerate primers composed of a mixture of two or more primers that differs in base composition at the positions placed in brackets.
PCR generates a 450-bp DNA fragment.
  –Nested primers:
    FP (GP5): 5′-TTT.GTT.ACT.GTG.GTA.GAT.AC-3′
    RP (GP6): 5′-GAA.AAA.TAA.ACT.GTA.AAT.CA-3′

PCR generates a 140-bp DNA fragment.

- Alternative primers for screening of the *HPV* genome:
  FP (L1C1): 5′-CGT.AAA.CGT.TTT.CCC.TAT.TTT.TTT-3′
  RP (L1C2-1): 5′-TAC.CCT.AAA.TAC.TCT.GTA.TTG-3′
  RP (L1C2-2): 5′-TAC.CCT.AAA.TAC.CCT.ATA.TTG-3′

PCR carried out in two phases using the forward primer and one of the two reverse primers. PCR generates a 250-bp DNA fragment.
For better results, it is recommended to screen with the three above-mentioned screening primer sets.

- Primers specific for *HPV types 2, 6, 11, 13, 16, 18, 30, 32 and 58*:
  FP: 5′-ATT.CAG.GAT.GGT.GAT.ATG.G-3′
  RP: 5′-CCC.CAG.CAA.ATG.CCA.TTA.TTG.TG-3′

PCR generates different DNA fragments according to virus type: 386 bp in HPV type 13; 392 bp in HPV type 2; 395 bp in types 6, 11, 16, 18, 32, 58 and 404 bp in HPV type 33.

- Primers specific for *HPV types 6b and 11*:
  FP: 5′-GTG.TTT.TGC.AGG.AAT.GCA.CTG.ACC.A-3′
  RP: 5′-CAG.CAT.AAT.TAA.AGT.GTC.TAT.ATT.G-3′

PCR generates a 157-bp DNA fragment.

- Alternative primers specific for *HPV types 6 and 11* (region E6/E7):
  FP: 5′-TAC.ACT.GCT.GGA.CAA.CAT.GC-3′
  RP: 5′-GTG.CGC.AGA.TGG.GAC.ACA.C-3′

PCR generates a 302-bp DNA fragment.

- Primers specific for *HPV type 16*:
  FP: 5′-ACC.GAA.ACC.GGT.TAG.TAT.AAA.AGC-3′
  RP: 5′-ATA.ACT.GTG.GTA.ACT.TTC.TGG.GTC-3′

PCR generates a 98-bp DNA fragment.

- Alternative primers specific for *HPV type 16* (region E6/E7):
  FP: 5′-CCC.AGC.TGT.AAT.CAT.GCA.TGG.AGA-3′
  RP: 5′-GTG.TGC.CCA.TTA.ACA.GGT.CTT.CCA-3′

PCR generates a 253-bp DNA fragment.

- Alternative primers specific for *HPV type 16*
  FP: 5′-TGA.GCA.ATT.AAA.TGA.CAG.CTC.AGA.G-3′
  RP: 5′-TGA.GAA.CAG.ATG.GGG.CAC.ACA.AT-3′

PCR generates a 212-bp DNA fragment.

- Primers specific for *HPV type 18*:
  FP: 5′-CGG.TCG.GGA.CCG.AAA.ACG.GTG-3′
  RP: 5′-CGT.GTT.GGA.TCC.TCA.AAG.CGC.GCC-3′

PCR generates an 80-bp DNA fragment.

- Alternative primers specific for *HPV type 18* (region E6/E7):
  FP: 5′-CGA.CAG.GAA.CGA.CTC.CAA.CGA-3′
  RP: 5′-GCT.GGT.AAA.TGT.TGA.TGA.TTA.ACT-3′

PCR generates a 201-bp DNA fragment.

- Alternative primers specific for *HPV type 18*:
  FP: 5′-GAC.CTT.CTA.TGT.CAC.GAG.CAA.TTA-3′
  RP: 5′-TGC.ACA.CCA.CGG.ACA.CAC.AAA.G-3′

PCR generates an 236-bp DNA fragment.

- Primers specific for *HPV type 26*:
  FP: 5′-CGA.AAT.TGA.CCT.ACG.CTG.CTA.CG-3′
  RP: 5′-TGG.CAC.ACC.AAG.GAC.ACG.TCT.TC-3′

PCR generates an 239-bp DNA fragment.

- Primers specific for *HPV type 31* (region L1):
  FP: 5′-ATG.GTG.ATG.TAC.ACA.ACA.CC-3′
  RP: 5′-GTA.GTT.GCA.GGA.CAA.CTG.AC-3′

PCR generates a 514-bp DNA fragment.

- Alternative primers specific for *HPV type 31*:
  FP: 5′-AGC.AAT.TAC.CCG.ACA.GCT.CAG.AT-3′
  RP: 5′-GTA.GAA.CAG.TTG.GGG.CAC.ACG.A-3′

PCR generates a 210-bp DNA fragment.

- Primers specific for *HPV type 33* (region E1/E2):
  FP: 5′-ATG.ATA.GAT.GAT.GTA.ACG.CC-3′
  RP: 5′-GCA.CAC.TCC.ATG.CGT.ATC.AG-3′

PCR generates a 456-bp DNA fragment.

- Alternative primers specific for *HPV type 33*:
  FP: 5′-ACT.GAC.CTA.(C/T)AC.TGC.TAT.GAG.CAA-3′
  RP: 5′-TGT.GCA.CAG.(C/G)TA.GGG.CAC.ACA.AT-3′

Both forward and reverse primers are degenerate primers composed of a mixture of two or more primers that differs in base composition at the positions placed in brackets.

PCR generates a 229-bp DNA fragment.

- Primers specific for *HPV type 35*:
  FP: 5′-CAA.CTG.ACC.TAT.ACT.GTT.ATG.AGC-3′
  RP: 5′-TGT.GAA.CAG.CCG.GGG.CAC.ACT.A-3′

PCR generates a 234-bp DNA fragment.

- Primers specific for *HPV type 39*:
  FP: 5′-TTG.TAT.GTC.ACG.AGC.AAT.TAG.GAG-3′
  RP: 5′-GAC.ACT.GTG.TCG.CCT.GTT.TGT.TTA-3′

PCR generates a 357-bp DNA fragment.

- Primers specific for *HPV type 45*:
  FP: 5′-GAC.CTG.TTG.TGT.TAC.GAG.CAA.TTA-3′
  RP: 5′-TGC.ACA.CCA.CGG.ACA.CAC.AAA.G-3′

PCR generates a 236-bp DNA fragment.

- Primers specific for *HPV type 51*:
  FP: 5′-GCT.ACG.AGC.AAT.TTG.ACA.GCT.CAG-3′
  RP: 5′-ATC.GCC.GTT.GCT.AGT.TGT.TCG.CA-3′

PCR generates a 242-bp DNA fragment.

- Primers specific for *HPV type 52*:
  FP: 5′-ACT.GAC.CTA.(C/T)AC.TGC.TAT.GAG.CAA-3′
  RP: 5′-CAG.CCG.GGG.CAC.ACA.ACT.TGT.AA-3′

The forward primer is a degenerate primer. PCR generates a 229-bp DNA fragment.

- Primers specific for *HPV type 53*:
  FP: 5′-ACC.TGC.AAT.GCC.ATG.AGC.AAT.TGA.A-3′
  RP: 5′-TTA.TCG.CCT.TGT.TGC.GCA.GAG.G-3′

PCR generates a 253-bp DNA fragment.

- Primers specific for *HPV type 56*:
  FP: 5′-ACC.TAC.A(A/G)T.GCA.ATG.AGC.AAT.TGG-3′
  RP: 5′-TGA.TGC.GCA.GAG.TGG.GCA.CGT.TA-3′

The forward primer is a degenerate primer. PCR generates a 244-bp DNA fragment.

- Primers specific for *HPV type 58*:
  FP: 5′-GCT.ATG.AGC.AAT.TAT.GTG.ACA.GCT-3′
  RP: 5′-TGT.GCA.CAG.(C/G)TA.GGG.CAC.ACA.AT-3′

The reverse primer is a degenerate primer. PCR generates a 219-bp DNA fragment.

- Primers specific for *HPV type 59*:
  FP: 5′-ACC.TTG.TGT.GCT.ACG.AGC.AAT.TAC-3′
  RP: 5′-GCT.GCA.CAC.AAA.GGA.CAC.ACA.AA-3′

PCR generates a 243-bp DNA fragment.

- Primers specific for *HPV type 66*:
  FP: 5′-ACC.TAC.A(A/G)T.GCA.ATG.AGC.AAT.TGG-3′
  RP: 5′-TGA.TGC.GCA.GAG.TGG.GCA.CGT.TA-3′

The forward primer is a degenerate primer. PCR generates a 244-bp DNA fragment.

- Primers specific for *HPV type 68*:
  FP: 5′-TTG.TAT.GTC.ACG.AGC.AAT.TAG.GAG-3′
  RP: 5′-GAT.TAC.TGG.GTT.TCC.GTT.GCA.CAC-3′

PCR generates a 258-bp DNA fragment.

- Primers specific for *HPV type 70*:
  FP: 5′-CAC.GAG.CAA.TTA.GAA.GAT.TCA.GAC.A-3′
  RP: 5′-TTC.CCG.ATG.CAC.ACC.AGG.GAC.A-3′

PCR generates a 237-bp DNA fragment.

- Primers specific for *HPV type 73*:
  FP: 5′-CTT.ACA.TGT.TAC.GAG.TCA.TTG.GAC-3′
  RP: 5′-GTT.TCT.GGA.ACA.GTT.GGG.GCA.C-3′

PCR generates a 221-bp DNA fragment.

- Primers specific for *HPV type 82*:
  FP: 5′-GCT.ACG.AGC.AAT.TTG.ACA.GCT.CAG-3′
  RP: 5′-CAT.TGC.CGA.TGT.TAG.TTG.GTC.GCA-3′

PCR generates a 240-bp DNA fragment.

3.7.3 Detection of Human Herpesviruses: Epstein–Barr Virus and Human Herpesvirus 8

The Herpesviridae virus family includes a large number of DNA viruses and is divided into alpha, beta and gamma subtypes. Eight members of the herpesvirus family are able to infect humans causing infectious inflammatory diseases or neoplastic changes. Both of Epstein–Barr virus (EBV; HHV-4) and human herpesvirus 8 (HHV-8) belong to the gamma subtype and described in association with different human neoplasia.

3.7.3.1 Epstein–Barr Virus

The Epstein–Barr virus (EBV) is one of the 8 human herpesviruses and one of the most common human viruses. It contains a 173 000-bp double-strand DNA gene, composed of five unique regions (U1–U5) separated by four internal repeats (IR1–IR4) and ended by terminal repeats (TR). Two types of EBV can be distinguished, type A (type 1) and type B (type 2) that differs at a number of genes and consequently shows different biological behavior, as EBV type A is found to be more efficient in immortalizing of B-lymphocytes. The virus is widely distributed within the adult population as a latent infection and the virus is maintained at low copy numbers in the memory B-cells and about 95% of healthy adults have antibodies to the virus antigens. In acute infectious mononucleosis caused by this virus, the virus infects mature B-lymphocytes by binding to the EBV receptor CR2 (CD21) on cell membrane causing the proliferation of infected cells. Infected cells synthesizing different EBV antigens including the *ly*mphocyte *d*etermined *m*embrane *a*ntigen (LYDMA) are recognized and destroyed by cytotoxic T-lymphocytes that selectively recognize the infected cells. After a primary infection, the EBV persists in the host throughout the whole life. In healthy virus carries, B-lymphocytes and mononuclear cells are the cellular reservoir of the virus.

The viral infection induces the synthesis of different viral antigens during different stages of the infection, which include the Epstein–Barr virus nuclear antigens (EBNAs: EBNA1, EBNA2, EBNA3A, B, C and EBNA-LP), the latent infection-associated membrane proteins (LMP1, LMP2A, LMP2B) and the terminal protein (TR). The following genes are found to play the most important roles during the course of the latent infection:

- *BZLF1* (ZEBRA) is the gene encoding an immediate early transcription activator responsible for the lytic infection phase.
- *BcLF1* gene, which encodes the major viral capsid protein.
- *BNLF1* gene encoding the *l*atent *m*embrane *p*rotein *1* (LMP1), which believed to be the only viral oncogene with obvious transforming properties playing a central role in the transformation of the B-lymphocytes and causing cell proliferation by activation of the bcl2 oncogene and the TNF/CD40 pathway.
- *BALF2* gene, encodes a single-strand-binding protein.
- *EBNA1* encodes a protein responsible for the maintenance of the EBV genome in an episomal state in latently infected cells. It mediates the replication and partitioning of the episome during the division of the host cells.
- *EBNA2* encodes a nuclear antigen with promoter activity functioning as the main viral transactivator.

EBV is able to infect different cell types including B- and T-lymphocytes, follicular dendritic cells, squamous epithelium of the upper respiratory tract and glandular epithelium of salivary glands, thyroid and stomach. In infected cells the viral genome presents as episomes, whereas the integration of the viral genes within the cell genome still controversial, how ever the viral infection causes the overexpression of different oncogenes by stimulation of cyclin D2 leading to the acceleration of the G1 phase. The EBV infection is associated with many reactive and neoplastic disorders such as Burkitt's lymphoma, nasopharyngeal (lymphoepithelial) carcinoma, Hodgkin's lymphoma, post-transplant lymphoproliferative disorders and different types of B- and T-cell lymphomas. EBV is detected in more than 50% of classic Hodgkin's lymphoma subtypes including mixed cellularity, lymphocyte depleted, nodular sclerosis and lymphocyte rich classic subtypes. EBV is rarely detected in lymphocyte predominance Hodgkin's lymphoma.

Because EBV persists after the primary infection, the molecular detection of the virus remains positive and it is not possible to discriminate a specific EBV-causing disorder from the background latent infection. However, recent studies show that the estimation of the viral load by real-time PCR can be a useful parameter to indicate a possible relapse in the case of nasopharyngeal carcinoma and Hodgkin's lymphoma. In order to classify the type of the viral infection it is essential to determine the transcriptional pattern of the viral genome by detecting the different viral transcripts using RT-PCR. Three major latent types of EBV infection are described:

- *Type I* latency, characterized by the expression of EBNA1 and LMP2A only. This transcriptional pattern found in circulating lymphocytes of healthy carriers, but also characteristic for Burkitt's lymphoma.
- *Type II* latency, characterized by the expression of both EBNA-1 and LMPs, usually associated with Hodgkin's lymphoma, lymphoepithelial (nasopharyngeal) carcinoma, peripheral T-cell lymphoma and NK/T-cell lymphoma.

- *Type III* latency, characterized by the expression of all EBNAs and LMPs. This transcriptional pattern is characteristic for the acute phase of infectious mononucleosis but also found to be associated with post transplant lymphoproliferative disorders.

3.7.3.2 **Human Herpesvirus-8**

Human herpesvirus-8 (HHV-8, Kaposi's sarcoma-associated virus) is another member of the human herpesvirus family described recently in association with different human malignancies. HHV-8 is a DNA virus with a 165 000-bp genome packed in a protein capsule. The natural cellular reservoirs for HHV-8 are CD19-positive B-cells. The viral genome persists in the infected cells as intranuclear viral episomes and includes many genes similar to the genes of the host cells encoding viral oncoproteins and cell cycle regulatory and signaling proteins stimulating the oncogenesis. The following viral oncoproteins play an important role in the viral oncogenesis and able to control the cellular regulatory pathways and use the cellular replication machinery for replication:

- *Viral cyclin D,* prevents the G1 arrest causing the cell cycle progression.
- *IL-6 homologue,* functions as a growth factor, inhibiting the normal apoptotic process.
- *Viral bcl-2,* a human bcl-2 homolog effecting as an inhibitor of apoptosis.
- *Fas-ligand inhibitory protein* (*viral FLIP*) also effects as an apoptosis inhibitor.
- *ORF73 gene* encodes the latency-associated nuclear antigen (LANA) permanently expressed in HHV-8-associated tumors. LANA interacts with different genes regulating the cell cycle such as p53 and the Rb genes.

In addition to these genes there are other genes playing important role in the viral oncogenesis like interferon regulatory factor, G protein-coupled receptor, as well as DNA synthesis proteins including dihydrofolate reductase, thymidine kinase, thymidylate synthetase and DNA polymerase.

The HHV-8 virus is found in association with Kaposi's sarcoma, rare forms of B-cell lymphoma, multiple myeloma, primary effusion lymphoma, Castleman's disease and sarcoidosis. HHV-8 has been detected in the majority of bone marrow samples taken from multiple myeloma patients. Recently HHV-8 was described in association with the pulmonary and extrapulmonary inflammatory pseudotumor (inflammatory myofibroblastic tumor).

Methods *In situ* hybridization (ISH) and PCR-based molecular methods are two highly sensitive molecular methods used to detect both viruses. Using ISH, we can study the viral persistence in preserved tissue structure. PCR-based methods allow to study different viral expression patterns and latent infection status. For PCR-based methods, viral DNA and/or mRNA is extracted from tissue or body fluids and followed by PCR or RT-PCR.

- Primers specific for the *EBV* gene (type A and B):
  FP (Gene 1): 5′-AGG. GAT.GCC.TGG.ACA.CAA.G-3′
  RP (Gene 2): 5′-GTG.CTG.GTG.CTG.CTG.GTG.G-3′

The product of this PCR is used as a template for the next two reactions with primers specific for both EBV subtypes A and B:

- Primers for *EBV subtype A* gene:
  FP: 5′-TCT.TGA.TAG.GGA.TCC.GCT.AGG.ATA-3′
  RP: 5′-ACC.GTG.GTT.CTG.GAC.TAT.CTG.GAT.C-3′

Reaction generates a 497-bp DNA fragment.

- Primers for *EBV subtype B* gene:
  FP: 5′-CAT.GGT.AGC.CTT.AGG.ACA.TA-3′
  RP: 5′-AGA.CTT.AGT.TGA.TGC.CCT.AG-3′

Reaction generates a 150-bp DNA fragment.

- Primers for the amplification of EBV gene fragment encoding a *major viral trans-activator*:
  FP: 5′-CTT.GGA.GAC.AGG.CTT.AAC.CAG -3′
  RP: 5′-CCA.TGG.CTG.CAC.CGA.TGA.AAG.TTA -3′

Reaction generates a 264-bp DNA fragment.

Primers specific for viral transcripts:

- *EBV-ENBA1* RT-PCR sequence-specific primers (nested PCR):
  –Outer primers:
  FP: 5′-CTT.AGG.AAG.CGT.TTC.TTG.AGC.TT-3′
  RP: 5′-GGG.TCT.CCG.GAC.ACC.ATC.TC-3′

Reaction generates a 386-bp DNA fragment.
  –Nested primers:
  FP: 5′-GAG.CGT.TTG.GGA.GAG.CTG.AT-3′
  RP: 5′-CAT.TTC.CAG.GTC.CTG.TAC.CT-3′

Reaction generates a 172-bp DNA fragment.

- *EBV-ENBA2* RT-PCR sequence-specific primers (nested PCR):
  –Outer primers:
  FP: 5′-AGA.GGA.GGT.GGT.AAG.CGG.TCC-3′
  RP: 5′-TGA.CGG.GTT.TCC.AAG.ACT.ATC.C-3′

Reaction generates a 381-bp DNA fragment.
  –Nested primers:
  FP: 5′-AGC.GGT.TCA.CCT.TCA.GGG.CC-3′
  RP: 5′-TGT.AGG.CAT.GAT.GGC.GGC.AG-3′

Reaction generates a 300-bp DNA fragment.

- *EBV-EBNA3A* RT-PCR sequence-specific primers:
  FP: 5′-CTT.AGG.AAG.CGT.TTC.TTG.AGC.TT-3′
  RP: 5′-TCT.TCC.ATG.TTG.TCA.TCC.AGG.G-3′

Reaction generates a 220-bp DNA fragment.

- *EBV-EBNA3B* RT-PCR sequence-specific primers:
  FP: 5′-CTT.AGG.AAG.CGT.TTC.TTG.AGC.TT -3′

RP: 5′-CAT.AAT.CTG.GTG.GGT.CCT.CGG-3′

Reaction generates a 287-bp DNA fragment.

- *EBV-EBNA3C* RT-PCR sequence-specific primers:

  FP: 5′-CTT.AGG.AAG.CGT.TTC.TTG.AGC.TT -3′

  RP: 5′-TTA.TCT.CCC.CGC.TCA.TTG.TCG-3′

Reaction generates a 253-bp DNA fragment.

- *EBV-LMP1* RT-PCR sequence-specific primers (nested PCR):

  –Outer primers:

  FP: 5′-ACA.CAC.TGC.CCT.GAG.GAT.GG-3′

  RP: 5′-ATA.CCT.AAG.(A/C)AA.GTA.AGC.A-3′

  –Nested PCR primers:

  FP: 5′-CTT.CAG.AAG.AGA.CCT.TCT.CT-3′

  RP: 5′-ACA.ATG.CCT.GTC.CGT.GCA.AA -3′

Reaction generates a 104-bp DNA fragment.

- *EBV-LMP2A* RT-PCR sequence-specific primers:

  FP: 5′-CAC.CGC.TTA.TGA.GGA.CCC.A-3′

  RP: 5′-CTG.CGG.CCA.GCA.ATG.AAA.C-3′

Reaction generates a 477-bp DNA fragment.

  –Nested PCR primers:

  FP: 5′-ATG.ACT.CAT.CTC.AAC.ACA.TA-3′

  RP: 5′-CAT.GTT.AGG.CAA.ATT.GCA.AAA-3′

Reaction generates a 280-bp DNA fragment.

- *EBV-BZLF1* RT-PCR sequence-specific primers:

  FP: 5′-ATT.GCA.CCT.TGC.CGG.CCA.CCT.TTG-3′

  RP: 5′-CGG.CAT.TTT.CTG.GAA.GCC.ACC.CGA-3′

Reaction generates a 608-bp DNA fragment.

  –Nested PCR primers:

  FP: 5′-GAC.CAA.GCT.ACC.AGA.GTC.TAT-3′

  RP: 5′-CAG.AAT.CGC.ATT.CCT.CCA.GCC.A-3′

Reaction generates a 442-bp DNA fragment.

- *EBV-BcLF1* RT-PCR sequence-specific primers:

  FP: 5′-TAT.GCC.CAA.TCC.CAA.GTA.CAC.G-3′

  RP: 5′-TGG.ACG.GGT.GGA.GGA.AGT.CTT.C-3′

  –Nested primers:

  FP: 5′-ACA.CAG.CAG.CTA.CCG.GTG.GA-3′

  RP: 5′-TGT.TGC.AGG.GAG.TAG.GTC.TC-3′

Reaction generates a 181-bp DNA fragment.

- Primers specific for the *HHV-8* virus (nested PCR):

  –Outer primers:

  FP: 5′-AGC.ACT.CGC.AGG.GCA.GTA.CG-3′

  RP: 5′-GAC.TCT.TCG.CTG.ATG.AAC.TGG-3′

Reaction generates a 571-bp DNA fragment.

–Nested primers (KS330 primers, minor capsid protein):
FP: 5′-AGC.CGA.AAG.GAT.TCC.ACC.AT-3′
RP: 5′-TCC.GTG.TTG.TCT.ACG.TCC.AG-3′
Reaction generates a 233-bp DNA fragment.

- Alternative primers specific for the *HHV-8* gene template (nested PCR):
  –Outer primers:
  FR: 5′-AGG.CAA.CGT.CAG.ATG.TGA.C-3′
  RP: 5′-GAA.ATT.ACC.CAC.GAG.ATC.GC-3′
  –Nested primers:
  FP: 5′-CAT.GGG.AGT.ACA.TTG.TCA.GGA.CCT.C-3′
  RP: 5′-GGA.ATT.ATC.TCG.CAG.GTT.GCC-3′

Nested PCR generates a 213-bp DNA fragment.

- Alternative primers specific for the *HHV-8* gene template (nested PCR):
  –Outer primers:
  FR: 5′-GGC.GAC.ATT.CAT.CAA.CCT.CAG.G-3′
  RP: 5′-ATA.TCA.TCC.TGT.GCG.TTC.ACG.AC-3′
  –Nested primers:
  FP: 5′-CGC.ATG.GAG.GAC.CTA.GTC.AAT.AAC-3′
  RP: 5′-GGT.TGT.AGT.CAT.TCT.CGT.CCA.GGG-3′

Nested PCR generates a 115-bp DNA fragment.

- Primers specific for the *HHV-8* virus (*LANA gene*):
  FP: 5′-TTG.GGA.AAG.GAT.GGA.AGA.CG -3′
  RP: 5′-AGT.CCC.CAG.GAC.CTT.GGT.TT-3′

Reaction generates a 346-bp DNA fragment.

Primers specific for viral transcripts:

- *HHV-8-IL-6* RT-PCR sequence-specific primers:
  FP: 5′-TCA.CGT.CGC.TCT.TTA.CTT.ATC.GTG-3′
  RP: 5′-CGC.CCT.TCA.GTG.AGA.CTT.CGT.AAC-3′

Reaction generates a 695-bp DNA fragment.

- *HHV-bcl-2* RT-PCR sequence-specific primers:
  FP: 5′-GCT.CAG.CCC.TAT.TAA.GCT.ATA.CAT.CAC-3′
  RP: 5′-TCA.TAC.GCA.TAT.ACA.GGT.AAA.ACG.G-3′

Reaction generates a 267-bp DNA fragment.

- *HHV-8-cyclin D* RT-PCR sequence-specific primers:
  FP: 5′-CAC.CCT.GAA.ACT.CCA.GGC-3′
  RP: 5′-GAT.CCG.ATC.CTC.ACA.TAG.CG-3′

Reaction generates a 115-bp DNA fragment.

4 Classification of Malignancies and the Status of Minimal Residual Cancer Disease (an Addition to the TNM Classification)

The histological type and the anatomical extension of malignant tumors considered being the most important prognostic factors on which depends the treatment strategy. The anatomical extension of tumors is principally classified by the clinical TNM or by the pathological pTNM classification. This classification includes the extension of the primary tumor within anatomical region (T; T1–T4), the involvement of locoregional lymph nodes (N; N0–N3), and the presence of distant metastases (M; M0 and M1). The additional R classification describes the presence of residual tumor structures in the margins after surgical resection of tumor (R1 for microscopic and R2 for macroscopic residual tumor).

During the last decade the widespread use of other supplementary morphologic diagnostic methods like immunohistochemistry and non-morphologic methods like molecular and genetic methods as well as flow cytometry delivered further important information concerning tumor classification and associated genetic abnormalities, tumor spread in addition to micrometastases and the presence of disseminated isolated tumor cells (ITC). The obtained results provide additional significant information concerning estimated clinical and biological behavior of tumors for further tumor therapy.

In order to integrate the results obtained by these unconventional new methods the following classification was suggested and offered to the UICC to include the results of the molecular and immunohistochemical methods in the TNM classification:

- *pN1 (mi):* Lymph node bearing micrometastasis (0.21–2.0 mm in diameter);
- *pN0:* Histologically no regional lymph node metastases, no special examination for isolated tumor cells (ITC);
- *pN0 (i–):* Histologically no regional lymph node metastases, negative morphologic finding for isolated tumor cells ITC (*i* for morphologic methods of examination including immunohistochemistry);
- *pN0 (i+):* Histologically no regional lymph node metastases, positive morphologic finding for ITC or tumor cell deposits smaller than 0.2 mm in diameter (nanometastases);

Phenotypic and Genotypic Diagnosis of Malignancies. Muin S.A. Tuffaha
Copyright © 2008 WILEY-VCH Verlag GmbH & Co. KGaA, Weinheim
ISBN: 978-3-527-31881-0

- *pN0 (mol–):* Histologically no regional lymph node metastases, negative non-morphologic finding for ITC (*mol* for non-morphologic methods including molecular techniques);
- *pN0 (mol+):* Histologically no regional lymph node metastases, positive non-morphologic finding for ITC.

For sentinel lymph nodes, the additional symbol (sn) may be used as proposed by the UICC, for example pN0 (i+)(sn). Sentinel lymph node is defined as the first draining lymph node from tumor region (mainly used in cases of breast cancer or melanoma).

If ITC are found in bone marrow, blood or other organs by non-morphologic methods the symbols *i* and *mol* are added to M0 in addition to the specification of the anatomical sites; e.g. M0 (i+, liver) or M0 (mol+, blood).

Molecular methods used for the detection of micrometastases and molecular staging are very sensitive and can detect up to one tumor cell in 10^6 background cells, which explain the high frequency of positive lymph nodes detected by these methods. Cases with pN0 (mol+) status may reflect an occult metastatic disease, and for many tumor types considered being an unfavorable prognostic factor associated with early relapse. Furthermore, recent studies shows that breast cancers with pN0 (i+) nodal status have an increased risk for metastatic relapse. Nevertheless we have to take in consideration that not all tumor with pN0 (i+) or pN0(mol+) nodal status or positive surgical margins detected by molecular methods give rise to local tumor recurrence or to lymph node metastases or cause tumor dissemination and should not be considered in the TNM classification as residual tumor (R1). Furthermore isolated disseminated tumor cells cannot be considered as micrometastases as they do not fulfill all the criteria of metastases, which include the arrest and implantation of tumor cells associated with proliferation and supported by stromal induction in addition to neovascularization.

Finally, there are no specific therapeutic protocols established for the pN0 (mol+) cases and it is not known if these patients benefit from the adjuvant therapy, keeping in mind that most micrometastases are mainly composed of nonproliferating cells in the G0 phase (dormant cells), in which standard cytotoxic chemotherapy is less effective. Nevertheless it is important to keep this category of patients under follow-up.

Bibliography

Molecular Detection of Synovial Sarcoma

1 Hiraga, H., Nojima, T., Abe, S. *et al.* (1998) Diagnosis of synovial sarcoma with the reverse transcriptase polymerase chain reaction: analyses of 84 soft tissue and bone tumors. *Diag Mol Pathol*, **7**, 102–110.

2 Argani, P., Zakowski, M.F., Klimstra, D. *et al.* (1998) Detection of the SYT-SSX chimeric DNA in paraffin embedded tissue and its application in problematic cases. *Mod Pathol*, **11**, 65–71. (1998) Erratum. *Mod Pathol*, **11**, 592.

3 Lasota, J., Jasinski, M. *et al.* (1998) Detection of the SYT-SSX fusion transcripts in formaldehyde-fixed, paraffin embedded tissue: a reverse transcriptase polymerase chain reaction amplification assay useful in the diagnosis of synovial sarcoma. *Mod Pathol*, **11**, 626–633.

4 kim, D.-H., Sohn, J.H., Lee, M.C. *et al.* (2000) Primary synovial sarcoma of the kidney. *Am J Surg Pathol*, **24**, 1097–1104.

5 van de Rijn, M., Barr, F.G., Xiong, Q.-B. *et al.* (1999) Poorly differentiated Synovial sarcoma. *Am J Surg Pathol*, **23**, 106–112.

6 Hisaoka, M., Hashimoto, H., Iwamasa, T. *et al.* (1998) Primary synovial sarcoma of the lung: report two cases confirmed by molecular detection of SYT-SSX fusion gene transcripts. *Histopathology*, **34**, 205–210.

7 Antonescu, C.R., Kawai, A., Leug, D.H. *et al.* (2000) Strong association of SYT-SSX fusion type and morphologic epithelial differentiation in synovial sarcoma. *Diag Mol Pathol*, **9**, 1–8.

8 Crew, A.J., Clark, J., Fischer, C. *et al.* (1995) Fusion of SYT to two genes, SSX1 and SSX2, encoding proteins with homology to the Kruppel associated box in human Synovial sarcoma. *EMBO J*, **14**, 2333–2340.

9 Argani, P., Faria, P., Epstein, J. *et al.* (2000) Primary renal Synovial sarcoma. Molecular and morphologic delineation of an entity previously included among embryonal sarcomas of the kidney. *Am J Surg Pathol*, **24**, 1087–1096.

10 Tsuji, S., Hisaoka, M., Morimitsu, Y. *et al.* (1998) Detection of SYT-SSX fusion transcripts in Synovial sarcoma by reverse transcriptase polymerase chain reaction using archival paraffin embedded tissues. *Am J Pathol*, **153**, 1807–1812.

11 de Leew, B., Balemans, M., Weghuis, D.O. *et al.* (1995) Identification of two alternative fusion genes, SYT-SSX1 and SYT-SSX2 in t (X; 18) (p11. 2; q11. 2)-positive Synovial sarcomas. *Hum Mol Genet*, **4**, 1097–1099.

12 Hashimoto, N., Myoti, A., Araki, N. *et al.* (2001) Detection of SYT-SSX fusion transcript gene in peripheral blood from a patient with Synovial sarcoma. *Am J Surg Pathol*, **25**, 406–410.

13 O'Connel, J.X., Browne, W., Gropper, P. *et al.* (1996) Intraneural biphasic Synovial sarcoma: an alternative "Glandular" tumor of peripheral nerve. *Mod Pathol*, **9**, 738–741.

14 van Kessel, A.G., de Brujin, D. *et al.* (1998) Masked t (X; 18)(p11; q11) in a biphasic

Phenotypic and Genotypic Diagnosis of Malignancies. Muin S.A. Tuffaha
Copyright © 2008 WILEY-VCH Verlag GmbH & Co. KGaA, Weinheim
ISBN: 978-3-527-31881-0

Synovial sarcoma revealed by FISH and RT-PCR. *Genes Chromosomes Cancer,* **23**, 198–201.

15 van de Rijn, M., Barr, F.G., Collins, M.H. *et al.* (1999) Absence of SYT-SSX fusion products in soft tissue tumors other than Synovial sarcoma. *Am J Surg Pathol,* **112**, 43–49.

16 Kawai, A., Woodruff, J., Healey, J.H. *et al.* (1998) SYT-SSX gene fusion as a determinant of morphology and prognosis in synovial sarcoma. *N Engl J Med,* **338**, 153–160.

17 Hashimoto, N., Myoui, A., Araki, N. *et al.* (2001) Detection of SYT-SSX fusion gene in peripheral blood from a patient with synovial sarcoma. *Am J Surg Pathol,* **25**, 406–410.

18 Nilsson, G., Skytting, B., Xie, Y. *et al.* (1999) The SYT-SSX1 variant of synovial sarcoma is associated with a high rate of tumor cell proliferation and poor clinical outcome. *Cancer Res,* **59**, 3180–3184.

19 Bijwaard, K.E., Fetsch, J.F., Przygodzki, R. *et al.* (2002) Detection of SYT-SSX fusion transcript in archival synovial sarcoma by Real-time reverse-transcriptase polymerase chain reaction. *J Mol Diag,* **4**, 59–64.

20 O'Sullivan, M.J., Kyriakos, M., Zhu, X. *et al.* (2000) Malignant peripheral nerve sheath tumors with t(X;18). A pathologic and molecular genetic study. *Mod Pathol,* **13**, 1336–1346; and correspondence: Ladanyi, M., Woodruff, J M., Scheithauer, B.W. *et al.* (2001) *Mod Pathol,* **14**, 733–737.

21 Coindre, J.-M., Hostein, I., Benhatter, J. *et al.* (2002) Malignant peripheral nerve sheath tumors are t(x; 18)-negative sarcomas. Molecular analysis of 25 cases occurring in neurofibromatosis type 1 patients, using different RT-PCR-based methods of detection. *Mod Pathol,* **15**, 589–592.

22 Tamborini, E., Agus, V., Perrone, F. *et al.* (2002) Lack of SYT-SSX fusion transcripts in malignant peripheral nerve sheath tumors on RT-PCR analysis on 34 archival cases. *Lab Invest,* **82**, 609–618.

23 dos Santos, N.R., de Bruijn, D.R., van Kessel, A.G. (2001) Molecular mechanisms underlying human synovial sarcoma development. *Genes Chromosomes Cancer,* **30**, 1–14.

24 Guillou, L., Coindre, J.-M., Gallagher, G. *et al.* (2001) Detection of the synovial sarcoma translocation t(X;18)(SYT; SSX) in paraffin-embedded tissue using reverse transcriptase-polymerase chain reaction: A reliable and powerful diagnostic tool for pathologists. *Hum Pathol,* **32**, 105–112.

25 Ladanyi, M., Antonescu, C.R., Leung, D. *et al.* (2002) Impact of SYT-SSX fusion type on the clinical behavior of synovial sarcoma: a multi-institutional retrospective study of 243 patients. *Cancer Res,* **62**, 135–140.

26 Thorson, J.A., Weigelin, H.C., Ruiz, R.E. *et al.* (2006) Identification of SYT-SSX transcripts from synovial sarcomas using RT-multiplex PCR and capillary electrophoresis. *Mod Pathol,* **19**, 641–647.

27 Amary, M., Berisha, F., Bernardi, F. *et al.* (2007) Detection of SS18-SSX fusion transcripts in formalin-fixed paraffin-embedded neoplasms: analysis of conventional RT-PCR, qRT-PCR and dual color FISH as diagnostic tools for synovial sarcoma. *Mod Pathol,* **20**, 1–15.

28 He, R., Patel, R.M., Alkan, S. *et al.* (2007) Immunostaining for SYT protein discriminates synovial sarcoma from other soft tissue tumors: analysis of 146 cases. *Mod Pathol,* **20**, 522–528.

Molecular Detection of Primitive Neuroectodermal Tumors and Ewing's Sarcoma

1 Adams, V., Hany, M., Schmid, M. *et al.* (1996) Detection of t(11; 22)(q24;q12) translocation breakpoint in paraffin embedded tissue of the Ewing's sarcoma family by nested of reverse-transcriptase polymerase chain reaction. *Diag Mol Pathol,* **5**, 107–113.

2 Scotland, K., Serra, M., Cristina, M. *et al.* (1996) Immunostaining of the p30/32 mic2 antigen and molecular detection of EWS rearrangements for the diagnosis of Ewing's sarcoma and peripheral neuroectodermal tumor. *Hum Pathol*, **27**, 408–416.

3 Jeon, I.-S., Davis, J.N., Braun, B.S. *et al.* (1995) A variant Ewing's sarcoma translocation (7; 22) fuses the EWS gene to the ETS gene ETV1. *Oncogene*, **10**, 1229–1234.

4 Seronsen, P.H.B., Shimada, H., Liu, X.F. *et al.* (1995) Biphenotypic sarcomas with myogenic and neural differentiation exoress the Ewing's sarcoma EWS/FLI1 fusion gene. *Cancer Res*, **55**, 1385–1392.

5 Thoner, P., Squire, J., Chilton-MacNeil, S. *et al.* (1996) Is the EWS/FLI-1 fusion transcript specific for Ewing's sarcoma and peripheral neuroectodermal tumor? *Am J Pathol*, **148**, 1125–1138.

6 Peter, M., Couturier, J., Pacquement, H. *et al.* (1997) A new member of the ES family fused to EWS in Ewing's tumours. *Oncogene*, **14**, 1159–1164.

7 Pagani, A., Masri, L., Rosolen, A. *et al.* (1998) Neuroendocrine differentiation in Ewing's sarcoma and primitive neuroectodermal tumors revealed by reverse-transcriptase polymerase chain reaction of Chromogranin mRNA. *Diag Mol Pathol*, **7**, 36–43.

8 de Alava, E., Lozano, M.D., Patino, A. *et al.* (1998) Ewing's family tumors: potential value of reverse-transcriptase polymerase chain reaction detection of minimal residual disease in peripheral blood samples. *Diag Mol Pathol*, **7**, 152–157.

9 Tsuji, S., Hisaoka, M., Morimitsu, Y. *et al.* (1998) Peripheral primitive neuroectodermal tumor of the lung: report of two cases. *Histopathology*, **33**, 396–374.

10 Parham, D.M., Roloson, G.J. *et al.* (2001) Primary malignant Neuroepithelial tumors, of the kidney. *Am J Surg Pathol*, **25**, 133–146.

11 Argani, P., Perez-Ordonez, B., Xiao, H. (1998) Olfactory neuroblastoma is not related to the Ewing's family of tumors. *Am J Surg Pathol*, **22**, 391–398.

12 Papotti, M., Croce, S. *et al.* (2000) Correlative immunohistochemical and reverse-transcriptase polymerase chain reaction analysis of somatostatin receptor type 2 in neuroendocrine tumours of the lung. *Diag Mol Pathol*, **9**, 47–57.

13 Liombart-Bosch, A., Pellin, A., Carda, C. *et al.* (2000) soft tissue Ewing's sarcoma-Peripheral primitive neuroectodermal tumor with atypical clear cell pattern shows type EWS.FEV fusion transcript. *Diag Mol Pathol*, **9**, 137–144.

14 Seronsen, P.H.B., Liu, X.F., Delattre, O. (1993) Reverse transcriptase PCR amplification of EWS/FLI1 fusion transcripts as a diagnostic test for peripheral primitive neuroectodermal tumours of childhood. *Diag Mol Pathol*, **2**, 147–157.

15 Zoubek, A., Dockhorn-Dworniczak, B., Delattre, O. (1996) Does expression of different EWS chimeric transcripts define clinically distinct risk groups of Ewing tumor patients. *J Clin Oncol*, **14**, 1245–1251.

16 Pfleiderer, C., Zoubek, A., Gruber, B. *et al.* (1995) Detection of tumour cells in peripheral blood and bone marrow from Ewing tumor patients by RT-PCR. *Int J Cancer*, **64**, 135–139.

17 Sandberg, A.A., Bridge, J.A. (2000) Updates on cytogenetic and molecular genetics of bone and soft tissue tumors: Ewing's sarcoma and peripheral primitive neuroectodermal tumors. *Cancer Genet Cytogenet*, **123**, 1–26.

18 Meier, V.S., Kühne, T., Jundt, G., Gudat, F. (1998) Molecular diagnosis of Ewing tumors: improved detection of EWS-FLI-1 and EWS-ERG chimeric transcripts and rapid determination of exon combinations. *Diag Mol Pathol*, **7**, 29–35.

19 Sarangarajan, R., Hill, D.A., Humphrey, A. *et al.* (2001) Primitive neuroectodermal tumors of the biliary and gastrointestinal

tracts: Clinicopathologic and molecular study of two cases. *Pediatr Dev Pathol*, **4**, 185–191.

20 Shing, D.C., McMullan, D.J., Roberts, P. *et al.* (2003) FUS/ERG Gene Fusions in Ewing's Tumors. *Cancer Res*, **63**, 4568–4576.

21 Qian, X., Jin, L., Shearer, B.M. *et al.* (2005) Molecular diagnosis of Ewing's sarcoma/primitive neuroectodermal tumor in formalin-fixed paraffin embedded tissues by RT-PCR and fluorescence in situ hybridization. *Diag Mol Pathol*, **14**, 23–28.

22 Stegmaier, S., Leuchner, I., Aakcha-Rudel, E. *et al.* (2004) Identification of various exon combinations of the ews/fli translocation:, an optimized RT-PCR method for paraffin embedded tissue. *Klin Paediatr*, **216**, 315–322.

23 Riggi, N., Cironi, L., Provero, P. *et al.* (2005) Development of Ewing's sarcoma from primary bone marrow-derived mesenchymal progenitor cells. *Cancer Res*, **65**, 11459–11468.

24 Mangham, D.C., Williams, A., McMullan, D.J. *et al.* (2006) Ewing's sarcoma of bone: the detection of specific transcripts in a large, consecutive series of formalin-fixed, decalcified, paraffin-embededd tissue samples using the reverse transcriptase-polymerase chain reaction. *Histopathology*, **48**, 363–376.

Molecular Detection of Myxoid Liposarcoma

1 Panagopoulos, I., Höglund, M., Mertens, F. *et al.* (1996) Fusion of the EWS and CHOP genes in myxoid liposarcoma. *Oncogene*, **12**, 489–494.

2 Tallini, G., Akerman, M., Dal Cin, P. *et al.* (1996) Combined morphologic and karyotypic study of 28 myxoid Liposarcoma. *Am J Surg Pathol*, **20**, 1047–1055.

3 Hisaoka, M., Tsuji, S., Morimitsu, Y. *et al.* (1998) Detection of TLS/FUS-CHOP fusion transcripts in myxoid and round cell Liposarcoma by nested reverse-transcriptase polymerase chain reaction using archival paraffin embedded tissue. *Diag Mol Pathol*, **7**, 96–101.

4 Hosaka, T., Nakashima, Y., Kusuzaki, K. *et al.* (2002) A novel type of EWS-CHOP fusion transcript gene in two cases of myxoid liposarcoma. *J Mol Diagn*, **4**, 164–171.

5 Nakanishi, H., Araki, N., Joyama, S. *et al.* (2004) Myxoid liposarcoma with adipocytic maturation: detection of TLS/CHOP fusion transcript. *Diag Mol Pathol*, **13**, 92–96.

6 Sandberg, A.A. (2004) Updates on the cytogenetics and molecular genetics of bone and soft tissue tumors: liposarcoma. *Cancer Genet Cytogenet*, **155**, 1–24.

7 Matsui, Y., Ueda, T., Kubo, T. *et al.* (2006) A novel type of EWS-CHOP fusion gene in myxoid liposarcoma. *Biochem Biophys Res Commun*, **22**, 437–444.

8 ten Heuvel, S.E., Hoekstra, H.J., van Ginkel, R. J. *et al.* (2007) Clinicopathologic prognostic factors in myxoid liposarcoma: a retrospective study of 49 patients with long-term follow-up. *Ann Surg Oncol*, **14**, 222–229.

Molecular Detection of Extraskeletal Myxoid Chondrosarcoma

1 Brody, R.I., Ueda, T., Hamelin, A. *et al.* (1997) Molecular analysis of the Fusion of EWS to an orphan nuclear receptor gene in extraskeletal myxoid chondrosarcoma. *Am J Pathol*, **150**, 1049–1058.

2 Okamoto, S., Hisaoka, M., Ishida, T. *et al.* (2001) Extraskeletal myxoid chondrosarcoma: a clinical, immunohistochemical, and molecular analysis of 18 cases. *Hum Pathol*, **32**, 1116–1123.

3 Panagopoulos, I., Mertens, F., Isaksson, M. *et al.* (2002) Molecular genetic characterization of the EWS/CHN and RBP56/CHN fusion genes in extraskeletal myxoid chondrosarcoma. *Genes Chromosomes Cancer*, **35**, 340–352.

4 Sjögren, H., Meis-Kindblom, J., Örndal, C. *et al.* (2003) Studies on the molecular pathogenesis of extraskeletal myxoid chondrosarcoma-cytogenetic, molecular genetic, and cDNA microarray analyses. *Am J Pathol*, **162**, 781–792.

5 Hisaoka, M., Okamoto, S., Yokoyama, K., Hashimoto, H. (2004) Coexpression of NOR1 and SIX3 proteins in extraskeletal myxoid chondrosarcomas without detectable NR4A3 fusion genes. *Cancer Genet Cytogenet*, **15**, 101–107.

6 Hisaoka, M., Hashimoto, H. (2005) Extraskeletal myxoid chondrosarcoma: updated clinicopathological and molecular genetic characteristics. *Pathol Int*, **55**, 453–463.

Molecular Detection of Desmoplastic Small Round Cell Tumor

1 Dorsey, B.V., Benjamin, L.E., Rauscher, F. *et al.* (1996) Intra-abdominal desmoplastic small round cell tumor: expression of the pathologic profile. *Mod Pathol*, **9**, 703–709.

2 Ladanyi, M., Gerald, W. (1994) Fusion of the EWS and WT1 genes in the desmoplastic small round cell tumor. *Cancer Res*, **54**, 2837–2840.

3 Ordi, J., de Alava, E., Torne, A. *et al.* (1998) Intraabdominal desmoplastic small round cell tumor with EWS/ERG fusion transcript. *Am J Surg Pathol*, **22**, 1026–1032.

4 Adsay, V., Cheng, J., Athanasian, E. *et al.* (1999) Primary desmoplastic small cell tumor of soft tissues and bone of the hand. *Am J Surg Pathol*, **23**, 1408–1413.

5 de Alava, E., Ladanyi, M., Rosai, J. *et al.* (1995) Detection of chimeric transcripts In desmoplastic small round cell tumor and related developmental tumours by reverse-transcriptase polymerase chain reaction. *Am J Pathol*, **147**, 1584–1591.

6 Antonescu, C.R., Gerald, W.L., Magid, M. S. *et al.* (1998) Molecular variants of the EWS-WT1 gene fusion in desmoplastic small round cell tumor. *Diag Mol Pathol*, **7**, 24–28.

7 Liu, J., Nau, M.M., Yeh, J.C. *et al.* (2000) Molecular heterogeneity and function of EWS-WT1 fusion transcripts in desmoplastic small round cell tumors. *Clin Cancer Res*, **6**, 3522–3529.

8 Hill, D.A., Pfeifer, J.D., Marley, E.F. *et al.* (2000) WT1 staining reliably differentiates desmoplastic round cell tumors from wing sarcoma/primitive neuroectodermal tumor. *Am J Clin Pathol*, **114**, 345–353.

9 Sandberg, A.A., Bridge, J.A. (2002) Updates on the cytogenetics and molecular genetics of bone and soft tissue tumors: desmoplastic round cell tumors. *Cancer Genet Cytogenet*, **138**, 1–10.

10 Gerald, W.L., Haber, D.A. (2005) The EWS-WT1 gene fusion in desmoplastic small round cell tumor. *Semin Cancer Biol*, **15**, 197–205.

Molecular Detection of Clear Cell Sarcoma

1 Pellin, A., Moteagudo, C. *et al.* (1998) New type of chimeric fusion product between the EWS and ATF1 genes in clear cell sarcoma (malignant melanoma of soft parts). *Genes Chromosomes Cancer*, **23**, 358–360.

2 Sonobe, H., Takeuchi, T., Taguchi, T. *et al.* (1999) Further characterization of the human clear cell sarcoma cell line HS-MM demonstrating a specific t(12;22)(q13;q12) translocation and hybrid EWS/ATF-1 transcript. *J Pathol*, **187**, 594–597.

3 Zucman, J., Delattre, O., Desmaze, C. *et al.* (1993) EWS and ATF-1 gene fusion induced by t(12;22) translocation in malignant melanoma of soft parts. *Nat Genet*, **4**, 341–345.

4 Langezaal, S.M., Van Roggen, J.F., Cleton-Jansen, A.M. *et al.* (2001) Malignant melanoma is genetically distinct from clear cell sarcoma of tendons and aponeurosis (malignant melanoma of soft parts). *Br J Cancer*, **84**, 535–538.

5 Antonescu, C.R., Tschernyavsky, S.J., Wodruff, J.M. *et al.* (2002) Molecular diagnosis of clear cell sarcoma. Detection of EWS-ATF1 and MITF-M transcripts and histopathological and ultrastructural analysis of 12 cases. *J Mol Diag*, **4**, 44–52.

6 Sandberg, A.A., Bridge, J.A. (2002) Updates on the cytogenetics and molecular genetics of bone and soft tissue tumors: clear cell sarcoma (malignant melanoma of soft parts). *Cancer Genet Cytogenet*, **138**, 1–7.

7 Coindre, L.M., Hostein, I., Terrier, P. *et al.* (2006) Diagnosis of clear cell sarcoma by real-time reverse transcriptase-polymerase chain reaction analysis of paraffin embedded tissues. *Cancer*, **107**, 1055–1064.

Molecular Detection of Rhabdomyosarcoma

A. Detection by Specific Gene Expression

1 Wang, N.P., Marx, J., McNutt, M.A. *et al.* (1995) Expression of Myogenic regulatory proteins (myogenin and MyoD1) in small blue round cell tumors of Childhood. *Am J Pathol*, **147**, 1799–1810.

2 Missias, A.C., Chu, G.C., Klocke, B.J. *et al.* (1996) Maturation of the acetylcholine receptor in skeletal muscle: regulation of the AChR γ-to-ε switch. *Dev Biol*, **179**, 223–228.

3 Frascella, E., Rosolen, A. (1998) Detection of the Myo D1 transcript in rhabdomyosarcoma cell lines and tumor samples by reverse-transcriptase polymerase chain reaction. *Am J Pathol*, **152**, 577–583.

4 Gattenloener, S., Vincent, A., Leuschner, I. *et al.* (1998) The fetal form of the acetylcholine receptor distinguishes in rhabdomyosarcoma from other childhood tumors. *Am J Pathol*, **152**, 437–447.

5 Gattenlöner, S., Müller-Hermelmik, H.-K., Marx, A. (1998) Comparison of the transcription of the fetal type acetylcholine receptor and myogenin in rhabdomyosarcomas. *Verh Dtsch Ges Pathol*, **82**, 195–201.

6 Gattenloener, S., Dockhorn-Dwornniczak, B., Leuschner, I. *et al.* (1999) A comparison of MyoD1and fetal acetylcholine receptor expression in childhood tumors and normal tissues. Implications for the molecular diagnosis of minimal disease in rhabdomyosarcomas. *J Mol Diag*, **1**, 23–31.

7 Michelagnoli, M.P., Burchill, S.A., Cullinane, C. *et al.* (2003) Myogenin – a more specific target for RT-PCR detection of rhabdomyosarcoma than MyoD1. *Med Pediatr Oncol*, **40**, 1–8.

8 Hostein, I., Andraud-fregeville, M., Guillou, L. *et al.* (2004) Rhabdomyosarcoma: value of myogenin expression analysis and molecular testing in diagnosing the alveolar type. *Cancer*, **101**, 2817–2824.

B. Detection by Specific Genetic Abnormalities (Alveolar Rhabdomyosarcoma)

1 Reichmuth, C., Markus, M., Hillemanns, M. *et al.* (1996) The diagnostic potential of the chromosome translocation t(2;13) in rhabdomyosarcoma: a PCR study of fresh-frozen and paraffin embedded tumour samples. *J Pathol*, **180**, 50–57.

2 Anderson, J., Renshaw, J., McManus, A. *et al.* (1997) Amplification of the t(2;13) and t(1;13) translocations of alveolar rhabdomyosarcoma in small formalin-fixed biopsies using a modified reverse-transcriptase polymerase chain reaction. *Am J Pathol*, **150**, 477–482.

3 Barr, F.G. (1997) Molecular genetics and pathogenesis of rhabdomyosarcoma. *J Pediatr Hematol Oncol*, **19**, 483–491.

4 Edwards, R.H., Chatten, J., Xiong, Q.-B. *et al.* (1997) detection of gene fusions in rhabdomyosarcoma by reverse-transcriptase polymerase chain reaction assay of archival samples. *Diag Mol Pathol*, **6**, 91–97.Errata. *Diag Mol Pathol*, **6**, 177–178.(1997)

5 Kelly, K.M., Wormer, R.B., Sorensen, P.H.B. *et al.* (1997) Common and variant gene fusions predict distinct clinical phenotypes in rhabdomyosarcoma. *J Clin Oncol*, **15**, 1831–1836.

6 Tobar, A., Avigad, S., Zoldan, M. *et al.* (2000) Clinical relevance of molecular diagnosis in childhood rhabdomyosarcoma. *Diag Mol Pathol*, **9**, 9–13.

7 Sorensen, P.H.B., Lynch, J.C., Qualman, S.J. *et al.* (2002) PAX3-FKHR and PAX7-FKHR gene fusions are prognostic indicators in alveolar rhabdomyosarcoma: a report from the children's oncology group. *J Clin Oncol*, **20**, 2672–2679.

8 Parham, D., Qualman, S., Teot, L. *et al.* (2007) Correlation between histology and PAX/FKHR fusion status in alveolar rhabdomyosarcoma: a report from the children's oncology group. *Am J Surg Pathol*, **31**, 895–901.

Molecular Detection of Congenital Fibrosarcoma and Mesoblastic Nephroma

1 Bourgeois, J.M., Knezevich, S.R., Mathers, J.A. *et al.* (2000) Molecular detection of the ETV6-NTRK3 gene fusion differentiates congenital fibrosarcoma from other childhood spindle cell tumors. *Am J Surg Pathol*, **24**, 937–946.

2 Knezevich, S.R., McFadden, D.E., Tao, W. *et al.* (1998) A novel ETV6-NTRK3 gene fusion in congenital fibrosarcoma. *Nat Genet*, **18**, 184–187.

3 Rubin, B.P., Chen, C.-J., Morgan, T.W. *et al.* (1998) Congenital mesoblastic nephroma t(12;15) is associated with ETV6-NTRK3 gene fusion. *Am J Pathol*, **153**, 1451–1458.

4 Knezevich, S.R., Garnett, M.J. *et al.* (1998) ETV6-NTRK3 gene fusion and trisomy 11 establish a histogenetic link between mesoblastic nephroma and congenital fibrosarcoma. *Cancer Res*, **58**, 5046–5048.

5 Argani, P., Fritsch, M. *et al.* (2000) Detection of the ETV6-NTRK3 chimeric RNA of infantile fibrosarcoma/cellular congenital mesoblastic nephroma in paraffin-embedded tissue: application to challenging pediatric renal stromal tumors. *Mod Pathol*, **13**, 29–36.

6 Sandberg, A.A., Bridge, J.A. (2002) Updates on the cytogenetics and molecular genetics of bone and soft tissue tumors: congenital (infantile) fibrosarcoma and mesoblastic nephroma. *Cancer Genet Cytogenet*, **132**, 1–13.

7 Anderson, J., Gibson, S., Sebire, N.J. (2006) Expression of ETV6-NTRK3 in classical, cellular and mixed subtypes of congenital mesoblastic nephroma. *Histopathology*, **48**, 748–753.

8 Tognon, C., Knezevich, S.R., Huntsman, D. *et al.* (2002) Expression the ETV6-NTRK3 gene fusion as a primary event in human secretory breast carcinoma. *Cancer Cell*, **2**, 367–376.

Molecular Detection of Dermatofibrosarcoma Protuberans and Giant Cell Fibroblastoma

1 Wang, J., Morimitsu, Y., Okamato, S. *et al.* (2000) COL1A1-PDGFB fusion transcripts in fibrosarcoma areas of six dermatofibrosarcoma protuberans. *J Med Diag*, **72**, 47–52.

2 Navarro, M., Simon, M.-P., Migeon, C. *et al.* (1998) COL1A1-PDGFB fusion in a ring chromosome 4 found in a dermatofibrosarcoma protuberans. *Genes Chromosomes Cancer*, **23**, 263–266.

3 Pedeutour, F., Simon, M.P. *et al.* (1996) Translocation, t (17;22)(q22;q13), in dermatofibrosarcoma protuberans: a new tumor associated chromosome rearrangement. *Cytogenet Cell Genet*, **72**, 171–174.

4 Terrier-Lacombe, M.J., Guillou, L., Maire, G. *et al.* (2003) Dermatofibrosarcoma protuberans, giant cell fibroblastoma, and hybrid lesions in children: clinicopathologic comparative analysis of 28 cases with molecular data. *Am J Surg Pathol*, **27**, 27–39.

5 Kashima, A., yamashita, A., Moriguchi, S. *et al.* (2006) Detection of COL1A1-PDGFB fusion transcripts and platelet-derivated factor α and β receptors in giant cell fibroblastoma of the postsacrococcygeal region. *Br J Dermatol*, **154**, 983–987.

6 Szollosi, Z., Scholtz, B., Egervari, K., Nemes, Z. (2007) Transformed dermatofibrosarcoma protuberans: real time polymerase chain reaction detection of COL1A1-PDGFB fusion transcripts in sarcomatous areas. *J Clin Pathol*, **60**, 190–194.

7 Takahira, T., Oda, Y., Tamiya, S. *et al.* (2007) Detection of COL1A1-PDGFB fusion transcripts and PDGFB/PDGFRB mRNA expression in dermatofibrosarcoma protuberans. *Mod Pathol*, **20**, 668–675.

Molecular Detection of Inflammatory Myofibroblastic Tumor

1 Ladanyi, M. (2000) Aberrant ALK tyrosine kinase signaling. Different cellular lineages, common oncogenic mechanisms. *Am J Pathol*, **157**, 341–345.
2 Lawrence, B., Perez-Atayde, A., Hibbard, M. K. *et al.* (2000) TMP3-ALK and TMP4-ALK oncogenes in inflammatory myofibroblastic tumors. *Am J Pathol*, **157**, 377–384.
3 Bridge, J.A., Kanamori, M., Ma, Z. *et al.* (2001) Fusion of the ALK gene to the clathrin heavy chain gene, CLTC, in inflammatory myofibroblastic tumor. *Am J Pathol*, **159**, 411–415.
4 Cook, J.R., Dehner, L.P., Collins, M.H. *et al.* (2001) Anaplastic lymphoma kinase (ALK) expression in the inflammatory myofibroblastic tumor. *Am J Surg Pathol*, **25**, 1364–1371.
5 Chan, J.K.C., Cheuk, W., Shimizu, M. (2001) Anaplastic lymphoma kinase expression in inflammatory pseudotumors. *Am J Surg Pathol*, **25**, 761–768.
6 Pulford, K., Lamant, L., Espinos, E. *et al.* (2004) The emerging normal and disease-related reles of anaplastic lymphoma kinase. *Cell Mol Life Sci*, **61**, 2939–2953.
7 Ma, Z., Hill, D.A., Collins, M.H. *et al.* (2003) Fusion of ALK to the Ran-binding protein 2 (RANBP2) gene in inflammatory myofibroblastic tumor. *Genes Chromosomes Cancer*, **37**, 98–105.
8 Tsuzuki, T., Magi-Galluzzi, C., Epstein, J.I. (2004) ALK-1 expression in inflammatory myofibroblastic tumor of the urinary bladder. *Am J Surg Pathol*, **28**, 1609–1614.
9 Panagopoulos, I., Nilsson, T., Domanski, H.A. *et al.* (2006) Fusion of the SEC31L1 and ALK genes in an inflammatory myofibroblastic tumor. *Int J Cancer*, **118**, 1181–1186.

Molecular Detection of Alveolar Soft Part Sarcoma

1 Ladanyi, M., Lui, M.L., Antonescu, C.R. *et al.* (2001) The der(17)t(X; 17)(p11; q25) of human alveolar soft part sarcoma fuses the TFE3 transcription factor gene to ASPL, a novel gene at 17q25. *Oncogene*, **20**, 48–57.
2 Argani, P., Antonescu, C.R., Illei, P.B. *et al.* (2001) Primary renal neoplasms with the ASPL-TFE3 gene fusion of alveolar soft part sarcoma. *Am J Pathol*, **159**, 179–192.
3 Ordonez, N.G. (1999) Alveolar soft part sarcoma: a review and update. *Adv Anatom Pathol*, **6**, 125–139.
4 Sandberg, A.A., Bridge, J.A. (2002) Updates on the cytogenetics and molecular genetics of bone and soft tissue tumors: alveolar soft part sarcoma. *Cancer Genet Cytogenet*, **136**, 1–9.
5 Huang, H.Y., Lui, M.Y., Ladanyi, M. (2005) Nonrandom cell-cycle timing of a somatic chromosomal translocation: t(X; 17) of alveolar soft part sarcoma occurs in G2. *Genes Chromosomes Cancer*, **44**, 170–176.
6 Aulmann, S., Longerich, T., Schirmacher, P. *et al.* (2007) Detection of the ASPSCR1–TFE3 gene fusion in paraffin-embedded alveolar soft part sarcomas. *Histopathology*, **50**, 881–886.

Molecular Detection of Endometrial Stromal Tumors

1 Koontz, J.I., Soreng, A.L., Nucci, M. *et al.* (2001) Frequent fusion of the JAZF1 and JJAZ1 genes in endometrial stromal tumors. *Proc Natl Acad Sci USA*, **98**, 6348–6353.
2 Huang, H.-Y., Ladanyi, M., Soslow, R.A. (2004) Molecular detection of JAZF1 and JJAZ1 gene fusion in endometrial stromal neoplasms with classic and variant histology, evidence for heterogeneity. *Am J Surg Pathol*, **28**, 224–232.

3 Hrzenjak, A., Moinfar, F., Tavassoli, F. *et al.* (2005) JAZF1/JJAZ1 gene fusion in endometrial stromal sarcomas. Molecular analysis by reverse transcriptase-polymerase chain reaction optimized for paraffin-embedded tissue. *J Mol Diagn*, **7**, 388–395.

4 Leunen, K., Amant, F., Debiec-rychert, M. *et al.* (2003) Endometrial stromal sarcoma presenting as postpartum hemorrhage: report of a case with a sole t(10; 17)(q22; p13) translocation. *Gynecol Oncol*, **91**, 265–271.

5 Micci, F., Panagopoulos, I., Bjerkehagen, B., Heim, S. (2006) Consistent rearrangement of chromosomal band 6p21 with generation of fusion genes JAFF1/PHF1 and EPC1/P HF1 in endometrial stromal sarcoma. *Cancer Res*, **66**, 107–112.

6 Nucci, M., Harburger, D., Koontz, J. *et al.* (2007) Molecular analysis of the JAZF1-JJAZ1 gene fusion by RT-PCR and fluorescence *in situ* hybridization in endometrial stromal neoplasms. *Am J Surg Pathol*, **31**, 65–70.

Molecular Detection of Follicular Lymphoma

1 Hostein, I., Menard, A., Soubeyran, I. *et al.* (2001) A 1-Kb Bcl-2-PCR fragment detection in a patient with follicular lymphoma and development of a new diagnostic PCR method. *Diag Mol Pathol*, **10**, 89–94.

2 Aiello, A., Delia, D., Giardini, R. *et al.* (1997) PCR analysis of IgH and bcl-2 gene rearrangement in the diagnosis of follicular lymphoma in lymph node fine-needle aspiration. *Diag Mol Pathol*, **6**, 154–160.

3 Natkunam, Y., Warnke, R.A., Zehnder, J.L. *et al.* (2000) Blastic/blastoid transformation of follicular lymphoma. *Am J Surg Pathol*, **24**, 525–534.

4 Soubeyran, P., Hostein, I., Debled, M. *et al.* (1999) Evolution of BCL-2/IgH hybrid gene RNA expression during treatment of t(14;18)-bearing follicular lymphomas. *Br J Cancer*, **81**, 860–869.

5 Segal, H.G., Jorgensen, D.T. *et al.* (1994) Optimal primer selection for clonality assessment: II follicular lymphoma. *Hum Pathol*, **25**, 1276–1282.

6 McCluggage, W.G., Catherwood, M., Alexander, H.D. *et al.* (2002) Immunohistochemical expression of CD10 and t(14; 18) chromosomal translocation may be indicators of follicle center origin in nodal diffuse large B-cell lymphoma. *Histopathology*, **41**, 414–420.

7 Gomez, M., Wu, X., Wang, Y.L. (2005) Detection or BCL-2-IGH using single-round PCR assays. *Diag Mol Pathol*, **14**, 17–22.

8 Goodlad, J.R., Batstone, P.J., Hamilton, H. *et al.* (2006) BCL2 gene abnormalities define distinct clinical subsets of follicular lymphoma. *Histopathology*, **49**, 229–241.

Molecular Detection of Anaplastic Large Cell Lymphoma

1 Cataldo, K.A., Jalal, S.M., Law, M.E. (1999) Detection of t(2;5) in anaplastic large cell lymphoma. Comparison of immuno-histochemical studies, FISH and RT-PCR in paraffin embedded tissue. *Am J Surg Pathol*, **23**, 1386–1392.

2 Ladanyi, M., Cavalchire, G. (1996) Detection of the NPM-ALK genomic rearrangement of Ki-1 lymphoma and isolation of the involved NPM and ALK introns. *Diag Mol Pathol*, **5**, 154–158.

3 Lamant, L., Dastugue, N., Pulford, K. *et al.* (1999) A new fusion gene TPM3-ALK in anaplastic large cell lymphoma created by a (1; 2) (q25; p23) translocation. *Blood*, **93**, 3088–3095.

4 Drexler, H.G., Gignac, S.M., von Waielewski, R. *et al.* (2000) Pathobiology of NPM-ALK and variant fusion genes in anaplastic large cell lymphoma and other lymphomas. *Leukemia*, **14**, 1533–1559.

5 Ma, Z., Cools, J., Marynen, P. *et al.* (2000) Inv(2)(p23q35) in anaplastic large-cell lymphoma induces constitutive anaplastic lymphoma kinase (ALK) tyrosine kinase activation by fusion to ATIC, an enzyme

involved in purine nucleotide biosynthesis. *Blood*, **95**, 2144–2149.

6 Trinei, M., Lanfrancone, L., Campo, E. *et al.* (2000) A new variant anaplastic lymphoma kinase (ALK)-fusion protein (ATIC-ALK) in a case of ALK-positive anaplastic large cell lymphoma. *Cancer Res*, **60**, 793–798.

7 Hernandez, L., Pinyol, M., Hernandez, S. *et al.* (1999) TRK-fused gene (TFG) is a new partner of ALK in anaplastic large cell lymphoma producing two structurally different TFG-ALK translocations. *Blood*, **94**, 3265–3268.

8 Stein, H., Foss, H.D., Dürkop, H. *et al.* (2000) $CD30^+$ anaplastic large cell lymphoma: a review of its histopathologic, genetic and clinical features. *Blood*, **96**, 3681–3695.

9 Tort, F., Pinyol, M., Pulford, K. *et al.* (2001) Molecular characterization of a new ALK translocation involving moesin (MSN-ALK) in anaplastic large cell lymphoma. *Lab Invest*, **81**, 419–426.

10 Colleoni, G.W.B., Bridge, J.A., Garicochea, B. *et al.* (2000) ATIC-ALT: A novel variant ALK gene fusion in anaplastic large cell lymphoma resulting from the recurrent cryptic chromosomal inversion. *Am J Pathol*, **156**, 781–789.

11 Downing, J.R., Shurtleff, S.A., Zielenska, M. *et al.* (1995) Molecular detection of the (2; 5) translocation of non-Hodgkin's lymphomas by reverse transcriptase-polymerase chain reaction. *Blood*, **85**, 3416–3422.

12 Ladanyi, M., Cavalchire, G., Morris, S.W. , *et al.* (1995) Reverse transcriptase polymerase chain reaction for the Ki-1 anaplastic large cell lymphoma-associated t(2; 5) translocation in Hodgkin's disease. *Am J Pathol*, **145**, 1296–1300. Erratum, *Am J Pathol*, **146**, 546.

13 Chikatsu, N., Kojima, H., Suzukawa, K. *et al.* (2003) ALK^+, $CD30^-$, $CD20^-$ large B-cell lymphoma containing anaplastic lymphoma kinase (ALK) fused to clathrin heavy chain gene (CKTC). *Mod Pathol*, **16**, 828–832.

14 Schumacher, J.A., Jenson, S.D. *et al.* (2004) Utility of linearly amplified RNA for RT-PCR detection of chromosomal translocations: validation using the t(2; 5) (p23; q35) NPM-ALK chromosomal translocation. *J Mol Diagn*, **6**, 16–21.

Molecular Detection of Mantle Cell Lymphoma

1 Molor, R.J., Meeker, T.C., Wittwer, C.T. (1994) Antigen expression and polymerase chain reaction amplification of mantle cell lymphoma. *Blood*, **83**, 1626–1631.

2 Imokh, R., Berger, F., Delsol, G. *et al.* (1994) detection of the chromosomal translocation t (11; 14) by polymerase chain reaction in mantle cell lymphoma. *Blood*, **83**, 1871–1875.

3 Fan, H., Gulley, M.L., Gascoyne, R.D. *et al.* (1998) Molecular detecting t(11;14) translocations in mantle cell lymphomas. *Diag Mol Pathol*, **7**, 209–214.

4 Lasota, J., Franssila, K., Koo, C.H.H. *et al.* (1996) Molecular diagnosis of mantle cell lymphoma in paraffin-embedded tissue. *Mod Pathol*, **9**, 361–366.

5 Falini, B., Mason, D.Y. (2002) Proteins encoded by genes involved in chromosomal alterations in lymphoma and leukemia: clinical value of their detection by immunohistochemistry. *Blood*, **99**, 4099–4126.

Molecular Detection of Chronic Myeloid Leukemia

1 Colleni, G.W.B., Jhanwar, S.C., Ladanyi, M. *et al.* (2000) Comparison of a multiplex reverse-transcriptase polymerase chain reaction for BCR-ABL to fluorescence *in situ* hybridization, southern blotting, and conventional cytogenetics in monitoring of patients with Ph1-positive leukemias. *Diag Mol Pathol*, **9**, 203–209.

2 Nogva, H.K., Evensen, S.A., Madshos, M. (1998) One-tube multiplex RT-PCR of BCR-ABL transcripts in analysis of patients with chronic myeloid leukemia and cute lymphoblastic leukaemia. *Scand J Clin Lab Invest*, **58**, 647–654.

3 Brandford, S., Hughes, T.P., Rudzki, Z. (1999) Monitoring chronic myeloid leukemia therapy by real-time quantitative PCR in blood is a reliable alternative to bone marrow cytogenetics. *Br J Haematol*, **107**, 587–599.
4 Bock, O., Reising, D., Kreipe, H. (2003) Multiplex RT-PCR for the detection of common BCR-ABL fusion transcripts in paraffin-embedded tissues from patients with chronic myeloid leukemia and acute lymphoblastic leukemia. *Diag Mol Pathol*, **12**, 119–2003.

Molecular Detection of Acute Lymphoblastic Leukemia

1 Scurto, P., Rocha, M.H., Kane, J.R. *et al.* (1998) A multiplex RT-PCR assay for the detection of chimeric transcripts encoded by the risk-stratifying translocations of pediatric acute lymphoblastic leukemia. *Leukemia*, **12**, 1994–2005.
2 Devaraj, P.E., Foroni, L., Sekhar, M. *et al.* (1994) E2A/HLF fusion cDNAs and the use of RT-PCR for the detection of minimal residual disease in t (17;19)(q22;p13) acute lymphoblastic leukemia. *Leukemia*, **8**, 1131–1138.
3 Rowley, J.D. (1999) The role of chromosome translocations in leukemogenesis. *Semin Hematol*, **36** (Suppl 7), 59–72.
4 Gleißner, B., Gökbuget, N., Bartram, C.R. *et al.* (2002) Leading prognostic relevance of the BCR-ABL translocation in adult acute B-lineage lymphoblastic leukemia: a prospective study of the German Multicenter Trial Group and confirmed polymerase chain reaction analysis. *Blood*, **99**, 1536–1543.
5 Satake, N., Kobayashi, H., Tsunemarsu, Y. *et al.* (1997) Minimal residual disease with TEL-AML1 fusion transcript in childhood acute lymphoblastic leukemia with t(12; 21). *Br J Haematol*, **97**, 607–611.
6 Pine, S.R., Yin, C., Matloub, Y.H. *et al.* (2005) Detection of central nervous system leukemia in children with acute lymphoblastic leukemia by real-time polymerase chain reaction. *J Mol Diagn*, **7**, 127–132.
7 Metzler, M., Mann, G., Monschein, U. *et al.* (2006) Minimal residual disease analysis in children with t(12;21)-positive acute lymphoblastic leukemia: comparison of Ig/TCR rearrangements and the genomic fusion gene. *Haematologica*, **91**, 683–686.

Molecular Detection of Acute Myeloid Leukemia

1 Longo, L., Pandolfi, P.P., Biondi, A. *et al.* (1990) Rearrangement and aberrant expression of the retinoic acid receptor α gene in acute promyelocytic leukemias. *J Exp Med*, **172**, 1571–1575.
2 Laczika, K., Mitterbauer, G., Korninger, L. *et al.* (1994) Rapid achievment of PML-RARα polymerase chain reaction (PCR)-negativity by combined treatment with Al-trans-retinoic acid and chemotherapy in acute promyelocytic leukemia: a pilot study. *Leukemia*, **8**, 1–5.
3 Biondi, A., Rambaldi, A., Pandolfi, P.P. *et al.* (1992) Molecular monitoring of the myl/retinoic acid aid receptor α fusion gene in acute promyelocytic leukemia by polymerase chain reaction. *Blood*, **80**, 492–497.
4 Huang, W., Sun, G.-L., Li, X.-S. *et al.* (1993) Acute promyelocytic leukemia: Clinical relevance of two major PML/RAR α isoforms and detection of minimal residual disease by retrotranscriptase polymerase chain reaction to predict relapse. *Blood*, **82**, 1264–1269.
5 Grimwalde, D., Howe, K., Langabeer, S. *et al.* (1996) Minimal residual disease detection in acute promyelocytic leukemia by reverse-transcriptase PCR: evaluation of PML-RAR α and RAR α-PML assessment in patients who ultimately relapse. *Leukemia*, **10**, 61–66.
6 Tobal, K., Yin, J.A.L. (1998) RT-PCR method with increased sensitivity shows persistence of PML-RAR fusion transcripts in patients in long-tem

remission of APL. *Leukemia*, **12**, 1349–1354.
7 Downing, J.R., Head, D.R., Curcio-Britnt, A.M. *et al.* (1993) An AML1/ETO fusion transcript is consistently detected by RNA-based polymerase chain reaction in acute myelogenous leukemia containing the (8; 21)(q22; q22) translocation. *Blood*, **81**, 2860–2865.
8 Kusec, R., Laczika, L., Knöbl, P. *et al.* (1994) AML1/ETO fusion mRNA can be detected in remission blood samples of all patients with t (8; 21) acute myeloid leukemia after chemotherapy or autologous bone marrow transplantation. *Leukemia*, **8**, 735–739.
9 Kwong, Y.L., Chan, V., Wong, K.F. *et al.* (1995) Use of the polymerase chain reaction in the detection of AML1/ETO fusion transcript in t (8; 21). *Cancer*, **75**, 821–825.
10 Marcucci, G., Livak, K.J., Bi, W. *et al.* (1998) Detection of minimal residual disease in patients with AML1/ETO-associated acute myeloid leukemia using a novel quantitative reverse transcription polymerase chain reaction assay. *Leukemia*, **12**, 1482–1489.
11 Wiemels, J.L., Xiao, Z., Buffler, P.A. *et al.* (2002) In utero origin of (8; 21) AML1-ETO translocations in childhood acute myeloid leukemia. *Blood*, **99**, 3801–3805.
12 Hebert, J., Cayuela, J.M., Daniel, M.-T. *et al.* (1994) Detection of minimal residual disease in acute myelomonocytic leukemia with abnormal eosinophils by nested polymerase chain reaction with allele specific amplification. *Blood*, **84**, 2291–2296.
13 Claxon, D.F., Liu, P., Hsu, H.B. *et al.* (1994) Detection of fusion transcripts generated by the inversion of 16 chromosome in acute myelogenous leukemia. *Blood*, **83**, 1750–1756.
14 Tobal, K., Johnson, P.R.E., Saundres, M.J. *et al.* (1995) Detection of CBFB/MYH11 transcripts in patients with inversion and other abnormalities of chromosome 16 at presentation and remission. *Br J Haematol*, **91**, 104–108.

Molecular Detection of B-Cell Lymphoma by Gene Rearrangement

1 Segal, S., Do, D., Jorgensen, J. *et al.* (1994) Optimal primer selection for clonality assessment by polymerase chain reaction analysis: I low grade B-cell lymphoproliferative disorders of nonfollicular type. *Hum Pathol*, **25**, 1269–1275.
2 Lombardo, J.A., Hwang, T.S., Maiese, R.U. *et al.* (1996) Optimal primer selection for clonality assessment by polymerase chain reaction analysis: III intermediate and high-grade B-cell neoplasms. *Hum Pathol*, **27**, 373–380.
3 Achille, A., Scarpa, A., Montresor, M. *et al.* (1995) Routine application of polymerase chain reaction in the diagnosis of monoclonality of B-cell lymphoid proliferations. *Diag Mol Pathol*, **4**, 14–24.
4 Arber, D.A. 2000; Review: molecular diagnostic approach to non-Hodgkin's lymphoma. *J Mol Diagn*, **2**, 178–190.and correspondence: PCR methods for determining B cell clonality. (2000) *J Mol Diagn*, **2**, 217–218.
5 Pongers-Willemse, M.J., Seriu, T., Stolz, F. *et al.* (1999) Primers and protocols for standardized detection of minimal residual disease in acute lymphoblastic leukemia using immunoglobulin and T cell receptor gene rearrangements and TAL1 deletions as PCR targets. *Leukemia*, **13**, 110–118.
6 Theriault, C., Galoin, S., Valmary, S. *et al.* (2000) PCR analysis of immunoglobulin heavy chain (IgH) and TCR-γ chain gene rearrangements in the diagnosis of lymphoproliferative disorders: results of a study of 525 cases. *Mod Pathol*, **13**, 1269–1279.
7 Gong, J.Z., Zheng, S., Chiarie, R. *et al.* (1999) Detection of immunoglobulin κ light chain rearrangements by polymerase chain reaction. An improved method for detecting clonal B-cell lymphoproliferative disorders. *Am J Pathol*, **155**, 355–363.

8 Hayashi, Y., Fukayama, M., Funata, N. *et al.* (1999) Polymerase chain reaction screening of immunoglobulin heavy chain and T cell receptor γ gene rearrangements: a practical approach to molecular DNA analysis of non-Hodgkin's lymphoma in a surgical pathology laboratory. *Pathol Int*, **49**, 110–117.

9 Greem, E., McConvile, C.M., Powel, J.E. (1998) Clonal diversity if Ig and T-cell receptor gene rearrangements identifies a subset of childhood B-precursor acute lymphoblastic leukemia with increased risk of relapse. *Blood*, **92**, 952–958.

10 Garcia-Sanz, R., Lopez-Perez, R., Langerak, A.W. (1999) Heteroduplex PCR analysis of rearranged immunoglobulin genes for clonality assessment in multiple myeloma. *Haematologica*, **84**, 328–335.

11 Dunn-Walters, D., Thiede, C., Alpen, B., Spencer, J. (2001) Somatic hypermutations and B-cell lymphoma. *Philos Trans R Soc Lond B*, **356**, 73–82.

12 Scrideli, C.A., Queiroz, R.P., Takayanagui, O.M. *et al.* (2003) Polymerase chain reaction on cerebrospinal fluid cells in suspected leptomeningeal involvement in childhood acute lymphoblastic leukemia: comparison to cytomorphological analysis. *Diagn Mol Pathol*, **12**, 124–127.

13 Noy, A., Verma, R., Glenn, M. *et al.* (2001) Clonotypic polymerase chain reaction confirms minimal residual disease in CLL nodular PR: results from a sequential treatment CLL protocol. *Blood*, **97**, 1929–1936.

14 Diss, T.C., Liu, H.X., Du, M.Q., Isaacson, P.G. (2002) Improvements to B cell clonality analysis using PCR amplification of immunoglobulin light chain genes. *J Clin Pathol Mol Pathol*, **55**, 98–101.

15 van Dongen, J.J., Langerk, A.W., Bürggeman, M. *et al.* (2003) Design and standardization of PCR primers and protocols for detection of clonal immunoglobulin and T-cell receptor gene recombinations in suspect lymphoproliferations. Report of the BIOMED-2 concerted action BMH4-CT98-3936. *Leukemia*, **17**, 2257–2317.

16 Pai, R.K., Chakerian, A.E., Binder, J.M. *et al.* (2005) B-cell clonality determination using an immunoglobulin κ light chain polymerase chain reaction method. *J Mol Diagn*, **7**, 300–330.

17 Evans, P.A.S., Pott, C., Groenen, P.J.T.A. *et al.* (2006) Significantly improved PCR-based clonality testing in B-cell malignancies by use of multiple immunoglobulin gene targets. Report of the BIOMED-2 Concerted Action BHM4-CT98-3936. *Leukemia*, **20**, 1–8.

18 Velden, V., Cazzaniga, G., Schrauder, A. (2007) Analysis of minimal residual disease by Ig/TCR gene rearrangements: guidelines for interpretation of real-time quantitative PCR data. *Leukemia*, **21**, 604–611.

Molecular Detection of T-Cell Lymphoma by Gene Rearrangement

1 Födinger, M., Winkler, K., Mannhalter, C. *et al.* (1999) Combined polymerase chain reaction approach for clonality detection in lymphoid neoplasms. *Diag Mol Pathol*, **8**, 80–91.

2 Benhatter, J., Delacretaz, F., Martin, P. *et al.* (1995) Improved polymerase chain reaction detection of clonal T-cell lymphoid neoplasms. *Diag Mol Pathol*, **4**, 108–112.

3 McCarthy, K.P., Sloane, J.P. *et al.* (1992) A simplified method of detection of clonal rearrangements of the T-cell receptor-γ chain gene. *Diag Mol Pathol*, **1**, 173–179.

4 Dadej, K., Gaboury, L., Lamarre, L. *et al.* (2001) The value of clonality in the diagnosis and follow-up of patients with cutaneous T-cell infiltrates. *Diag Mol Pathol*, **10**, 1078–1088.

5 Cave, H., Guidal, C., Rohrlich, P. *et al.* (1994) Prospective monitoring and quantitation of residual blasts in childhood acute lymphoblastic leukemia by polymerase chain reaction study of δ and

γ T-cell receptor genes. *Blood*, **83**, 1892–1902.

6 Arber, D.A., Braziel, R.M., Bagg, A. *et al.* (2001) Evaluation of T cell receptor testing in lymphoid neoplasms. Results of a multicenter study of 29 extracted DNA and paraffin-embedded samples. *J Mol Diagn*, **3**, 133–140.

7 Delabesse, E., Burtin, M.-L., Millein, C. *et al.* (2000) Rapid multifluorescent TCRG Vγ and Jγ typing: application to T cell acute lymphoblastic leukemia and to the detection of minor clonal populations. *Leukemia*, **14**, 1143–1152.

8 Kröber, S.M., Greschniok, A., Kaiserling, E., Horny, H.-P. (2000) Acute lymphoblastic leukaemia: correlation between morphological/immunohistochemical and molecular biological findings in bone marrow. *J Clin Pathol Mol Pathol*, **53**, 83–87.

9 Assaf, C., Hummel, M., Steinhoff, M. *et al.* (2005) Early TCR-β and TCR-γ PCR detection of T-cell clonality indicates minimal tumor disease in lymph nodes of cutaneous T-cell lymphoma: diagnostic and prognostic implications. *Blood*, **105**, 503–510.

Molecular Detection of Carcinoembryonic Antigen-Expressing Tumors

1 Gerhard, M., Juhl, H. *et al.* (1994) Specific detection of carcinoembryonic antigen expressing tumor cells in bone marrow aspirates by polymerase chain reaction. *J Clin Oncol*, **12**, 725–729.

2 Jonas, S., Windeatt, S., O-Boateng, A. *et al.* (1996) Identification of carcinoembryonic antigen producing cells circulating in the blood of patients with colorectal carcinoma by reverse-transcriptase polymerase chain reaction. *Gut*, **39**, 717–721.

3 Funaki, N.O., Tanaka, J., Kasamatsu, T. *et al.* (1996) Identification carcinoembryonic antigen m RNA in circulating peripheral blood of pancreatic and gastric carcinoma patients. *Life Sciences*, **59**, 2187–2199.

4 Mori, M., Mimori, K. *et al.* (1995) Detection of cancer metastases in lymph nodes by reverse-transcriptase polymerase chain reaction. *Cancer Res*, **55**, 3417–3420.

5 Yon, K.O., Klinz, M. *et al.* (1998) Limitations of the reverse-transcription-polymerase chain reaction method for the detection of carcinoembryonic antigen-positive tumor cells in peripheral blood. *Clin Cancer Res*, **4**, 2141–2146.

6 Neumaier, M., Gerhard, M., Wagner, C. (1995) Diagnosis of micrometastases by the amplification of tissue-specific genes. *Gene*, **159**, 43–47.

7 Castaldo, G., Tomaiuolo, R., Saduzi, A. *et al.* (1997) Lung cancer metastatic cells detected in blood by reverse-transcriptase polymerase chain reaction and dot-blot. *J Clin Oncol*, **15**, 3388–3393.

8 Mori, M., Mimori, K., Ueo, H. *et al.* (1996) molecular detection of circulating solid carcinoma in the peripheral blood: the concept of early systemic disease. *Int J Cancer*, **68**, 739–743.

9 Uchikura, K., Takao, S., Nakajo, A. *et al.* (2002) Intraoperative molecular detection of circulating tumor cells by reverse transcription-polymerase chain reaction in patients with biliary-pancreatic cancer is associated with hematogenous metastasis. *Ann Surg Oncol*, **9**, 364–370.

10 To, E.M.C., Chan, W.-Y., Chow, C. *et al.* (2003) Gastric cancer cell detection in peritoneal washing: cytology versus RT-PCR for CEA transcripts. *Diag Mol Pathol*, **12**, 88–95.

11 Wang, J.-Y., Wu, C.-H., Lu, C.-Y. *et al.* (2006) Molecular detection of circulating tumor cells in the peripheral blood of patients with colorectal cancer using RT-PCR: significance of the prediction of postoperative metastasis. *World J Surg*, **30**, 1007–1013.

Molecular Detection of Squamous Cell Carcinoma

1 Stenman, J., Lintula, S. *et al.* (1997) Detection of squamous cell carcinoma antigen expressing tumour cells in blood by reverse-transcriptase polymerase chain reaction in cancer of the uterine cervix. *Int J Cancer Pred Oncol*, **1997**, 74–85.

2 Hamakawa, H., Fukizumi, M., Bao, Y. *et al.* (1999) Genetic diagnosis of micrometastasis based on SC antigen mRNA in cervical lymph nodes of head and neck cancer. *Clin Exp Metastasis*, **17**, 593–599.

3 Kano, M., Shimada, Y., Kaganoi, J. *et al.* (2000) Detection of lymph node metastasis of esophageal cancer by RT-nested PCR for SCC antigen mRNA. *Br J Cancer*, **82**, 429–435.

4 Cataltepe, S., Gornstein, E.R., Schick, C. *et al.* (2000) Co-expression of the squamous cell carcinoma antigens 1 and 2 in normal adult human tissues and squamous cell carcinomas. *J Histochem Cytochem*, **48**, 113–122.

Molecular Detection of Prostate Cancer

1 Katze, A.E., Seamen, E. *et al.* (1994) Molecular staging of prostate cancer with the use of an enhanced reverse-transcriptase-PCR assay. *Urology*, **43**, 765–775.

2 Gao, C.L., Maheshwari, S., Dean, R.C. *et al.* (1999) Blinded evaluation of reverse-transcriptase polymerase chain reaction prostate specific antigen peripheral blood assay for molecular staging of prostate cancer. *Urology*, **53**, 714–721.

3 Wood, D.P., Jr, Banerjee, M. (1997) Presence of circulating prostate cells in the bone marrow of patients undergoing radical prostatectomy is predictive of disease-free survival. *J Clin Oncol*, **15**, 3451–3457.

4 Sokoloff, M.H., Tso, C.-L., Kaboo, R. *et al.* (1996) Quantitative polymerase chain reaction does not improve preoperative prostate cancer staging: a clinicopathological molecular analysis of 121 patients. *J Urol*, **156**, 1560–1566.

5 Ghossein, R.A., Scher, H.I., Gerald, W.L. *et al.* (1995) Detection of circulating tumor cells in patients with localized and metastatic prostatic carcinoma: clinical implants. *J Clin Oncol*, **13**, 1195–1200.

6 Seiden, M.V., Kantoff, P.W., Krithivas, K. *et al.* (1994) Detection of circulating tumor cells in men with localized prostate cancer. *J Clin Oncol*, **12**, 26340–2639.

7 Corey, E., Corey, M. (1998) Detection of disseminated prostate cells by reverse-transcriptase polymerase chain reaction (RT-PCR): technical and clinical aspects. *Int J Cancer*, **7**, 655–673.

8 Corey, E., Arfman, E.W., Oswin, M.M. *et al.* (1997) Detection of circulating prostate cells by reverse-transcriptase polymerase chain reaction of human glandular kallikrein (hk2) and prostate specific antigen (PSA) messages. *Urology*, **50**, 184–188.

9 Ferrari, A.C., Stone, N.N., Eyler, J.N. *et al.* (1997) Prospective analysis of prostate specific markers in pelvic lymph nodes of patients with high-risk prostate cancer. *J Nat Cancer Inst*, **89**, 1498–1504.

10 Ennis, E., Katz, A.E., de Vries, G.M. *et al.* (1997) Detection of circulating prostate carcinoma cells via an enhanced reverse-transcriptase polymerase chain reaction assay in patients with early stage prostate carcinoma. *Cancer*, **79**, 2402–2408.

11 Ghossein, R.A., Carusone, L., Bhattacharya, S. (1999) Review: polymerase chain, prostate and thyroid carcinomas. *Diag Mol Pathol*, **8**, 165–175.

12 Sun, Y., Lin, J., Katz, A.K. *et al.* (1997) Human prostatic carcinoma oncogene PTI-1 is expressed in human tumor cell lines and prostate carcinoma patient blood samples. *Cancer Res*, **57**, 18–23.

13 Israeli, R.S., Miller, W.H. *et al.* (1995) Sensitive detection of prostatic hematogenous tumor cell dissemination using prostate specific antigen and prostate specific membrane-derived primers in the polymerase chain reaction. *J Urol*, **153**, 573–577.

14 Straus, B., Müller, M., Krause, H. *et al.* (2001) Reverse transcriptase polymerase chain reaction for prostate specific antigen in the molecular staging of pelvic surgical margins after radical prostatectomy. *Urology*, **57**, 1006–1011.

15 Patel, K., Whelan, P.J., Prescott, S. *et al.* (2004) The use of real-time reverse transcription-PCR for prostate-specific antigen mRNA to discriminate between blood samples from healthy volunteers and from patients with metastatic prostate cancer. *Clin Cancer Res*, **10**, 7511–7519.

16 Yu, L., Wu, G., Wang, L. *et al.* (2006) Transient reduction of PTI-1 expression by short interfering RNAs inhibits the growth of human prostate cancer cells. *Tohoku J Exp Med*, **209**, 141–148.

Molecular Detection of Neuroendocrine Tumors and Neuroblastoma

1 Kuroda, T., Saeki, M. *et al.* (1997) Clinical application of minimal residual neuroblastoma cell detection by reverse-transcriptase polymerase chain reaction. *J Pediatr Surg*, **32**, 69–72.

2 Miyajima, Y., Kato, K. *et al.* (1995) detection of neuroblastoma cells in bone marrow and peripheral blood at diagnosis by the reverse-transcriptase polymerase chain reaction for tyrosine hydroxylase mRNA. *Cancer*, **75**, 2757–2761.

3 Yanagisawa, T.Y., Sasahara, Y., Fujie, H. *et al.* (1998) Detection of the PGP9. 5 and tyrosine hydroxylase mRNA for minimal residual neuroblastoma cells in bone marrow and peripheral blood. *J Exp Med*, **184**, 229–240.

4 Lioyd, R.V., Jin, L., Kulig, E. *et al.* (1992) Molecular approaches for the analysis of chromogranins and secretogranins. *Diag Mol Pathol*, **1**, 2–15.

5 Shono, K., Tajiri, T., Fujii, Y., Suita, S. (2000) Clinical implication of minimal disease in the bone marrow and peripheral blood in neuroblastoma. *J Pediatr Surg*, **35**, 1415–1420.

6 Cheung, I.Y., Barber, D. *et al.* (1998) Detection of microscopic neuroblastoma in marrow by histology, immunocytology, and reverse transcriptase-PCR of multiple molecular markers. *Clin Cancer Res*, **4**, 2801–2805.

7 Cheung, I.Y., Cheung, N.-K.V. (2001) Quantitation of marrow disease in neuroblastoma by real-time reverse transcription-PCR. *Clin Cancer Res*, **7**, 1698–1705.

8 Ito, R., Asami, S., Kagawa, S. *et al.* (2004) Usefulness of tyrosine hydroxylase mRNA for diagnosis and detection of minimal residual disease in neuroblastoma. *Biol Pharm Bull*, **27**, 315–318.

9 Corrias, M.V., Faulkner, L.B., Pistorio, A. *et al.* (2004) Detection of neuroblastoma cells in bone marrow and peripheral blood by different techniques: accuracy and relationship with clinical features of patients. *Clin Cancer Res*, **10**, 7978–7985.

Molecular Detection of Hepatocellular Carcinoma

1 Hillaire, S., Barbu, V., Boucher, E. *et al.* (1994) Albumin messenger RNA as a marker of circulating hepatocytes in hepatocellular carcinoma. *Gastroenterology*, **106**, 239–242.

2 Komeda, T., Fukuda, Y., Sando, T. *et al.* (1995) Sinsitive detection of circulating Hepatocellular carcinoma cells in peripheral blood. *Cancer*, **75**, 2214–2219.

3 Matsumura, M., Niwa, Y., Kato, N. *et al.* (1994) Detection of alpha fetoprotein mRNA, an indicator of hematogenous spreading hepatocellular carcinoma, in the circulation: a possible predictor of metastatic hepatocellular carcinoma. *Hepatology*, **20**, 1418–1425.

4 Sheen, I.S., Jeng, K.S., Shih, S.C. *et al.* (2004) Does surgical resection of hepatocellular carcinoma accelerate cancer dissemination? *World J Gastroenterol*, **10**, 31–36.

5 Yang, S.-Z., Dong, J.-H., Li, K. *et al.* (2005) Detection of AFP-mRNA and melanoma antigen gene-1mRNA as markers of disseminated hepatocellular carcinoma cells in blood. *Hepatobil Pancreat Dis Int*, **4**, 227–233.

6 Wu, W., Yao, D.-F., Yuan, Y.-M. *et al.* (2006) Combined serum hepatoma-specific alpha-fetoprotein and circulating alpha-fetoprotein mRNA in diagnosis of hepatocellular carcinoma. *Hepatobil Pancreat Dis Int*, **5**, 538–544.

Molecular Detection of Malignant Melanoma

1 Battayani, Z., Grob, J.J. *et al.* (1995) Polymerase chain reaction detection of circulating melanocytes as a prognostic marker in patients with melanoma. *Arch Dermatol*, **131**, 443–447.
2 Fathmann, B., Eberle, J., Krasagkis, K. *et al.* (2005) RT-PCR for tyrosine-mRNA-positive cells in peripheral blood: evaluation strategy and correlation with know prognostic markers in 123 melanoma patients. *J Invest Dermatol*, **110**, 263–267.
3 Foss, A.J.E., Guille, M.J., Occleston, N.L. *et al.* (1995) The detection of melanoma cells in peripheral blood by reverse-transcriptase polymerase chain reaction. *Br J Cancer*, **72**, 155–159.
4 Mellado, B., Colomer, D., Castel, T. *et al.* (1996) Detection of circulating neoplastic cells by reverse-transcriptase polymerase chain reaction in malignant melanoma: association with clinical stage and prognosis. *J Clin Oncol*, **14**, 2091–2097.
5 Hoon, D.S.B., Wang, Y., Dale, P.S. *et al.* (1995) Detection of occult melanoma cells in blood with a multiple – marker polymerase chain reaction assay. *J Clin Oncol*, **13**, 2109–2116.
6 Curry, B.J., Myers, K., Hersey, P. (1998) Polymerase chain reaction of melanoma cells in the circulation: Relationb to clinical stage, surgical treatment and recurrence from melanoma. *J Clin Oncol*, **16**, 1760–1769.
7 Ghossein, R.A., Coit, D., Brennan, M. *et al.* (1998) Prognostic significance of peripheral blood and bone marrow tyrosinase messenger RNA in malignant melanoma. *Clin Cancer Res*, **4**, 419–428.
8 Kunter, U., Buer, J., Probst, M. *et al.* (1996) Peripheral blood tyrosinase messenger RNA detection and survival in malignant melanoma. *J Nat Cancer Inst*, **88**, 590–594.
9 Buzaid, A.C., Balch, C.M. (1996) Polymerase chain reaction for detection of melanoma in peripheral blood: too early to accesses clinical value. *J Nat Cancer Inst*, **88**, 569–570.
10 Bieligk, S.C., Ghossein, R., Bhattacharya, S. *et al.* (1999) Detection of mRNA by reverse-transcriptase polymerase chain reaction in melanoma sentinel nodes. *Ann Surg Oncol*, **6**, 232–240.
11 Sarantou, T., Chi, D.D.J., Garrison, D.A. (1997) Melanoma associated antigens as mRNA detection markers for melanoma. *Cancer Res*, **57**, 1371–1376.
12 Cheung, I.Y., Cheung, N.V., Ghossein, R. A. *et al.* (1999) Association between molecular detection of GAGE and survival in patients with malignant Melanoma: a retrospective cohort study. *Clin Cancer Res*, **5**, 2042–2047.
13 De Plaen, E., Arden, K., Travesari, C. *et al.* (1994) Structure, chromosomal localization, and expression of 12 genes of the MAGE family. *Immunogenetics*, **40**, 360–369.
14 Gutzmer, R., Kaspari, M., Broderson, J.P. *et al.* (2002) Specificity of tyrosinase and HMB45 PCR in the detection of melanoma metastases in sentinel lymph node biopsies. *Histopathology*, **41**, 510–518.
15 Davids, V., Kidson, S.H., Hanekom, G.S. (2003) Accurate molecular detection of melanoma nodal metastases: an assessment of multimarker assay specificity, sensitivity, and detection rate. *J Clin Pathol Mol Pathol*, **56**, 43–51.
16 Wlodzimierz, R., Rutkowski, P., Nowecki, Z. *et al.* (2004) Detection of melanoma cells in the lymphatic drainage after lymph node dissection in melanoma patients by using two-marker reverse transcriptase-polymerase chain reaction assay. *Ann Surg Oncol*, **11**, 988–997.

17 Baruch, A.C., Shi, J., Feng, Y. *et al.* (2005) New developments in the staging of melanoma. *Cancer Invest*, **23**, 561–567.

Molecular Detection of Thyroid Carcinoma

A. Detection by Specific Gene Expression

1 Ditkoff, B.A., Marvin, M.R., Yemul, S. *et al.* (1996) Detection of circulating thyroid cells in peripheral blood. *Surgery*, **120**, 959–965.
2 Tallini, G., Ghossein, R.A., Emanuel, J. *et al.* (1998) Detection of thyroglobulin, thyroid peroxidase and RET/PTC1 mRNA transcripts in the peripheral blood of patients with thyroid disease. *J Clin Oncol*, **16**, 1158–1166.
3 Wingo, S.T., Ringel, M.D., Anderson, J.S. (1999) Quantitative reverse-transcriptase PCR measurement of thyroglobulin mRNA in peripheral blood of healthy subjects. *Clin Chem*, **45**, 785–789.
4 Katoh, R., Miyagi, E., Nakamura, N. *et al.* (2000) Expression of thyroid transcription factor-1 (TTF-1) in human C cells and medullary thyroid carcinoma. *Hum Pathol*, **31**, 386–393.
5 Bojunga, J., Dragan, C., Schumm-Draeger, P.M. *et al.* (2001) Circulating calcitonin and carcinoembryonic antigen m-RNA detected by RT-PCR as tumor markers in medullary thyroid carcinoma. *Br J Cancer*, **85**, 154–1550.

B. Detection by Specific Genetic Abnormalities

1 Nikiforov, Y.E., Rowland, J.M., Bove, K.E. *et al.* (1997) Distinct pattern of RET oncogene rearrangements in morphologic variants of radiation-induced and sporadic thyroid papillary carcinomas in children. *Cancer Res*, **57**, 1690–1694.
2 Nikiforov, Y.E., Ericson, L.A., Nikiforova, M.N. *et al.* (2001) Solid variant of papillary thyroid carcinoma. *Am J Surg Pathol*, **25**, 1478–1484.
3 Chua, E.L., Wu, W.M., Tran, K.T. *et al.* (2000) Prevalence and distribution of ret/ptc 1, 2, 3 in papillary thyroid carcinoma in New Caledonia and Australia. *J Clin Endocrinol Metabol*, **85**, 2733–2739.
4 Nikiforov, Y.E. (2002) RET/PCT rearrangement in thyroid tumors. *Endocrine Pathol*, **13**, 3–16.
5 Kroll, T.G., Sarraf, P., Pecciarini, L. *et al.* (2000) PAX8-PPARγ1 fusion oncogene in human thyroid carcinoma. *Science*, **289**, 25.
6 Nikiforova, M.N., Biddinger, P.W., Caudill, C.M. *et al.* (2002) PAX8-PPARγ (rearrangement in thyroid tumors. *Am J Surg Pathol*, **26**, 1016–1023.
7 Pierotti, M.A., Vigneri, P., Bongarzone, I. (1998) Rearrangements of RET and NTRK1 tyrosine kinase receptors in papillary thyroid carcinoma. *Rec Res Cancer Res*, **154**, 237–247.
8 Rabes, H.M., Demidchik, E.P., Sidorow, J.D. (2000) Pattern of radiation-induced RET and NTRK1 rearrangements in 191 post-Chernobyl papillary thyroid carcinoma: biological phenotypic and clinical implications. *Clin Cancer Res*. **6**.1093–1103.
9 Marques, A., Espadinha, C., Catarino, A. *et al.* (2002) Expression of PAX8 PPARγ1 rearrangements in both follicular thyroid carcinomas and adenomas. *J Clin Endocrinol Metab*, **87**, 3947–3952.
10 Cheung, L., Messina, M., Gill, A. *et al.* (2003) Detection of the PAX8-PPARγ fusion oncogene in both follicular thyroid carcinomas and adenomas. *J Clin Endocrinol Metab*, **88**, 354–357.
11 Lacroix, L., Lazar, V., Michiels, S. *et al.* (2005) Follicular thyroid tumors with the PAX8-PPARγ1 rearrangement display characteristic genetic alterations. *Am J Pathol*, **167**, 223–231.
12 Castro, P., Rebocho, A.P., Soares, R.J. *et al.* (2006) PAX8-PPARγ rearrangement is frequently detected in the follicular variant of papillary thyroid carcinoma. *J Clin Endocrinol Metab*, **91**, 213–220.

Molecular Detection of Breast Cancer and Other Adenocarcinomas by the Molecular Detection of CK19, MUC-1 and Mammaglobin

1 Bostick, P.J., Chatterjee, S., Chi, D.D. *et al.* (1998) Limitations of specific reverse-transcriptase polymerase chain reaction markers in the detection of metastases in the lymph nodes and blood of breast cancer patients. *J Clin Oncol*, **16**, 2632–2640.

2 Vannucchi, A.M., Bosi, A., Glinz, S. *et al.* (1998) Evaluation of breast tumour cell contamination in bone marrow and leukapheresis collection by RT-PCR for cytokeratin-19 mRNA. *Br J Haematol*, **103**, 610–617.

3 Noguchi, S., Aihara, T., Motomura, K. *et al.* (1996) Detection of breast cancer micrometastases in axillary lymph nodes by means of reverse-transcriptase polymerase chain reaction. *Am J Pathol*, **148**, 649–656.

4 Traystman, M.D., Cochram, G.T., Hake, S.J. *et al.* (1997) Comparison of molecular cytokeratin 19 reverse-transcriptase polymerase chain reaction (CK19 RT-PCR) and immunocytochemical detection of micrometastatic breast cancer cells in hematopoietic harvests. *J Hematother*, **6**, 551–561.

5 Fields, K.K., Elfenbein, G.J., Trudeau, W.L. (1996) Clinical significance of bone marrow metastases as detected using the polymerase chain reaction in patients with breast cancer undergoing high dose chemotherapy and autologous bone marrow transplantation. *J Clin Oncol*, **14**, 1868–1876.

6 Datta, Y.H., Adams, P.T., Drobyski, W.R. *et al.* (1994) Sensitive detection of occult breast cancer by the reverse-transcriptase polymerase chain reaction. *J Clin Oncol*, **12**, 475–482.

7 Schoenfeld, A., Luqmani, Y., Smith, D. *et al.* (1994) Detection of breast cancer metastases in axillary lymph nodes by using polymerase chain reaction. *Cancer Res*, **54**, 2986–2990.

8 Min, C., Tafra, L., Verbanac, K.M. (1998) Identification of superior markers for polymerase chain reaction detection of breast cancer metastases in sentinel lymph nodes. *Cancer Res*, **58**, 4581–4584.

9 Noguchi, S., Aihara, T., Motomura, K. *et al.* (1996) Histologic characteristics of breast cancers with occult lymph node metastases detected by keratin 19 mRNA reverse-transcriptase polymerase chain reaction. *Cancer*, **78**, 1235–1240.

10 Liefers, G.J., Tollenaar, R.A.E.M., Cleton-Jansen, A.M. (1999) Molecular detection of minimal residual disease on colorectal and breast cancer, *Histopathology*, **34**, 385–390.

11 Moscinski, L.C., Trudeau, W.L., Fields, K.K. *et al.* (1996) High sensitivity detection of minimal residual breast carcinoma using the polymerase chain reaction for cytokeratin 19. *Diag Mol Pathol*, **5**, 173–180.

12 Mori, M., Mimori, K., Inoue, H. *et al.* (1995) Detection of cancer micrometastases in lymph nodes by reverse-transcriptase polymerase chain reaction. *Cancer Res*, **55**, 3417–3420.

13 Ruud, P., Fodstad, O., Hovig, E. (1999) Identification of a novel cytokeratin 19 pseudogene that may interfere with reverse-transcriptase polymerase chain reaction assays used to detect micrometastatic tumor cells. *Int J Cancer*, **80**, 119–125.

14 Burchill, S., Bradbury, M., Pittman, K. *et al.* (1995) Detection of epithelial cancer cells in peripheral blood by reverse-transcriptase polymerase chain reaction. *Br J Cancer*, **71**, 271–281.

15 Peck, K., Sehr, Y.-P. *et al.* (1998) Detection and quantitation of circulating cancer cells in the peripheral blood of lung cancer patients. *Cancer Res*, **58**, 2761–2765.

16 Salerno, C.T., Frizelle, S., Niehans, G.A. *et al.* (1998) Detection of occult micrometastases in non-small cell lung carcinoma by reverse transcriptase-polymerase chain reaction. *Chest*, **113**, 1526–1532.

17 Stimpfl, M., Schmid, B.C., Schiebel, I. *et al.* (1999) Expression of mucins and cytokeratins in ovarian cancer cell lines. *Cancer Lett*, **145**, 133–141.

18 de Cremoux, P., Extra, J.M., Denis, M.G. *et al.* (2000) Detection of MUC1-expressing mammary carcinoma cells in the peripheral blood of breast cancer patients by real-time polymerase chain reaction. *Clin Cancer Res*, **6**, 3117–3122.

19 Zach, O., Kasparu, H., Kriege, O. *et al.* (1999) Detection of circulating mammary carcinoma cells in the peripheral blood of breast cancer patients via a nested reverse-transcriptase polymerase chain reaction assay for mammaglobin mRNA. *Clin Oncol*, **17**, 2015–2019.

20 Leygue, E., Snell, L., Dotzlaw, H. *et al.* (1999) Mammaglobin, a potential marker of breast cancer nodal metastases. *J Pathol*, **189**, 28–33.

21 Watson, M.A., Fleming, T.P. (1996) Mammaglobin, a mammary specific member of the uteroglobin family, is overexpressed in human breast cancer. *Cancer Res*, **56**, 860–865.

22 Ooka, M., Sakita, I., Fujiwara, Y. *et al.* (2000) Selection of mRNA markers for detection of lymph node micrometastases in breast cancer patients. *Oncology Rep*, **7**, 561–566.

23 Watson, M., Darrow, C., Zimonjic, D. *et al.* (1998) Structure and transcriptional regulation of the human mammaglobin gene family localized to chromosome 11q13. *Oncogene*, **16**, 817–824.

24 Masuda, N., Tamaki, Y., Sakita, I. *et al.* (2000) Clinical significance of micrometastases in axillary lymph nodes assessed by reverse transcription-polymerase chain reaction in breast cancer patients. *Clin Cancer Res*, **6**, 4176–4185.

25 Zehentner, B.K., Dillon, D.C., Jiang, Y. *et al.* (2002) Application of a multigene reverse transcriptase PCR assay for the detection of Mammaglobin and complementary transcribed genes in breast cancer lymph nodes. *Clin Chem*, **48**, 1225–1231.

26 Ooka, M., Tamaki, Y., Sakita, I. *et al.* (2001) Bone marrow metastases detected by RT-PCR for mammaglobin can be an alternative prognostic factor of breast cancer. *Breast Cancer Res Treat*, **67**, 169–175.

27 Janku, F., Kleibl, Z., Novotny, J. *et al.* (2004) Mammaglobin A, a novel marker of minimal residual disease in early stages breast cancer. *Neoplasm*, **51**, 204–208.

28 Reinholz, M.M., Nibbe, A., Jonart, L.M. *et al.* (2005) Evaluation of a panel of tumor markers for molecular detection of circulating cancer cells in women with suspected breast cancer. *Clin Cancer Res*, **11**, 15.

29 Brown, N.M., Stenzel, T.T., Friedman, P.N. *et al.* (2005) Evaluation of expression based markers for the detection of breast cancer cells. *Breast Cancer Res Treat* DOI 10.1007/s10549-005-9085-8, 41–47.

30 Roncella, S., Ferro, F., Bacigalupo, B. *et al.* (2005) Human mammaglobin mRNA is a reliable molecular marker for detecting occult breast cancer cells in peripheral blood. *J Exp Cancer Res*, **24**, 2.

31 Jäger, D., Stockert, E., Güre, A. *et al.* (2001) Identification of a tissue-specific putative transcription factor in breast tissue by serological screening of a breast cancer library. *Cancer Res*, **61**, 2055–2061.

32 Saintigny, P., Coulon, S., Kambouchner, M. *et al.* (2005) real-time RT-PCR detection of CK19, CK7 and MUC1 mRNA of diagnosis of lymph node micrometastases in non small cell lung carcinoma. *Int J Cancer*, **115**, 777–782.

33 Uen, Y.-H., Lin, S.-R., Wu, C.-H. *et al.* (2006) Clinical significance of Muc1 and c-met RT-PCR detection of circulating tumor cells in patients with gastric carcinoma. *Clin Chim Acta*, **367**, 55–61.

Molecular Detection of Colorectal Adenocarcinoma

A. Detection of CK 20

1 Gunn, J., McCall, J.L., Yun, K. *et al.* (1996) Detection of micrometastases in colorectal

cancer patients by K19 and K20 reverse-transcriptase polymerase chain reaction. *Lab Invest*, **75**, 611–616.

2 Soeth, E., Vogel, I., Röder, C. *et al.* (1997) Comparative analysis of bone marrow and venous blood isolates from gastrointestinal cancer patients for the detection of disseminated tumor cells using reverse transcriptase PCR. *Cancer Res*, **57**, 3106–3110.

3 Futamura, M., Takagi, Y., Koumura, H. *et al.* (1998) Spread of colorectal cancer micrometastases in regional lymph nodes by reverse-transcriptase polymerase chain reactions for carcinoembryonic antigen and CK20. *J Surg Oncol*, **68**, 34–40.

4 Liefers, G.J., Tollenaar, R.A.E.M., Cleton-Jansen, A.M. (1999) Molecular detection of minimal residual disease in colorectal and breast cancer. *Histopathology*, **34**, 385–390.

5 Dorudi, S., Kinrade, E., Marshall, N.C. *et al.* (1998) Genetic detection of lymph node micrometastases in patients with colorectal cancer. *Br J Surg*, **85**, 98–100.

6 Yamamato, N., Kato, Y., Yanagisawa, A. *et al.* (1997) Predictive value of genetic diagnosis for cancer micrometastases. *Cancer*, **80**, 1393–1398.

7 Weitz, J., Kienle, P., Lacroix, J. *et al.* (1998) Dissemination of tumor cells in patients undergoing surgery for colorectal cancer. *Clin Cancer Res*, **4**, 343–348.

8 Funaki, N., Tanaka, J., Ohshio, G. *et al.* (1998) Cytokeratin 20 mRNA in peripheral venous blood of colorectal carcinoma patients. *Br J Cancer*, **77**, 1327–1332.

9 Wyld, D.K., Selby, P., Perren, T.J. *et al.* (1998) Detection of colorectal cancer cells in peripheral blood by reverse-transcriptase polymerase chain reactions for cytokeratin 20. *Int J Cancer Pred Oncol*, **79**, 288–293.

10 Jung, R., Petersen, K., Krüger, W. *et al.* (1999) Detection of micrometastasis by cytokeratin 20 RT-PCR is limited due to stable background transcription in granulocytes. *Br J Cancer*, **81**, 870–873.

11 Miake, Y., Yamamoto, H., Fujiwara, Y. *et al.* (2001) Extensive micrometastases to lymph nodes as a marker for rapid recurrence of colorectal cancer: a study for lymphatic mapping. *Clin Cancer Res*, **7**, 1350–1357.

12 Vlems, F., Soong, R., Diepstra, H. *et al.* (2002) Effect of blood sample handling and reverse transcriptase-polymerase chain reaction assay sensitivity on detection of CK20 expression in healthy donor blood. *Diag Mol Pathol*, **11**, 90–97.

13 Vlems, F.A., Diepstra, J.H.S., Cornelissen, I.M.H.A. *et al.* (2002) Limitations of cytokeratin 20 RT-PCR to detect disseminated tumour cells in blood and bone marrow of patients with colorectal cancer: expression in controls and downregulation in tumour tissue. *J Clin Pathol Mol Pathol*, **55**, 156–163.

14 Zhang, Y.-L., Feng, J.-G., Gou, J.-M. *et al.* (2005) Detection of CK20mRNA in peripheral blood of pancreatic cancer and its clinical significance. *World J Gastroenterol*, **11**, 1023–1027.

B. Detection of Guanylyl Cyclase C and CDX-2

1 Bonhomme, C., Duluc, I., Martin, E. *et al.* (2003) The Cdx2 homeobox gene has a tumour suppressor function in the distal colon in addition to a homeotic role during gut development. *Gut*, **52**, 1465–1471.

2 Fava, T.A., Desnoyers, R., Schulz, S. *et al.* (2001) Ectopic expression of guanylyl cyclase C in CD34+ progenitor cells in peripheral blood. *J Clin Oncol*, **19**, 3951–3959.

3 Bustin, S., Siddiqi, S., Ahmed, S. *et al.* (2004) Quantification of cytokeratin 20, carcinoembryonic antigen and guanylyl cyclase C mRNA levels in lymph nodes may not predict treatment failure in colorectal cancer patients. *Int J Cancer*, **108**, 412–417.

4 Conzelmann, M., lineman, U., Berger, M.R. (2005) Molecular detection of clinical colorectal cancer metastasis: how should multiple markers be put to use. *Int J Colorectal Dis*, **20**, 137–146.

5 Schulz, S., Hyslop, T., Haaf, J. *et al.* (2006) A validated quantitative assay to detect occult micrometastases by reverse transcriptase-polymerase chain reaction of guanylyl cyclase C in patients with colorectal cancer. *Clin Cancer Res*, **12**, 4545–4552.

6 Birbe, R., Palazzo, J.P., Waltersa, R. *et al.* (2005) Guanylyl cyclase C is a marker of intestinal metaplasia, dysplasia, and adenocarcinoma of the gastrointestinal tract. *Hum Pathol*, **36**, 170–179.

Molecular Detection of Transitional Cell Carcinoma

1 Wu, X.-R., Lin, J.-H., Walzg, T. *et al.* (1994) Mammalian uroplakins: a group of highly conserved urothelial differentiation-related membrane proteins. *J Biol Biochem*, **269**, 13716–13724.

2 Lobban, E.D., Smith, B.A., Hall, G.D. *et al.* (1998) Uroplakin gene expression by normal and neoplastic human urothelium. *Am J Pathol*, **153**, 1957–1967.

3 Yuasa, T., Yoshiki, T., Isono, T. *et al.* (1999) Expression of transitional cell-specific genes, uroplakin Ia and II, in bladder cancer: detection of circulating cancer cells in the peripheral blood of metastatic patients. *Int J Urol*, **6**, 286–292.

4 Lu, J.-J., Kakehi, Y., Takahashi, T. *et al.* (2000) Detection of circulating cancer cells by reverse transcription-polymerase chain reaction for uroplakin II in peripheral blood of patients with urothelial cancer. *Clin Cancer Res*, **6**, 3166–3171.

5 Gazzaniga, P., Gandini, O., Giuliani, L. *et al.* (2001) Detection of epidermal growth factor receptor mRNA in peripheral blood: a new marker of circulating neoplastic cells in bladder cancer patients. *Clin Cancer Res*, **7**, 577–583.

6 Liang, F.-X., Riedel, I., Deng, F.-M. *et al.* (2001) Organization of uroplakin subunits: transmembrane topology, pair formation and plaque composition. *Biochem J*, **355**, 13–18.

7 Kurahashi, T., Hara, I., Oka, N. *et al.* (2005) Detection of micrometastases in pelvic lymph nodes in patients undergoing radical cystectomy for focally invasive bladder cancer by real-time reverse transcriptase-PCR for cytokeratin19 and uroplakin II. *Clin Cancer Res*, **11**, 3773–3777.

Hormone Receptor Status, HER-2 and Epidermal Growth Factor Receptor

A. Hormone Receptors

1 Remmele, W., Stegner, H.E. (1987) Vorschlag zur einhheilichen Definition einer Immunreaktiven score (IRS) für den immunhistochemischen Östrogenrereptor-Nachweis (ER-ICA) in Mammakarzinomgewebe. *Pathologe*, **8**, 138–140.

2 McCarty, K.S., Jr, Miller, L.S., Cox, E.B., Konath, J. *et al.* (1985) Estrogen receptor analyses: correlation of biochemical and immunohistochemical methods using monoclonal antireceptor antibodies. *Arch Pathol Lab Med*, **109**, 716–721.

3 Remmele, W. (1993) Immunohistochemical determination of estrogen and progesterone receptor content in human breast cancer. *Pathol Res Pract*, **189**, 862–866.

4 Grünewald, K., Haun, M., Urbanek, M. *et al.* (2000) Mammaglobin gene expression: a superior marker of breast cancer cells in peripheral blood in comparison to epidermal-growth-factor receptor and cytokeratin-19. *Lab Invest*, **80**, 1071–1077.

5 Span, P.N., Waanders, E., Manders, P. *et al.* (2004) Mammaglobin is associated with low grade, steroid receptor-positive breast tumors from postmenopausal patients, and has independent prognostic value for relapse-free survival time. *J Clin Oncol*, **22**, 691–698.

6 Agoff, S.N., Swanson, P.E., Linden, H. *et al.* (2003) Androgen receptor expression in estrogen receptor-negative breast cancer. *Am J Clin Pathol*, **120**, 725–731.
7 Umemura, S., Kurosumi, M., Moriya, T. *et al.* (2006) Immunohistochemical evaluation for hormone receptors in breast cancer: a practically useful evaluation system and handling protocol. *Breast Cancer*, **13**, 232–235.
8 Henriksen, K., Rasmussen, B., Lykkesfeldt, A. *et al.* (2007) Semi-quantitative scoring of potentially predictive markers for endocrine treatment of breast cancer: a comparison between whole sections and tissue microarrays. *J Clin Pathol*, **60**, 397–404.

B. HER-2

1 Lebeau, A., Deimling, D., Kaltz, C. *et al.* (2001) HER/2 neu analysis in archival tissue samples of human breast cancer: Comparison of immunohistochemistry and fluorescence in situ hybridization. *J Clin Oncol*, **19**, 354–363.
2 Koeppen, H.K.W., Wright, B.D., Burt, A.D. *et al.* (2001) Overexpression of HER/neu in solid tumours: an immunohistochemical survey. *Histopathology*, **38**, 96–104.
3 Konecny, G.E., Pegram, M.D., Venkatesan, N. *et al.* (2006) Activity of the dual kinase inhibitor lapatinib (GW572016) against HER-2-overexpressing and trastuzumab-treated breast cancer cells. *Cancer Res*, **66**, 1630–1639.
4 Dowsett, M., Hanna, W.M., Kockx, M. (2007) Standardization of HER2 testing: results of an international proficiency-testing ring study. *Mod Pathol*, **20**, 584–591.

C. Epidermal Growth Factor Receptor

1 Putti, T.C., To, K.F., Hsu, H.C. *et al.* (2002) Expression of epidermal growth factor receptor in head and neck cancers correlates with clinical progression: a multicentre immunohistochemical study in the Asia-Pacific region. *Histopathology*, **41**, 144–151.
2 Reichert, J.M., Valge-Archer, V.E. (2007) Development trends for monoclonal antibody cancer therapeutics. *Drug Discov*, **6**, 349–356.

CD117 (c-kit) Expression

1 Smithey, B.E., Pappo, A.S., Hill, D.A. (2002) c-kit Expression in pediatric solid tumors. *Am J Surg Pathol*, **26**, 486–492.
2 Sandberg, A.A., Bridge, J.A. (2002) Updates on cytogenetic and molecular genetics of bone and soft tissue tumors: gastrointestinal stromal tumors. *Cancer Genet Cytogenet*, **135**, 1–22.
3 Rubin, B.P., Singer, S., Tsao, C. *et al.* (2001) Kit activation is a ubiquitous feature of gastrointestinal stromal tumors. *Cancer Res*, **61**, 8118–8121.
4 Markku, M., Jerzy, L. (2005) KIT (CD117): a review on expression in normal and neoplastic tissues, and mutations and their clinicopathologic correlation. *Appl Immunohistochem Mol Morphol*, **13**, 205–220.

Molecular Detection of Human Papilloma Virus

1 Husnjak, K., Grce, M., Magdic, L. *et al.* (2000) Comparison of five different polymerase chain reaction methods for detection of human papillomavirus in cervical cell specimens. *J Virol Methods*, **88**, 125–134.
2 Poblet, E., Alfaro, L., Fernander-Segoviano, P. *et al.* (1999) Human papillomavirus-associated penile Squamous cell carcinoma in HIV-positive patients. *Am J Surg Pathol*, **23**, 1119–1123.
3 Kuroki, T., Tusukibashi, O., Saito, M. *et al.* (2000) Design of a unique PCR primer for detection of oral HPV infection. *J Oral Sci*, **42**, 189–193.
4 Gheit, T., Landi, S., Gemignani, F. *et al.* (2006) Development of a sensitive and

specific assay combining multiplex PCR and DANN microarray primer extension to detect high-risk mucosal human papillomavirus types. *J Clin Microbiol*, **44**, 2025–2031.

Molecular Detection of Epstein–Barr Virus and HHV-8

1 Menet, A., Speth, C., Larcher, C. *et al.* (1999) Epstein–Barr virus infection of human astrocyte cell lines. *J Virol*, **73**, 7722–7733.
2 Emile, J.-F., Sebagh, M., Marchadier, E. *et al.* (2000) Hepatocellular carcinoma with lymphoid stroma: a tumor with good prognosis after liver transplantation. *Histopathology*, **37**, 523–529.
3 Hsiao, C.-H., Su, I.-J. *et al.* (1999) Epstein–Barr virus-associated intravascular lymphomatosis within Kaposi's sarcoma in an AIDS patient. *Am J Surg Pathol*, **23**, 482–487.
4 Jäger, M., Prag, N., Mitterer, M. *et al.* (1996) Pathogenesis of chronic Epstein–Barr virus infection: Detection of a virus strain of high rate of lytic replication. *Br J Haematol*, **95**, 626–636.
5 Gomez-Roman, J.J., Ocejo-Viyals, G., Sanchez-Velasco, P. *et al.* (2000) Presence of human herpisvirus-8 DNA sequences and overexpression of human IL-6 and cyclin D1 in inflammatory myofibroblastic tumor (inflammatory pseudotumor). *Lab Invest*, **80**, 1121–1126.
6 Gomez-Roman, J.J., Sanchez-Velasco, P., Ocejo-Vinyals, G. *et al.* (2001) Human herpisvirus-8 genes are expressed in pulmonary inflammatory myofibroblastic tumor (inflammatory pseudotumor). *Am J Surg Pathol*, **25**, 624–629.

Classification of Tumors and Minimal Residual Cancer Disease

1 Hermanek, P., Hutter, R.V.P., Sobin, L.H., Wittekind, C. (1999) Classification of isolated tumor cells and micrometastasis. A communication of the UICC. *Cancer*, **86**, 2668–2673.
2 Wittekind, C., Henson, D.E., Hutter, R.V.P., Sobin, L.H. (2001) *TNM Supplement. A Commentary on Uniform Use.* 2nd edn., Wiley–Liss, New York.
3 Sobin, L.H., Wittekind, C. (2002) (eds) *TMN Classification of Malignant Tumours*, 6th edn, Wiley–Liss, New York.
4 Singletary, S.E., Allred, C., Ashley, P. *et al.* (2002) Revision of the american joint committee on cancer staging system for breast cancer. *J Clin Oncol*, **20**, 3628–3636.
5 Colleoni, M., Rotmensz, N., Peruzzotti, G. *et al.* (2005) Size of breast cancer metastases in axillary lymph nodes: clinical relevance of minimal lymph node involvement. *J Clin Oncol*, **23**, 1379–1389.
6 Singletary, S.E., Connolly, J.L. (2006) Breast cancer staging: working with the sixth edition of the AJCC cancer staging manual. *CA Cancer J Clin*, **56**, 37–47.
7 Querzoli, P., Pedriali, M., Rinaldi, R. *et al.* (2006) Axillary lymph node nanometastases are prognostic factors for disease-free survival and metastatic relapse in breast cancer patients. *Clin Cancer*, **12**, 6997–6701.

Nucleic Acid Techniques

1 Sambrook, A., Russell, B. (2000) *Molecular Cloning: a Laboratory Manual*, 3rd edn, Cold Spring Harbor Laboratory Press, New York.
2 Ausubel, F.M. *et al.* (1987) *Current Protocols in Molecular Biology*, John Wiley and Sons, New York.
3 Alberts, B. *et al.* (1994) *Molecular Biology of the Cell*, 3rd edn, Garland Publishing, New York.
4 Wilfinger, W.W., Mackey, M., Chomczynski, P. (1997) Effect of pH and ionic strength on the spectrophotometric assessment of nucleic acid purity. *BioTechniques*, **22**, 474.
5 Lewin (1997) *Genes VI*, Oxford University Press, Oxford.

6 Kraft, A.M., Duncan, B.W., Bijwaard, K.E. *et al.* (1997) Optimization of the isolation and amplification of RNA from formalin-fixed, paraffin-embedded tissue: The armed forces institute of pathology experience and literature review. *Mol Diagn*, **2**, 217–230.

7 Godfrey, T.E., Kim, S.-H., Chavira, M. *et al.* (2000) Quantitative mRNA expression analysis from formalin-fixed, paraffin-embedded tissues using 5′ nuclease quantitative reverse transcription-polymerase chain reaction. *J Mol Diag*, **2**, 84–91.

8 Bock, O., Kreipe, H., Lehmann, U. (2001) One-step extraction of RNA from Archival biopsies. *Anal Biochem*, **295**, 116–117.

9 Chen, J., Byrne, G., Lossos, I. (2007) Optimization of RNA extraction from formalin-fixed, paraffin-embedded lymphoid tissues. *Diagn Mol Pathol*, **16**, 61–72.

10 Garci-Sanz, R., Lopez-Perez, R., Langerak, A.W. (1999) Heteroduplex PCR analysis of rearranged immunoglobulin genes for clonality assessment in multiple myeloma. *Haematologica*, **84**, 328–335.

Index

Phenotypic and Genotypic Diagnosis of Malignancies. Muin S.A. Tuffaha
Copyright © 2008 WILEY-VCH Verlag GmbH & Co. KGaA, Weinheim
ISBN: 978-3-527-31881-0

d